DARWINIAN PSYCHIATRY

DARWINIAN PSYCHIATRY

MICHAEL McGUIRE

ALFONSO TROISI

New York   Oxford
OXFORD UNIVERSITY PRESS
1998

Oxford University Press

Oxford   New York
Athens   Auckland   Bangkok   Bogota   Bombay
Buenos Aires   Calcutta   Cape Town   Dar es Salaam
Delhi   Florence   Hong Kong   Istanbul   Karachi
Kuala Lumpur   Madras   Madrid   Melbourne
Mexico City   Nairobi   Paris   Singapore
Taipei   Tokyo   Toronto   Warsaw

and associated companies in
Berlin   Ibadan

Published by Oxford University Press, Inc.
198 Madison Avenue, New York, New York 10016

Oxford is a registered trademark of Oxford University Press

Library of Congress Cataloging-in-Publication Data
McGuire, Michael T., 1929–
Darwinian psychiatry / Michael McGuire, Alfonso Troisi.
p.   cm.
Includes bibliographical references and index.
ISBN 0-19-511673-9
1. Mental illness—Genetic aspects.   2. Mental illness—Etiology.
3. Genetic psychology.   I. Troisi, Alfonso.   II. Title.
[DNLM: 1. Psychiatry.   2. Behavior.   3. Mental Disorders.   WM 100 M47832d   1998]
RC455.4.G4M34   1998
616.89—dc21   98-2504
DNLM/DLC for Library of Congress

3 5 7 9 9 8 6 4 2

Printed in the United States of America
on acid-free paper

For Giles Mead, who from the beginning said, "Yes!"

For Antonio Troisi, who taught me the value of freedom

Preface

In what follows we will argue that evolutionary theory can and should serve as the general theoretical framework for explaining and treating mental disorders. The idea is not new. It has been a possibility since Darwin, and one that, during the middle of their careers, both Freud and Jung considered in their efforts to unmask the hidden workings of the mind. *Totem and Taboo* (1912–1913) is perhaps the clearest example of Freud's attempts to integrate evolutionary and psychoanalytic ideas (Sulloway, 1979). As for Jung, evolutionary and psychoanalytic concepts commingle in his formulations of archetypes and the collective unconscious (1961, 1966). Yet, what might have developed to become a fruitful marriage never fully took place, and by the mid-1920s, both Freud and Jung had largely abandoned their early links with evolutionary theory and turned their interests elsewhere.

From the 1930s until the late 1960s, evolutionary explanations of disorders appeared only sporadically in the psychiatric literature. Examples can be found in the writings of the American psychiatrist Adolph Meyer (1948–1952) and the Australian psychiatrist Aubrey Lewis (1936). Meyer suggested that some disorders, as well as some features of disorders, represent adaptations. Lewis postulated that depression, because it often initiates caretaking behavior by others, is an evolved trait. Three decades separate Lewis's suggestion from the publication of "The Dominance Hierarchy and the Evolution of Mental Illness" (1967) by the English psychiatrist John Price. The origins of Darwinian psychiatry—psychiatric disorders viewed in evolutionary context—can be traced to these pioneers, along with Freud and Jung.

But this is getting ahead of the story. From the time Darwin first published his ideas about evolution in 1859 until the 1920s, most scholars were skeptical of evolutionary explanations of human behavior, normal or otherwise. Among evolutionists, signs of change began to appear in the early 1930s in the works of R. A. Fisher (1930), J. B. S. Haldane (1932), and S. Wright (1930), three of the architects of modern evolutionary thinking. Their ideas not only spawned a host of investigative studies but also charted the course of much of evolutionary theorizing for the three decades that followed. By the 1960s, evolutionary theory had evolved into a complex system of theoretical, often mathematically couched, formulations that could accommodate diverse sets of data, incorporate and integrate findings from archaeology, anthropology, genetics, natural history, physiology, and ethology, as well as serve as a source of testable hypotheses.

The 1960s brought further changes. What Fisher, Haldane, and Wright had begun some 30 years earlier, W. Hamilton, G. Williams, and R. Trivers would largely complete. In 1964, Hamilton published his theory of kin selection, which accounts for the preferential selection of investment and altruistic behavior toward kin. Williams (1966) spelled out the evolutionary principle of self-interest, the likely conditions favoring its selection, and its critical importance in explaining a broad array of behavior. And Trivers (1971) provided an evolutionary explanation for the selection of nonkin altruism: Self-interest and investment in nonkin can coexist when the probability of reciprocation is high.

The 1960s also mark an important period in the history of psychiatry. Psychoanalysis, which had dominated psychiatric thinking and practice for six decades, began to lose its place at psychiatry's center stage. Alternative theories of disorders, developed by investigators working in genetics, biochemistry, physiology, psychology, and sociology, began to gain prominence. And as the tools of modern science began to focus on mental disorders, reports of new research methodologies and findings, as well as new treatment techniques became weekly, not yearly, occurrences. It was a period of optimism, imbued with the sense that in the near future, mental disorders might become a thing of the past.

Three decades have passed since the optimism of the 1960s. In the interim, evolutionary theory has undergone significant change. Its hypotheses have become increasingly refined, specific, and testable, and its capacity for explaining much of human behavior cannot be easily ignored. From a distance, a similar story might seem to apply to psychiatry. Thousands of experimental and clinical studies have been conducted. Thousands of research reports have been published. New research and treatment techniques have been developed. And literally hundreds of disorder-related hypotheses have been put forth. Still, all is not well in psychiatry. In the words of one investigator:

> What other recourse does psychiatry have to surmount the present and continuing explosion of its data? . . . Our libraries are filled with volumes of old research results from played out areas, which no one consults or remembers. There are reams of data on cerebrospinal fluid (CSF) and plasma metabolites; urinary excreta; electroencephalographic (EEG) asymmetries; evoked potential peaks and troughs; brain lesions; cortical, electrical and chemical stimulation; unit recording; and a dozen other activities. The data were expensive to obtain and now are not worth replicating or even retrieving, because there was *little or no theory* [italics added] to specify what to measure, in what ancillary conditions, and with what expected outcomes. This is not the kind of chaos of which we can be proud. (Freeman, 1992, p. 1080)

To say that psychiatry is in a state of chaos is not to say that it remains in the Middle Ages. Compared to the 1960s, today's psychiatry has a far better understanding of the brain's anatomy and physiology, as well as of the causes of disorders. Drugs that rapidly ameliorate debilitating symptoms have been developed, and their worth has been established. And the conditions under which different treatments are likely to work can be more precisely specified. Certainly, these are signs of progress. Yet, for almost every sign of progress, there are other signs that make one uneasy. Schools of psychiatry with their familiar patterns of causal explanations and speculations continue to flourish. Physiological, genetic, intrapsychic, social, and learning explanations—each can often be used to explain the same disorder. With time and new findings, the composition of the schools changes, along with the explanations most cherished by each school's advocates. Nevertheless, the schools persist, and at best, their causal formulations explain only bits and pieces of disorders. As to the hope that disorders would become a thing of the past, some disorders have declined in frequency while others have increased. Currently, the estimated lifetime prevalence of mental disorders in the United States approaches 50%, which means that one out of every two persons will suffer from a mental disorder during his or her lifetime (Kessler et al., 1994). And much of the optimism of the 1960s has faded. As it has, psychiatry and its related professions have become uncertain about where to turn next.

This book addresses in part the reasons for psychiatry's uncertainty about where to turn. In it, we emphasize that psychiatry operates without a theory of behavior that can explain both nondisordered and disordered states, that can organize and prioritize its findings, that can provide novel and testable hypotheses about the causes of disorders, that can guide its research, and that can focus clinical interventions. The book builds on a theory of behavior based on evolutionary ideas. This theory serves as a framework for integrating many features of psychiatry's prevailing models into an evolutionary context. The book also considers arguments that are seldom encountered in psychiatry and its related professions. For example, it might seem that the honing process of evolution (natural selection) would have largely eliminated disorders, especially those disorders that by almost any criteria appear to be minimally adaptive (e.g., residual schizophrenia). Yet, to reason this way is to misread the workings of evolution, which do not promise species perfection, only that some genes and traits are more likely to be preserved than others. Moreover, those features that are preserved are often far from ideal, at least from a psychiatric perspective. Diseases and disorders of all types are everywhere apparent in nature, and the vast majority of species (>95%) become extinct. Said another way, it is an error to assume that *Homo sapiens* has been selected to be mentally healthy. Such an assumption is riddled with contradictions, with misinterpretations about how evolution works, and with misunderstandings of its products. This is also a book with which readers may disagree. Nothing surprising here. Dispute has been psychiatry's bedfellow since the 16th century (Hunter and Macalpine, 1963).

Most important, this book has been written to introduce an evolutionary perspective to those who study and are involved in the alleviation of mental suffering. As currently understood, mental health has relatively little to do with the view that traits evolve and that some traits may be adaptive while others are not. Instead, psychiatry and the related professions are deeply immersed in activities such as the definition and characterization of signs and symptoms, the classification of disorders, the identification of proximate causes of behavior (e.g., genes, physiological states, cognitive processes), and the assessment of intervention outcomes. It is not that these activities are unimportant; rather, there is a price to pay when they dominate psychiatric thinking to the exclusion of a theory dealing with why people behave as they do. The price can become enormous if the species that clinicians and investigators try to understand and treat is out of focus.

We turn now to disclaimers. A book that addresses both nondisordered and disordered human behavior will inevitability overlook many important findings and insights developed by others. We apologize for these oversights, and we especially apologize for using ideas the sources of which we have long forgotten. Further, we have not attempted to take into account all of the possibly useful approaches to characterizing, explaining, and treating disorders (e.g., Dubrovsky, 1995). Our aim is to *explore* the implications of evolutionary theory for mental disorders. Some will find this aim grandiose. But note that we have said *explore,* which means that our discussion will leave much to be said and debated: what follows is not the final word on Darwinian psychiatry. Moreover, the book is not primarily a recipe for treatment. While there are numerous implications for treatment throughout the book, including a chapter that discusses an evolutionary approach to interventions, developing a framework for understanding and explaining mental disorders is the book's first order of business.

As to references, we have taken the approach of providing one or two relevant references for key findings, not a comprehensive set. This point applies particularly to findings in studies of disorder-physiology relationships. Many of these findings will be outdated by the time the book is published. A related issue has to do with psychiatric research findings that are not predicted by the views developed here: Should these research findings be cited? Our answer is yes. When we are aware of findings that differ from those predicted by evolutionary theory, we discuss them. Identifying such findings turns out to be a more difficult task than might be imagined, however. For example, is a research finding such as "Alzheimer's disease is associated with the presence of amyloid plaques in XX% of the cases thus far autopsied" predicted by evolutionary theory? To our knowledge, the presence of plaques has not been predicted. Yet, it doesn't follow that the presence of plaques refutes evolutionary explanations. Rather, the question remains unanswered largely because it has not been addressed. The point is that evolutionary theory is only *now* turning its focus on mental illness. Thus, its explanations are far from comprehensive, the limits of its explanations remain to be explored, and in most instances, the prevailing explanations of disorders are not developed in ways that either support or refute evolutionary explanations.

Organization of the Book

The book has four parts. Part I (chapters 1 and 2) sets the stage for subsequent parts by providing an overview of the approach we have taken (chapter 1) and an analysis of how psychiatry currently approaches the diagnosis of a disorder and its explanation (chapter 2). Several conclusions emerge, the most important being that psychiatry's current practices of diagnosing and explaining disorders are limited in scope, insufficient for the tasks psychiatry faces, and begging for revision.

Part II (chapters 3 to 6) begins with a review of evolutionary concepts important to psychiatry (chapter 3). It then turns to a theory of behavior and a discussion of function (chapter 4). Chapters 5 and 6 elaborate details of the theory and focus on mechanisms, emotions, moods, symptoms, affects, and information processing. Again, several conclusions emerge. Evolutionary theory offers several potential remedies for psychiatry's current chaos by providing a theory of behavior; specifying the kinds of data that need to be collected to explain disorders; introducing novel, yet powerful and testable, causal hypotheses; providing a theoretical framework that permits the consideration of some disorders, as well as features of disorders, as adaptive; and incorporating and integrating features of the prevailing-model hypotheses.

Part III (chapters 7 to 14) focuses on specific disorders. Chapter 7 addresses sexual selection, the 15% principle, alternative disorder classifications, and evolutionary models of depression. Chapter 8 offers a review of regulation-dysregulation theory. Chapters 9 to 14 interpret personality conditions, anorexia nervosa, schizophrenia, phobias, and other conditions from an evolutionary perspective.

Part IV (chapter 15) takes findings and concepts from parts I to III and applies them to intervention strategies.

Part V (chapter 16) summarizes key points from the preceding chapters and asks what psychiatry needs to do to integrate itself within evolutionary biology.

As the book unfolds, readers might wonder why we have not devoted large sections to genetics and basic evolutionary biology or used evolutionary concepts to systematically review disorders as they are currently classified (e.g., learning disorders, mood disorders, somatoform disorders). Our reasons will become clear in subsequent chapters, but briefly, they are as follows: Although there are many highly suggestive gene-disorder relationships, very few disorders can be compellingly tied to *known* genetic information, and many of those that can be are discussed; reviewing basic evolutionary theory, while important, would require several volumes, and others have already published such reviews (e.g., Wilson, 1975); and systematically interpreting disorders from an evolutionary perspective by using psychiatry's current taxonomic categories as a guide is inconsistent with key ideas and findings discussed throughout the book. To put the latter point another way, we might have taken two very different approaches in writing this book. One is to accept psychiatry's current classification system and explanatory models and attempt to accommodate evolutionary reasoning and data *within* these frameworks. This approach would give conceptual priority to the prevailing explanatory models and secondary priority to evolutionary models. This approach has been tried for at least three decades without much success; that is, it has failed to convince psychiatry of the value of evolutionary thinking. The second approach is to reverse the relationship between evolutionary theory and the prevailing-model explanations and use evolutionary theory as the basic conceptual framework both for explaining disorders and for evaluating the explanatory relevance of the prevailing models. This is the approach we have taken. For several reasons, it is the riskier approach: It assumes that evolutionary theory, not prevailing theories, is the basic model within which disorders are optimally explained; it assumes that the prevailing models will have greater explanatory power if they are incorporated and amplified by evolutionary concepts and findings; and it asks readers to think about familiar material in novel ways.

Once we had chosen the second approach, a number of strategic problems arose, for example: What vocabulary should be used, that most familiar to evolutionary biologists or to clinicians? Because the book is aimed at clinicians, we have opted for the vocabulary of clinical psychiatry and psychology and, when relevant, the translation of evolutionary concepts into this idiom. Another problem concerned what disorders should be discussed. We have elected to discuss a wide variety of disorders, partly because some support and partly because some raise questions about the framework we develop. A technical point with which we struggled concerned whether to capitalize the names of disorders that appear in various editions of the *Diagnostic and Statistical Manual of Mental Disorders* (*DSM;* American Psychiatric Association, 1980, 1987, 1994). With the exception of direct quotations from one of the editions of the *DSM*, we have *not* capitalized disorder names. This decision reflects our uneasiness about the validity of psychiatry's taxonomic categories, a topic that we address in detail in chapter 2.

Finally, a point about what is new and what is not new in evolutionary explanations. Individuals often respond to evolutionary explanations with comments such as "What's new about that?" Such responses are not surprising. If the influence of evolution is as pervasive as we will argue, persons should be aware of many of its effects. However, explanations of the same behavior frequently differ among nonevolutionists and evolutionists. For example, the ubiquitous presence of an easily recognizable trait, such as mother-infant bonding, has both evolutionary and nonevolutionary explanations: "Mothers bond with their offspring because they are predisposed to act in ways that increase the probability of the replication of their genes" is an example of an evolutionary explanation. "Mothers bond with their offspring because that's the natural thing to do" is an example of a nonevolutionary explanation. Both explanations begin from the same set of observations, that is, that mothers and infants bond. Mothers who believe that they bond "because it's the natural thing to do" often have little trouble accepting evolutionary explanations of their behavior, and when they do, they sometimes ask, "What's new about that?" What's new is that the evolutionary explanation of why mothers bond is only one example of the more general evolutionary proposition that traits have been selected because they either directly or indirectly enhance genetic replication in subsequent generations. Saying that bonding is "the natural thing to do," while perhaps descriptively correct, not only is far less informative and amenable to both theoretical analysis and empirical testing but also offers minimal information about possible evolved systems that may be responsible for the behavior. In our view, identifying, understanding, and explaining both the evolved systems and their behavioral manifestations are steps essential to progress in psychiatry. As we go about this task, from time to time expect to ask, "What's new about that?"

Acknowledgments

In a variety of ways, the following people and organizations have contributed to this book: Kent Bailey, John Beahrs, Nicholas Blurton Jones, Nancy Brown, William

Charlesworth, Toby-Ann Cronin, Lynn A. Fairbanks, Michael Fisher, Jennifer Levitt, Kevin MacDonald, Roger Masters, Giles Mead, Detlev Ploog, Michael Raleigh, David Torigoe, Arthur Yuwiler, Louis Jolyon West, students who read and criticized an earlier version of the book, several anonymous reviewers, and the Human Behavior and Evolution Society. Paul Gilbert, Kevin MacDonald, Marsden McGuire, Randolph Nesse, Tyge Schelde, and Robert Trivers were especially helpful in their critiques, and much of the book reflects points they suggested. We deeply appreciate their help. In addition, support has been provided by the Giles W. and Elise G. Mead Foundation, Harry Frank Guggenheim Foundation, The Gruter Institute for Law and Behavioral Research, the University of California at Los Angeles, the Veterans Administration, and the Oxford University Press. Again, our thanks.

Los Angeles, California   M. M.
Rome, Italy   A. T.
October 1997

Contents

IV. Treatment in Evolutionary Context

V. Conclusion

PART I

INTRODUCTION

1

Darwinian Psychiatry:
The Context

Mrs. M

Mrs. M arrived for her first appointment late in the afternoon. She began slowly: "I am Mrs. M....Dr. X referred me....I think my problem is anxiety....It's there all day....At times I can't stand it....My body—my whole body—it's everywhere....I seem to jump at every little thing....I'm too sensitive....I hate everyone—it's senseless....It won't seem to go away. ...It's like something foreign inside me....I'm not myself....I can't sleep." Fifteen minutes into the appointment she began to cry. "I always cry....It began last year when I lost my job....At times I scream uncontrollably....I've become impossible to live with....My son must hate me....I sometimes want to die....I need help....Dr. X said you would prescribe a drug."

Mrs. M was 37 years old, divorced, and the mother of a teenage son. Her mother died when Mrs. M was 3. By Mrs. M's own account, her emotional problems began soon thereafter and centered on her relationship with her father, whom, from her earliest memories, she viewed as excessively controlling and insensitive about her mother's death. From age 5 to age 16, she and her father fought almost daily over Mrs. M's behavior. At age 12, she first noticed feelings of inferiority. A year of psychotherapy at age 16 failed to change her feelings about herself or to improve her relationship with her father. An older brother, whom she liked, and an older sister, whom she disliked, viewed their father as decent, caring, and responsible. Following her departure for college, Mrs. M maintained minimal contact with her family.

Mrs. M performed well in college and was graduated at age 21. She met her husband-to-be the same year. They were married when she was 22. Their only child was born two years later. At age 25, she entered law school, and three years later, she received a law degree. She divorced her husband when she was 35 after discovering that he was having an active sexual relationship with one of her "close friends." Her anger toward her ex-husband and her former friend had not subsided.

Until nine months before seeking psychiatric help, Mrs. M had been employed by a large law firm. A downsizing of the firm had left her without a job. Within weeks, she had begun to experience periods of intense worry, anxiety, and anger. She had stopped socializing and refused to visit with even her closest friends. Unprovoked outbursts of anger toward her son and periods of uncontrollable crying had followed. Three members of Mrs. M's family, including her sister, suffered from symptoms of anxiety and depression, but all to a lesser degree than Mrs. M.

Mrs. M's clinical profile is not atypical of a woman in her mid-30s who for the first time experiences debilitating emotional, cognitive, and motivational symptoms. Nor was her treatment atypical of the times: During the 1980s (as well as currently), the use of drugs to reduce symptoms such as anxiety and depression was the preferred type of intervention among many psychiatrists. Following several weeks of medication, her symptoms began to decline, and at three months, her worrying, crying, and anxiety had largely disappeared. She began seeing close friends and socializing, but not with the pleasure of previous years. Although less frequent, her angry outbursts toward her son continued.

Evaluating Mrs. M

How do clinicians evaluate persons like Mrs. M? Descriptive questions come first: In what ways is Mrs. M suffering? How long has she suffered? Is there a prior history of anxiety, anger, and depression? What situations increase and decrease her suffering? Do her symptoms compromise her ability to function? Were there precipitating events? What is her current mental state? Have other members of her family suffered in similar ways? Does Mrs. M suffer from any medical illness? Has she been treated for her symptoms and, if so, what was the outcome? What are her current living conditions? Are there others who will assist her? Is she psychologically sophisticated? Is she likely to comply with treatment recommendations? The list of questions is long, and it differs from clinician to clinician and patient to patient.

For the moment, we will assume that Mrs. M has a diagnosable disorder. Once a diagnosis is made, or even if it is deferred, clinicians turn to possible causes: How is Mrs. M's suffering best explained?

Because psychiatry's prevailing causal models differ, its explanations of disorders also differ. Biomedical models emphasize genetic and physiological explanations. Mrs. M may be genetically predisposed to anxiety or depression, or both. Her predispositions may manifest themselves in the dysfunction of one or more physiological systems, which, in turn, may precipitate her signs and symptoms. The history of family members with similar symptoms is consistent with this interpretation, as is, perhaps, her positive response to antianxiety and antidepression medications. Biomedical models do not preclude the possibility that social or psychological events can contribute to dysfunctional physiological states, but the presence of such events is not required, nor are they usually thought to be the primary causes of disorders.

Psychoanalytic models view intrapsychic conflicts and distortions as the basis of disorders. The premature death of Mrs. M's mother, her frequent fights with her father, and her dislike of her sister are possible contributing factors to her conflicts and her response to life events. Intrapsychic conflicts or distortions may explain her lingering anger toward her husband and her former close friend, her response to the loss of her job, her outbursts at her son, and her symptoms. Psychoanalytic models do not preclude the possibility that persons may be genetically predisposed to disorders or that dysfunctional physiological systems are present when persons suffer from severe symptoms, but in most instances, these possibilities are viewed as secondary, not primary. Psychological events are antecedent to physiological changes.

Behavioral explanations are built on the view that persons suffering from disorders have learned the wrong things or they have not fully learned what they need to learn in order to cope adequately. Mrs. M's response to her mother's death and her preoccupations with her father's behavior may have led to a failure to learn that effective socialization requires tolerance and cooperation; that maturation requires the development of psychological techniques for dealing with both successes and failures; and that there are methods other than anger for resolving social conflicts. Mrs. M's responses to stressful and frustrating situations, her difficulty in adjusting to the loss of her job, her withdrawal from her social milieu, and her angry outbursts at her son are consistent with behavioral explanations. Behavioral models of disorders downplay the possibility that genetic predispositions, intrapsychic conflicts, or physiological change contribute to disorders. Atypical learning is primary.

Sociocultural explanations look to the environment for the precipitants of disorders. Interpersonal stress, social and economic discrimination, poverty, and social isolation are viewed as the mediators of signs and symptoms. In the sociocultural model, persons like Mrs. M are victims of adverse circumstances. The loss of her mother, the behavior of her father and her husband, her divorce, the loss of her job, and the responsibilities of being a single parent with a teenage son are all possible adverse factors that may have contributed to her condition. Mrs. M may have predispositions to certain disorders; physiological changes, dysfunctional learning, or intrapsychic conflicts may develop in response to adverse events; but as in the psychoanalytic and behavioral models, they are not considered primary.

Psychiatry's Causal Explanations

What is to be made of the many possible explanations of Mrs. M's disorder? The four models discussed above (biomedical, psychoanalytic, behavioral, and sociocultural) are currently the models clinicians most frequently use in explaining disorders. Yet these are only the most prominent among a far larger group of models that view disorders as, for example, consequences of inadequate nutrition (e.g., vitamin deficiency), viral infections, or the absence of will—a shortage of causal hypotheses has never been psychiatry's Achilles' heel. Moreover, for the vast majority of disorders, it is unclear whether one explanation is more valid than another, or whether seemingly different explanations simply describe different features of the same disorder. What *is* clear is that clinicians spend considerable time and energy debating both the supposed virtues of the causal models they prefer and the shortcomings of those they reject.

Debates about causes can be productive or unproductive, and for psychiatry and its related professions, both outcomes are apparent. On the productive side, they have required advocates of models to be more specific about their hypotheses, to specify when and how their explanations apply, and to muster supporting evidence. It is the unproductive debates and their consequences that are of the most immediate concern, however, for they are a major contributor to psychiatry's current state of confusion, its conceptual pluralism, its far from successful attempts either to disconfirm or to integrate its prevailing models, and its misconceptions about the value of new knowledge and new research techniques. These consequences deserve a closer look.

Conceptual Pluralism

Conceptual pluralism is the most prominent of psychiatry's current trademarks. It is present when more than one model is used to explain similar clinical or experimental findings. It has been defended by the argument that multiple models are an unavoidable by-product of the early stages of a science and that, at some future date, a unified model, or at least a few basic models, will prevail. From this perspective, pluralism is a sign of scientific health.

Should psychiatry's conceptual pluralism be viewed as a sign of health? A cursory look at its history invites skepticism. Models of mental disorders come and go, and they have done so for centuries (Hunter and Macalpine, 1963). The present century has witnessed the rise and partial fall of the psychoanalytic, sociocultural, and behavioral models of disorders; the rise of the biomedical model; and more recently, the rise and questionable relevance of general systems theory (Bertalanffy, 1974; Grinker, 1975; Marmor, 1983) and chaos theory (e.g., Mandell, 1982; Ehlers, 1995; Gottschalk, Bauer, and Whybrow, 1995). How long these models will remain prominent is uncertain. For example, there are growing indications that the biomedical model, currently psychiatry's most influential model, can explain only a limited number of disorder features (Maas and Katz, 1992). Chaos theory has yet to demonstrate that it can develop more compelling explanations of disorders than those offered by prevailing models. And general systems theory has not established that it can meaningfully integrate different models or their key elements. Viewed dispassionately, each of psychiatry's four prevailing models has fallen short of explaining critical details of disordered behavior. Yet models endure, and along the way, they gather supporters who offer different explanations of disorders, engage in different types of treatment, edit and publish their own journals, and attempt to convince the uninitiated of the superiority of their explanations (McGuire, 1978). It's unlikely that everyone can be right. More important, today's psychiatry is not in a position to decide.

Conceptual pluralism is plagued by yet other problems. Two of the most important are the accumulation of low-utility data and the absence of agreed-upon methods for testing hypotheses (Colby and McGuire, 1981). On the subject of low-utility data, physiological and psychological measures, self-reports of childhood and adult experiences, family pedigrees, patients' fantasies, signs and symptoms, the weather, others' behavior, and diet are only a few examples of findings that have found a special place in psychiatry's database. Once there, they are difficult to dislodge. More than two and a half decades ago, Kuhn (1970) identified a major consequence of conceptual pluralism, and in doing so, he may well have had psychiatry in mind: "In the absence of a paradigm . . . all the facts that could possibly pertain to the development of a given science are likely to seem equally relevant" (p. 15). As to agreed-upon methods for testing hypotheses, one must look to the future. To be sure, today's investigators use hypothesis-testing methods in both clinical and laboratory settings. But this does not mean that clinicians or investigators agree on which methods are most appropriate for answering the same question, let alone which questions are most in need of answers.

It might be supposed that much of the explanatory confusion that accompanies conceptual pluralism would disappear if it could be agreed that there are "multiple-psychiatries," that is, that there are multiple causal factors and each factor explains its own set of data. Different causal explanations could then be yoked to specific sets of findings (McHugh and Slavney, 1986). On first reading, the idea is appealing. But a closer look invites concern. The multiple-psychiatries approach is likely to do little more than refocus existing debates on such topics as which data belong to which data set, and which causal explanations best explain which sets of data. A new form of pluralism would be upon us. For example, applied to Mrs. M, the findings of anxiety, anger, and depression, coupled with a statistically atypical physiological measure (e.g., low CNS serotonin activity), does not disqualify other possible explanations of her condition. Her atypical physiological state could be the primary cause of her disorder; it could be a secondary consequence of events explained by behavioral, sociocultural, or psychoanalytic models; it could be noise; it could be a response to changes occurring in other neurochemical systems; or it could be a harbinger of clinical improvement.

Escaping from conceptual pluralism requires at least two conditions: the presence of a theory from which hypotheses can be deduced, and agreed-upon methods for hypothesis disconfirmation (Popper, 1969). Put another way, without a theory with deductive and testable properties as well as the use of methodologies that facilitate the testing and rejection of hypotheses, more data will be collected, more causal explanations will be generated, more methodological debates will occur, and a resolution to psychiatry's conceptual confusion will remain a distant goal.

Failure of Model Integration

Attempts to integrate models and findings (e.g., systems theory) have their own special histories. Many of the reasons for these histories are clear. For example, psychiatry's explanatory models are cast within different metaphysical systems, each of which is built on different assumptions and uses different explanatory logics. The biomedical model is built on mechanistic assumptions and applies mechanistic logic. Psychoanalytic models are built on formistic, mechanistic, and organismic assumptions and apply their respective logics (McGuire, 1979b; Beahrs, 1986). Depending on which system of assumptions and logic one selects, the same datum can be explained differently (Pepper, 1942). Further, different theories of truth are associated with each metaphysical system. This means that there are different rules for deciding if an explanation is valid (Pepper, 1942). When they are applied to prevailing models, it is not difficult to predict the consequences of the preceding points: models will remain separate; causal events may be obscured; and efforts to refute or integrate models will get bogged down in the rhetoric of professional politics, funding opportunities, and philosophical debates.

New Knowledge and New Research Techniques

New research techniques leading to knowledge that will finally clarify which explanatory model(s) psychiatry should adopt have been advocated as a cure for conceptual pluralism. Psychiatry's current enthusiasm over molecular biology provides a convenient example. Here again, caution is advisable. Even if one assumes that the primary cause of disorders is to be found in genes, the steps between gene and phenotype are many: gene → protein → enzyme → interactions among enzymes → neuroproteins → interactions among neuroproteins → behavior. Each step may be influenced by a variety of factors (e.g., intrauterine environment, early experiences) over which genes and their products exert varying but far-from-complete control. Thus, for example, it comes as no surprise that the adult phenotypes of two persons who begin life with the same genetic makeup (monozygotic twins) may differ.

Applied to disorders, the preceding points mean that genetic profiles may offer only a partial explanation of disorder causes and phenotypes. Yet advocates of the molecular approach often suggest otherwise (e.g., Gershon et al., 1987, 1990; Baron, 1991, 1994a, 1994b; cf. Harris and Schaffner, 1992). Such suggestions are not without precedent. Literally hundreds of medical diseases have well-established genetic causes. For example, Type-1 diabetes (insulin-dependent diabetes) is known to be influenced by genetic information. However, for mental disorders, findings of disorder-related DNA profiles are more the exception than the rule. Moreover, even for diseases like Type-1 diabetes, both the age at which clinical features become apparent and the clinical features themselves differ across persons. Usually, there is an abrupt onset during the first two decades of life, but the disease may develop as late as age 40; in some instances, the disease is associated with early death, while in other instances it is not. Such differences may or may not be genetically influenced.

Still, it is likely that clear genotype-phenotype relationships will emerge for some disorders. This point can be granted, but it should be put in perspective: As the majority of disorders that clinicians encounter are brought into focus, the hope that clear genotype-phenotype relationships will emerge becomes little more than wishful thinking. Further, many currently classified disorders are likely to have statistically normal molecular and physiological profiles. The fact that most of the physiological measures of persons with disorders are within normal limits (usually one standard deviation from the mean) suggests as much. The opposite possibility holds as well: Some statistically atypical molecular and physiological profiles may be associated with normal phenotypes. We could discuss here the fact that much of the current confusion dealing with molecular and physiological interpretations of clinical findings is a consequence of how disorders are classified. However, we will wait until chapter 2. For the present, what needs to be underscored is that findings from molecular biology and other new research techniques, such as positron emission tomography (PET) and autoradiography, while they will inform our understanding of disorders, are likely to fall short of satisfactorily explaining the majority of the features of the majority of disorders.

The thrust of the preceding discussion is not that psychiatrists should stop debating their causal hypotheses. Nor should they discontinue their attempts to integrate models or forgo the use of new research techniques. These efforts are as valuable as they are inevitable. Rather, the point is that psychiatry has spent much of its time and energy in activities other than coming to terms with its need to embrace a theoretical framework that can facilitate the selection, organization, prioritization, and explanation of its findings, as well as direct its research. As a result, psychiatry lacks an *integrated* set of concepts that can facilitate the development of testable hypotheses, address issues of causal sequence and feedback, and foster studies that will purge low-utility data. Thus, much of the time and effort devoted to diagnosing, studying, and explaining disorders turns out to be unproductive.

What This Book Is About

This book is about evolutionary theory and how it can inform psychiatry. It is a book that, in the words of one author, is devoted to exploring "Darwin's dangerous idea" (Dennett, 1995) and its implications for clarifying our understanding of disorders. It is a book about an alternative framework for thinking about, investigating, and treating disorders. We will argue that evolutionary theory is positioned to explain many behaviors and features of disorders that are of key importance in psychiatry. We will also argue that the theory offers a framework within which causal hypotheses can be developed and many of the features of psychiatry's prevailing causal models can be integrated. Many symptoms (e.g., anxiety, depression, and delusions), signs (e.g., social withdrawal and deception), and disorders classified by the prevailing models not only turn out to be more understandable, but also take on a different character when they are explained in evolutionary context. Furthermore, developing interventions in an evolutionary context broadens treatment options through the recognition of the often adaptive utility of states and traits normally viewed as signs, symptoms, or undesirable personality features. In essence, evolutionary theory provides a framework that psychiatry might well use to exit from conceptual pluralism, to purge low-utility data, to facilitate model integration, and deflate its belief that findings from new research techniques will finally explain disorders.

Central to the preceding claims is the fact that psychiatry lacks a theory of behavior, that is, a theory that explains why people behave as they do. The consequences of this are not easily dismissed given both psychiatry's interest in the causes of disorders and its attempts to alter behavior through interventions. Evolutionary theory is *now* giving rise to a theory of behavior, and it is that theory on which this book builds, attempting to construct an answer to the question: Why do people behave as they do in both ordered and disordered states?

At a more detailed level, evolutionary theory is, in part, a theory of motivations; in part, a theory of the systems responsible for information processing and initiating behavior (infrastructures); in part, a theory of capacities and their uses; in part, a theory of function; and in part, a theory of environment-behavior-gene interactions. It is also, in part, a theory of traits, trait variation, and the cross-generational influence of genetic information on traits. Studies of traits leave little doubt that a host of socially important behaviors—those associated with reproduction, personal survival, social navigation, information processing, and the investment of time and energy in both kin and nonkin—are genetically influenced. Trait predispositions range from strong (e.g., withdrawal from a painful stimulus) to weak (e.g., color preference). The genetic mixing that occurs at each conception, when combined with such factors as the trait-refining effects of different upbringing environments, ensures that the same trait will differ across individuals (within-trait variation); for example, some persons empathize easily, while others do so only with great difficulty. The same point applies to features of physiology, such as CNS neurotransmitter baseline activity. Some persons have low—and others have high—baseline activities.

One point about traits deserves special emphasis. Like many others before us (e.g., Jung, 1972; Rowan, 1990; Coon, 1992), we view individuals as mosaics of independent and semi-independent traits (e.g., capacities) that, in different combinations, combine in carrying out specific behaviors. A good analogy is returning a tennis shot. A well-hit shot requires one to track a moving object, to determine where to locate oneself in space and time to return the opponent's shot, to plan a return shot, to translate one's planned shot into behavior, and so forth. Within-trait variation may apply to any of these capacities; for example, one may be accomplished at tracking a moving object but may have only an average capacity to execute a planned shot, in which case the shot is unlikely to be well hit. At first glance, the mosaic characterization may seem to do little more than stir already murky waters, largely, we suspect, because we often view others holistically. Nonetheless, not only does evidence strongly support the mosaic view, but the view also opens the door to a host of new insights into the possible causes of disorders.

There are, of course, many phenotypic features that invite nontrait explanations. Values, aesthetic and ideological preferences, special knowledge of the environment, and the language one speaks are examples. Nontrait features are sometimes explained by theories that assume that the brain is uninfluenced by past evolutionary events. At times, sociological and psychological theories have built on this assumption. Yet such explanations are not compelling. The idea that the brain is not a product of its history is as untenable as it is uninteresting. A more prudent view is that *Homo sapiens* has evolved capacities for learning, for refining the use of information, and for organizing and executing behavior, and that these capacities were selected in the past because they helped solve past adaptive problems. But this is not to say that what one learns, how one adjusts, and degree of behavioral plasticity are uninfluenced by experience. Far from it. For example, the profound grief and subsequent effects that result from the death of a child may never be experienced unless the event actually happens. At times, the influence of experience is more important, and at times less important, than the influence of evolved traits. These points will be addressed in subsequent chapters. For the present, the point to emphasize is that an in-depth understanding of behavior and the systems responsible for behavior requires that traits, within-trait variation (e.g., physiological stability), learning, and social context all are a part of the explanatory equation. Only then will reductionistic explanations be avoided.

An example involving social context and social signals illustrates the importance of taking a nonreductionistic approach to behavior. Up and down the animal kingdom, animals of the same and different species signal one another, and they do so for a variety of reasons. Birds signal their interest in mates and threaten intruders that enter their territories. Nonhuman primates solicit grooming from conspecifics and threaten one another in dominance competitions. And so forth. In principle, humans with and without disorders are no different. Individuals signal their wants, their likes and dislikes, and their status, and such signals often alter others' behavior. A key feature of others' signals often goes unnoticed, however: Signals may have a significant influence on the receivers' physiological systems—ranging from the release of peripheral hormones to the release of CNS neurotransmitters—and, in turn, on their behavior. It follows that attempts to explain behavior outside the social context in which it occurs require settling for less than a full understanding of such behavior.

Taken together, the preceding points suggest a multidimensional picture. In one dimension, there are predisposed or genetically-influenced traits. In a second dimension, there are cross-person differences in the strength of trait predispositions. In a third dimension, there are ontogenetic contingencies that influence the expression, refinement, plasticity, and use of traits. In a fourth dimension, there are social and physical environments that have their own influences, options, and constraints. And in a fifth dimension, there is the behavior of others. Embedded in this dimensionality are both ordered and disordered behavior. In our view, the only available theory capable of giving meaning to and facilitating the integration of this dimensionality is evolutionary theory.

Our focus on evolutionary theory asks readers to step back from their currently preferred models of disorders and to view *Homo sapiens* in an evolutionary context. Biomedically oriented clinicians concern themselves primarily with physiological mechanisms and possible disorder-predisposing genes; psychoanalytic clinicians concern themselves primarily with intrapsychic distortions and conflicts; and so forth. Often, both the complexity and the consequences of multidimensionality are put aside in favor of single-cause explanations (e.g., Engle, 1980). In our assessment, not only are prevailing model explanations of disorders too narrow, but they need to be integrated into a broader theoretical framework if their utility is to be optimized (e.g., Buss, 1995a, 1995b). The infusion of evolutionary thinking into psychiatry calls for a more expansive strategy of inquiry than is currently fashionable, as well as a willingness to consider explanations that go beyond what has been demonstrated empirically.

When one looks to evolutionary biology for explanations of disorders, one's interest quickly turns to those parts of evolutionary theory that have a central role in the explanation of behavior. Traits, trait variation, learning, and the social environment have already been mentioned. Ultimate and proximate causation, sexual selection, kin selection, reciprocal altruism, ontogeny, the information-processing characteristics of infrastructures, and function can be added to the list. In treating Mrs. M, a clinician with an evolutionary orientation not only would ask many of the questions listed at the beginning of this chapter but would also assess Mrs. M's motivational priorities, her interactions with kin, the functionality of her information-processing systems, and her capacity to engage in behaviors integral to social navigation, such as reading others' behavior rules, developing and carrying out novel behavior strategies, and monitoring her own behavior.

We will assume that readers are familiar with many of the basic concepts of evolutionary biology. Thus, our discussions of theory will be limited to those parts of the theory most relevant to psychiatry. As noted in the preface, much of evolutionary theory is common knowledge. Thus, some of what is said will amount to little more than a rephrasing of what is already known. More comprehensive and general discussions of the theory and its empirical basis can be found elsewhere (e.g., Williams, 1966; Tiger, 1969; Tiger and Fox, 1971; Wilson, 1975; Dawkins, 1976, 1987; Daly and Wilson, 1978; Alexander, 1979; Symons, 1979; Barash, 1982; Trivers, 1985; Ploog, 1992; Nesse and Williams, 1994). Books by White (1974), McGuire and Fairbanks (1977), Wenegrat (1984, 1990), Marks (1987), Gilbert (1992), and Stevens and

Price (1996) focus primarily on psychiatric disorders and address many of the points discussed in the following pages.

At this stage, we should add what this book is *not* about. It is not a defense of evolutionary theory. The theory is taken in its generally accepted form. Like all theories, it has limitations. There are features of the theory that are disputed, and the theory continually undergoes refinement. Disputed features that are important in the context of this book will be discussed, however. This book is also not a review of other interesting and potentially useful theories that might help explain features of disorders, for example, semantic theory. Nor is it a review of psychiatry's current state of knowledge. Thus, much of the material found in standard psychiatric textbooks will not be repeated. On the other hand, material seldom found in these textbooks will take on importance.

For us, psychiatry's most pressing need is to embrace evolutionary theory and to begin the process of identifying its most important data and of testing novel explanations of disorders. Attempts to explain behavior, normal or otherwise, without having an in-depth understanding of the species one is studying invite misinterpretation.

Finally, to get an overview of where we are headed, it may be helpful to read Chapter 16, which summarizes the main points of the book.

2

Diagnosing and Explaining
Mental Conditions

In the preceding chapter, both the questions asked of Mrs. M and her answers serve as the building blocks of clinical evaluations. While clinicians differ in how they carry out their evaluations, most have the same aims in mind: to diagnose, explain, and treat disorders. Before turning to the details of Darwinian psychiatry, it will be helpful to take a closer look at these aims and how they shape psychiatry's thinking. Following some introductory comments about medical and psychiatric models, diagnostic and explanatory practices are discussed. (Treatment is the topic of chapter 15.) The chapter then turns to an analysis of two explanatory systems and asks if they can accommodate psychiatry's prevailing causal models.

This chapter also addresses two key points reexamined in later chapters: why an evolutionary-based theory of behavior holds the promise of accelerating our understanding of mental conditions and improving intervention effectiveness, and what the central place of behavior in classification, explanation, and treatment is. We will use the terms *disorder* and *condition* differently: *disorder* when we are referring to disorders specifically as they are described in the fourth edition of the American Psychiatric Association's *Diagnostic and Statistical Manual of Mental Disorders* (*DSM-IV*, 1994) and *condition,* in a more general sense, when we are referring to *DSM-IV* disorders, syndromes, atypical behavior, or isolated signs or symptoms. The terms *signs* and *symptoms* are used as they are in the psychiatric literature.

Perhaps the best place to start our inquiry into diagnostic and explanatory practices is with the medical approach to disease. Over the past three centuries, medicine has moved through four stages in its efforts to diagnose and explain diseases: from (1) the recognition of signs and symptoms to (2) definitions of syndromes to (3) the identification of tissue pathology to (4) the development of hypotheses about the causes of tissue pathology and associated phenotypes. The process begins with clinical observations and descriptions. It ends with postulates about the internal and external causes of diseases.

Are the same four stages applicable to mental disorders, or, said another way, can disorders be understood through the medical model? An answer to this question depends in part on how disorders are defined. As most readers will be aware, there is no generally accepted definition. For the present, we will not dwell on this point but will settle for the definition used in *DSM-IV* (American Psychiatric Association [APA], 1994):

> Each of the mental disorders is conceptualized as a clinically significant behavioral or psychological syndrome or pattern that occurs in an individual and that is associated with present distress (e.g., a painful symptom) or with a significantly increased risk of suffering death, pain, disability, or an important loss of freedom. (p. xxi)

Given this definition, an answer to the question above hinges on two points: which disorder is being discussed and which of psychiatry's prevailing models is used to explain the disorder. For example, consider biomedical explanations for the class of disorders referred to as *dementias*: Syndromes have been recognized and defined; there are often identifiable types of tissue pathology; and specific types of tissue pathology are associated with characteristic phenotypes. Thus, a reasonable case can be made that all four stages in the medical model apply. But what about the use of the biomedical model to explain the majority of disorders in which there is no evidence of tissue pathology? The first two stages (the recognition of signs and symptoms and the definition of syndromes) are present. The third stage (the identification of tissue pathology) is absent. And the fourth stage (hypotheses about the causes of tissue pathology) consists of postulates about dysfunctional genetic, anatomic, or physiological systems that may or may not apply.

When the same question is asked of the psychoanalytic, behavioral, and sociocultural models, different answers emerge. For those disorders in which evidence clearly points to either tissue pathology (e.g., substance abuse) or disorder-related genetic information (atypical DNA profiles) as critical causal factors, causal explanations are seldom offered, although contributing factors (e.g., why a person might abuse substances) may be considered. For disorders in which evidence of tissue pathology is absent, the first two stages apply (the recognition of signs and symptoms and the definition of syndromes); the third (the identification of tissue pathology) is not relevant; and the fourth (hypothesis development) consists of process hypotheses (e.g., intrapsychic distortions, dysfunctional learning) that do not require the assumption of tissue pathology or atypical DNA profiles.

The fact that the biomedical model is the one psychiatric model that achieves a close fit with the medical model might be taken as an indication that psychiatry is adopting the medical approach to mental disorders and leaving behind some of its seemingly low-utility explanatory models. More is involved, however. There is compelling evidence that some disorders are due to dysfunctional learning; that adverse upbringing environments can disrupt maturation; and that the information systems which are responsible for thinking, emotions, and behavior can be in conflict. None of these findings presupposes tissue pathology or atypical DNA profile explanations.

Moreover, they raise questions such as: Is the biomedical model applicable to only a small percentage of currently classified disorders?

Apart from the points above, what is meant by the term *model* is often unclear. In its normal usage, the term implies specificity. Yet models are usually less specific than might be assumed. In part, the lack of specificity is a consequence of factors influencing model development and choice. One such factor deals with psychiatry's boundaries, which remain poorly defined despite a century of serious effort at self-definition. These boundaries overlap with those of the disciplines of genetics, virology, physiology, psychology, biochemistry, and cultural anthropology. The positive side of overlap is that it fosters cross-fertilization and facilitates external checks on causal hypotheses. The negative side is that it contributes to territorial disputes over such matters as what the disorder-defining criteria are, where explanations begin and end, and which discipline's explanations should prevail. A second factor is that new information (some valid, some not) continually alters the makeup of models. A third factor is the tendency of nonpsychiatric medical specialties to take difficult-to-explain conditions and drop them into psychiatry's lap only to reclaim them if tissue pathology or atypical DNA profiles are identified. This is the ownership issue over which medical disciplines often fuss. The upshot is that psychiatry's boundaries remain porous and subject to change. In turn, its models are in a continual state of flux, and what they are attempting to model is often unclear.

Although the preceding points are troublesome, they are potentially resolvable by the adoption of an alternative approach to model development, a point to which we return in subsequent chapters. For the moment, it is more important to take a closer look at psychiatry's prevailing practices of diagnosis and explanation and at the ways in which they contribute to some of the conceptual and interpretive problems under discussion (Kendell, 1975, 1984; Hafner, 1987; Pam, 1990; Yuwiler, 1995).

How Are Conditions Diagnosed?

To diagnose conditions, clinicians usually go through two steps: First, they identify a condition; then, they classify it.

Identifying the Condition

At least four criteria are used in identifying conditions: suffering, statistical variance, tissue pathology, and functional impairment.

Suffering

People seek medical advice because they feel ill. The apparent triviality of this statement hides an idea that is deeply rooted in human experience, namely, that unpleasant experiences, ranging from discomfort and malaise to severe pain, are often signs of diseases and conditions. Still, discomfort is not a necessary requirement for identification. Diseases can be diagnosed when persons are not suffering, as in the early stages of diabetes mellitus, cancer, hypertension, kidney failure, and caries. The same point holds for persons with personality disorders, who often seem unaware that it is they, not others, who are behaving atypically. Further, some forms of epilepsy, grandiose delusions, hypomania, and substance-induced states are associated with pleasurable feelings. The fact that people often feel ill even though they are in good health adds a further twist. Dentition, menstruation, the first trimester of pregnancy, labor, overeating, and overexercise often coexist with suffering, although these states are seldom thought of as diseases or mental conditions.

Moreover, the degree to which persons actually suffer is often unclear. Suffering is private, and devising reliable assessment tools has been difficult. Cross-person differences in the ability to tolerate suffering are likely, as are differences in the degree to which persons reveal their suffering. At times, persons with seemingly mild symptoms, such as intermittent periods of low-intensity anxiety, claim their suffering is unbearable, while persons with severely debilitating disorders (e.g., schizophrenia, conversion reactions) say the opposite. There is also the problem of nonspecific signs and symptoms; for example, low self-esteem, depressed mood, and anxiety are present across a variety of conditions, while only a few signs and symptoms, such as auditory hallucinations and amnesia, are specific to one or a few conditions. Nonspecific signs and symptoms have only moderate diagnostic utility, much as high temperature, muscle pain, sweating, and fatigue are common to many medical diseases and have only moderate diagnostic utility. There is also the fact that clinicians differ in their views about the degree to which human nature, even in its optimal state, is associated with temporary periods of suffering (e.g., pain, malaise, anxiety, depression) that are often reported by persons seeking psychiatric help and are often considered manifestations of conditions.

Where do these observations leave us? Clearly, suffering is frequently used in identifying conditions. In many languages, the word for disease and suffering share the same etymological root. *Maladie* in French and *malattia* in Italian derive from the Latin *male habitus* ("in a bad state"). Equally clearly, when suffering is intense and prolonged, a disease or a condition may be present. Yet, on occasion, suffering is a bellwether of improvement, as when one attempts to discontinue an addiction, tries to constrain an impulse, or experiences the end phase of a time-limited depression. Thus, suffering alone is neither sufficient nor necessary for identifying conditions, and the degree of suffering and the severity of conditions do not consistently correlate.

Statistical Variance

Statistical variance is a second criterion. Behavioral, psychological, physiological, or genetic measures that vary from age- and sex-characteristic norms are often taken as indices of specific conditions. At first glance, statistical assessments appear to require little more than the systematic measurement of specific attributes and their separation into statistically normal and nonnormal categories. But such assessments turn out to be more complicated than is usually imagined. Changes in disorder definition are partly responsible. For example, homosexuality was a disorder in *DSM-II* (APA, 1968), but only ego-dystonic homosexuality was a disorder in *DSM-III* (APA, 1980). The fifth edition of the *Comprehensive Textbook of Psychiatry* (Kaplan and Sadock, 1989a) states that malingering and adult antisocial behavior, which were disorders in *DSM-III-R* (APA, 1987), are "not attributable to a mental disorder" (p. 1396). And

DSM-IV (APA, 1994) has deleted or subsumed into other diagnostic categories a number of disorders listed in *DSM-III,* including overanxious disorder of childhood, avoidant disorder of childhood, and passive-aggressive personality disorder.

Concepts of normality are also critical. For example, highly social behaviors, such as overgenerosity and unexplained optimism, have sometimes been viewed as signs of conditions. (See J. Harris, Birley, and Fulford, 1993, for an amusing discussion of why happiness qualifies as a disorder if the usual statistical criteria are applied.) Similar points can be made about cultural attitudes toward behavior. Intense grieving up to 18 months following the loss of a loved one is viewed as normal in some cultures, while in the United States, 18 months is usually seen as excessive; and social tolerance of, for example, of eccentric behavior, mild to moderate dissociative states, and excessive emotional fluctuation varies across cultures (J. M. Murphy, 1978; Kirmayer, 1989, 1991; Konner, 1989). Often, these behaviors are not viewed as disorders. There are also unresolved questions about the meaning of the term *normality:* Some disorders are thought to be as common as not. This is the case for prevalence estimates in *DSM-IV* (APA, 1994); if they are to be believed, somewhere between 44% and 60% of preadolescents suffer from one or more disorders of development (pp. 37–123). These estimates are consistent with recently published 6-month and lifetime prevalence estimates of mental disorders for adults, which are in the ranges of 20% and 50%, respectively (Kessler et al., 1994). We may say all the above without addressing the still unsolved technical issue of measurement in psychiatry (Snaith, 1991).

A related issue is continuous and discontinuous variation and its use in identifying conditions. *Continuous variation* refers to attributes that can be measured on a continuum, such as mild, moderate, and intense anxiety. *Discontinuous variation* refers to attributes that may or may not be present among persons, such as the presence or absence of a particular enzyme. Different measurement and statistical techniques apply to these two types of variance. Statistical assessments of physiological measures pose yet another type of problem. For example, concentrations of cerebrospinal fluid 5-hydroxyindoleacetic acid (CSF 5-HIAA) are know to be low among some individuals with depression, as well as among some males who are impulsive arsonists (Linnoila et al., 1983). Yet CSF 5-HIAA is also low among some individuals who do not suffer from either condition. Thus, a finding of low CSF 5-HIAA, while a correlate of a number of conditions, is not sufficient to identify a condition. Other information is required, a point illustrated in the following passage from *DSM-IV* (APA, 1994) dealing with the diagnosis of mental retardation:

> Thus, it is possible to diagnose Mental Retardation in individuals with IQs between 70 and 75 who exhibit significant deficits in adaptive behavior. Conversely, Mental Retardation would not be diagnosed in an individual with an IQ lower than 70 if there are no significant deficits or impairments in adaptive functioning. (pp. 39–40)

The utility of statistical assessments for identifying conditions turns out to be similar to the utility of suffering. Statistical criteria are useful. However, they are not foolproof, and they are often applied idiosyncratically. Moreover, in many instances, they do not distinguish between deviations from the norm that are neutral, that may be beneficial, and that are harmful (Kendell, 1975). The *when* and *how* of their use frequently requires additional information, and cultural as well as other factors often determine the importance of such information.

Tissue Pathology

Most clinicians and investigators agree that the identification of tissue pathology (specific anatomic, physiological, or molecular features) is the best criterion for defining a condition. As noted, tissue pathology or atypical DNA profiles turn out to be acceptable identifying criteria provided they can be consistently tied to specific phenotypes. Alzheimer's disease seemingly meets these criteria. Neuritic plaques and neurofibrillary tangles, which are thought to be responsible for many of the signs of this disorder, are observed in postmortem analyses of the brains of persons who have received a diagnosis of Alzheimer's disease. Yet, even here, questions remain, for plaques and tangles are also found in the brains of persons who exhibited no detectable signs of dementia while they were alive.

There is also the issue of reversibility. For example, studies of persons with obsessive-compulsive disorder using positron emission tomography (PET) indicate that atypical glucose metabolism rates in the caudate nucleus can normalize following either drug or behavior therapy (Baxter et al., 1992). Such findings, which are developed further in chapter 8, are consistent with two important clinical observations: A percentage of conditions remit spontaneously, and different treatments often result in similar clinical outcomes. In effect, tissue pathology may not be a cut-and-dried affair. (The possibility of reversibility may underlie attempts by Kraepelinian-oriented clinicians to distinguish between disorders that worsen and those that get better.)

To restate the points in the preceding paragraphs in the context of the earlier discussed medical model of disease, the sequence of (1) the recognition of signs and symptoms, (2) the definition of syndromes, (3) the identification of tissue pathology, and (4) the development hypotheses about tissue pathology or atypical DNA profiles does not uniformly work in reverse. *Tissue pathology or atypical DNA profiles may be present, but disorders may not be, and in some instances, tissue pathology reverses itself.* One can take exception to this view by arguing that for many disorders, it is only a matter of time (e.g., looking in the right place, developing new investigative techniques) before evidence of tissue pathology or atypical DNA profiles is found. This assumption has led some investigators to suggest that at the heart of every psychiatric condition is a specific organic abnormality waiting to be uncovered (e.g., Guze, 1989). But this and similar views are no more than speculations. In addition, other points, such as the degree of fit between tissue pathology, atypical DNA profiles, and phenotypes, require refinement. Even for those disorders in which tissue pathology or atypical DNA profiles are clearly implicated as causal factors, the clinical manifestations of the same disorder often differ significantly across persons of the same age and sex. Thus, multiple factors are likely to influence the timing, form, and severity of conditions. A similar point may be made about postulated physiological contributions to conditions; for example, over time, measures of physiological variables often shift significantly with no associated change in clinical conditions.

The utility of tissue pathology or atypical DNA profiles is similar to the utility of the two preceding condition-identifying categories: Tissue pathology is a useful identifying criterion, but there are limits, and when mental conditions are considered as a whole, there are few convincingly documented one-to-one relationships between tissue pathology or atypical DNA profiles and specific phenotypes.

Functional Impairment

Functional impairment may also serve as a condition-identifying criterion. Impairment measures are frequently used in medicine, and they are often tied to before-and-after comparisons, such as lung capacity before and after developing emphysema. Impairment assessments are applicable to disorders when there are clear differences between before-and-after states, as often occurs when a disorder has an abrupt onset. Acute schizophrenia, panic disorder, and amnesia often qualify. Still, in many disorders, such as most personality disorders, somatoform disorders, and developmental disorders, it is often difficult to distinguish before-and-after states.

The usual way of assessing functional impairment is for a clinician to compare the behavior of an individual with that of age- and sex-matched persons not suffering from a condition. Like the use of statistical variance, this practice introduces an opportunity for idiosyncratic usage: The inability to easily acquire a foreign language or to master differential equations is seldom viewed as a functional deficit, while the inability to read often is. In addition, the usual procedure for evaluating function is more global than detailed. Seldom are the moment-to-moment features of behavior or their specific functions closely examined, even though studies show that moment-to-moment assessments lead to novel findings and favor different explanatory hypotheses from those that currently prevail (e.g., Polsky and McGuire, 1980; Rosen, Mueser, et al., 1981; Rosen, Tureff, et al., 1981; Troisi, Pasini, et al., 1991).

Two Related Issues

We next turn to two issues related to the discussion of diagnosis: one dealing with disorders as natural classes, the other dealing with epiphenomena.

Natural Classes

A natural class is a grouping existing in nature that can be described without reference to other such groupings. Natural classes are relatively easy to establish when different species are the topics of interest. Elephants can be described as elephants without reference to beavers. However, when the natural class concept is applied to variants within the same species, as when disorders such as schizophrenia or bipolar disorder are viewed as examples of natural classes, an interesting set of interpretive issues arises. In part, the interpretive problems are a consequence of the ambiguities associated with disorder classification, a topic to be addressed later in this chapter. For the present, we will assume that classification practices are free of ambiguities and will explore the disorder-identifying implications of the natural class concept.

If disorders can be viewed as examples of natural classes, identification is in principle a straightforward, although perhaps tedious, exercise: Data must be collected and analyzed in ways that facilitate the characterization and measurement of attributes specific to each natural class or disorder. The belief that such characterizations are possible has at times fueled the search for biological markers.

Philosophers more than psychiatrists have concerned themselves with whether disorders make philosophical sense (e.g., Sadler, Wiggins, and Schwartz, 1994). And one of the questions philosophers have asked is whether more than one natural class can exist within the same species. In literally all instances, they have concluded that they cannot (Reznek, 1987). From a philosophical perspective, tissue pathology, statistically atypical behavior, and statistically atypical physiological measures are best viewed as variants of the normal. Unlike the characteristics that distinguish elephants from beavers, within-species deviations are inexorably yoked, either explicitly or implicitly, to concepts of normality. This point applies even to instances of discontinuous variation (e.g., absence of an enzyme).

Similar reasoning applies to DNA profiles, particularly to the view that specific profiles represent biological markers of disorders. There is, first of all, an unusual amount of DNA diversity among humans (Cavalli-Sforza, 1991). Biologists often refer to cross-person DNA differences as *biological markers*. In this usage, only cross-person differences, not separate classes, are implied. Not only is DNA diversity the norm, but it is also so prevalent that with the possible exception of monozygotic twins, all persons are likely to have different DNA profiles. (Pushed to its logical extreme, this point might lead to the view that every person, or every member of every species, should be viewed as his or her own natural class.) Second, DNA differences do not necessarily translate into protein differences. That is, putative biological markers may have no intervening variable or phenotypic consequences. The preceding points may be granted, yet given our current understanding of the gene, it is reasonable to assume that specific DNA profiles are associated with a percentage of disorders. We agree with this assumption, provided the following caveat is included: Even for those disorders in which specific DNA profiles can or will be consistently associated with specific phenotypes, it has been, or will be, essential to match variant DNA profiles to normal profiles and to distinguish between normal variants and disorder-related variants. One consequence of the preceding points is that different types of identification and statistical problems must be addressed: those that apply at the DNA level and those that apply at the phenotype level. Another is that the definition of normality becomes critically important.

Concordance rates for disorders among monozygotic twins pose yet other problems for the natural class concept: Rates rarely exceed 50% even when genetic information is thought to be a significant contributing factor (e.g., to residual schizophrenia). For such disorders, putative DNA markers could qualify as indices only if the methods used for classifying disorders are changed, for example, classifying at the DNA level irrespective of phenotype. The interpretive complications do not stop here, however, for it is likely that some persons classified as having the same disorder will have DNA profiles that differ from those that are thought to be responsible for many instances of a disorder. This is an implication of common final pathway constraints on the phenotypic expression of genetic information (Yuwiler, 1995).

It follows that disorders are best viewed as statistical variants of the normal (although, as noted in the statistical variance section, statistical assessments are not necessarily straightforward undertakings). It also follows that thinking about disorders as separate classes introduces distinctions that are difficult to defend philosophically; that may invite misinterpretations of both data and potential disorder-contributing variables; and that often overlook a more likely prospect: In many disorders, tissue pathology and specific genetic profiles are only two of many possible causal factors.

Epiphenomena

In psychiatry, the term *epiphenomena* most often refers to phenotypic characteristics that are assumed to be products of one or more underlying dysfunctional processes, such as tissue pathology and intrapsychic distortions. For example, the excess production of thyroid hormone in hyperthyroidism is viewed as an instance of somatic dysfunction, and the associated behavioral and cognitive-emotional features are viewed as consequences—epiphenomena—of excess thyroid production. Or in the case of monoamine oxidase (MAO-A) deficiency, reports point to an increased frequency of aggressiveness and mental retardation, which are viewed as consequences of the absence of a specific enzyme that modulates neurotransmitter concentrations (Brunner et al., 1993). Because interventions such as reducing thyroid production or altering dopamine activity in persons with schizophrenia are often associated with a reduction of signs and symptoms, there is a ready model supporting the epiphenomena concept.

As usually applied, the epiphenomena concept assumes that events occur in a linear fashion: A causes B, B causes C, and so forth. This assumption may apply to certain simple machines, and it may be useful for making first approximations of tissue-pathology-phenotype relationships. Nevertheless, literally all available evidence suggests that biological systems cannot be accurately characterized as linear, primarily because they have numerous positive and negative feedback systems as well as self-correcting features. The current understanding of biological systems is more like A causes B, and B causes C, but C modifies A and B, and so forth. Understandably, it is often difficult to interpret clinical phenomena when they are the products of multidirectional and offsetting events, although in the future, this issue will need to be addressed. Here, another point requires emphasis: Thinking about and explaining either conditions or features of conditions primarily as epiphenomena ensures that other possible functions of events at A, B, or C will be overlooked. For example, persons who are depressed or who are diagnosed as schizophrenic often withdraw from social participation. The epiphenomena view of such behavior is that it is a consequence of events at A, such as dysregulation of the CNS norepinephrine or dopamine systems. In an evolutionary context, social withdrawal is as likely to reflect an *adaptive strategy* to reduce the undesirable effects of depression and schizophrenia (e.g., avoid the costs of interacting with others) as it is to reflect an epiphenomenal event.

One might contest the preceding view in the following way: Despite feedback systems and the fact that persons act to reduce their pain, A-type events are known to occur (e.g., chromosome breakage), and when they do, conditions are often present. We would agree, but we would also add two critical points and a clarification. First, in an ecological model of conditions, A-type events may be environmental, such as loss of an important other, a decline in social status, or social ostracism. Second, A-type events may initiate B-type events (e.g., alterations in receptor number in response to changes in neurotransmitter concentrations) that counterbalance the effects of A-type events. This second possibility is a strong candidate for explaining fluctuations in the clinical course of conditions as well as spontaneous reversal of conditions. As to the clarification, evolutionary theory introduces novel explanatory hypotheses, some of which conflict with the epiphenomena view. For example, sexual jealousy may reflect mate-guarding strategies; anxiety may serve as information that an ongoing strategy will fail and may increase the possibility that an individual will develop alternative strategies; narcissism may reflect the fact that one is highly skeptical of the value of altruistic strategies; and so forth. From this perspective, moods, much of the material of consciousness, and behavior may be indications of strategies or may reflect responses to environmental contingencies, neither of which imply underlying psychopathology.

To return to the discussion of disorder identification, continuing debates both within and outside psychiatry show that consensus on how to define, diagnose, and classify conditions has not as yet been reached (e.g., Scadding, 1967, 1988; Kendell, 1975; Reznek, 1987; Costello, 1993a, 1993b). These debates are hampered by psychiatry's continuing lack of agreement on critical terms, such as the definition of *mental illness* (Wiggins and Schwartz, 1994). Definitions that are closer to the definition we will use have been suggested by a number of authors. For example, Scadding (1967) characterized disease as the sum of the abnormal phenomena displayed by a group of living organisms in association with a specified common characteristic or set of characteristics by which they differ from the norm for their species in such a way as to place them at a biological disadvantage. Klein (1978) expressed the same idea more succinctly: Disorders are the result of things that have gone wrong with evolved structures that allow for adequate functioning. And Wakefield (1992) used a similar definition, to which he added a cultural judgment factor dealing with the social undesirability of disorders.

To bring this section to a close, we have touched on some of the ambiguities encountered in identifying conditions. Suffering is a relatively good, but not a consistently applicable, criterion. The use of statistical criteria introduces other issues, the most troublesome of which is the use of different statistical criteria for different conditions. Tissue pathology and atypical DNA profiles are known to apply to only a small percentage of *DSM*-classified disorders, and instances of tissue pathology reversibility, as well as the nonexpression of atypical genetic information, are known to happen. Compromised functionality applies to most but not all conditions, but before this criterion can be usefully incorporated into psychiatry, more precise and detailed techniques for assessing behavior and its functions need to be routinely put to use. For example, rather than devoting so much effort to dissecting and classifying moods, researchers might pay more attention to the functions of moods and the behaviors they influence. We have also reviewed some of the conceptual problems that arise when disorders are viewed as members of natural classes or instances of epiphenomena. Many of these problems are not insurmountable. Nevertheless, the preceding discussion clearly suggests that psychiatry could benefit from a revamping of many of its condition-identification practices, and once done, the revamping could be expected to reconcile many of the seemingly unsettled points raised here. Finally, our analysis should not be taken to imply that identification is unimportant. How conditions are defined and diagnosed has numerous implications. Illness implies that there will be attempts at treatment, and diagnostic categories are associated with constraints on treatment options; for example, if an illness is not present, treatment is unethical.

How Conditions Are Classified

Condition-identification practices are not easily separated from condition-classification practices. Once a condition is identified, how it will be classified is largely a function of two factors: the taxonomic system that is adopted and the system's classification rules. The history of *DSM-III* (APA, 1980) and *DSM-IV* (APA, 1987) illustrates these factors at work.

From its inception, the *DSM-III* classification had three primary aims: to describe disorders accurately using a language free of theoretical biases (accurate descriptions); to develop a system based on a vocabulary that would achieve a high degree of cross-clinician agreement (reliability among multiple users); and to devise a system that was consistent with the empirical trend in psychiatry (Akiskal, 1989). On an initial reading of *DSM-III*, these aims appear to have been accomplished. The descriptions consist of sets of signs, symptoms, historical information, test scores (e.g., reading ability), and functional assessments. Only those clusters of attributes that occur relatively frequently or, if infrequently (e.g., amnesia), with some measure of within-cluster consistency are accorded disorder status. And clinicians increasingly use *DSM* diagnostic categories when discussing conditions.

A closer reading of the several versions of *DSM* leads to a somewhat different impression. Concerns about *DSM*-type classification have been present from the beginning, starting with *DSM-II* (Colby and McGuire, 1981; van Praag, 1989; Gert, 1992; Mathis, 1992). At times, critics have focused on the limitations of taxonomies designed to partition and characterize human behavior. At other times, they have questioned the reasons for preferring one taxonomic system over another. Taxonomists may have an intuitive sense of what they are trying to classify. A system may be preferred because it is consistent with prior systems. It may be chosen because it is understandable to those whom it classifies or because natural classes are assumed to exist (e.g., J. M. Murphy, 1978). Elements of the system may reflect current research interests. Or a system may be preferred for reasons that are not obviously related to taxonomic objectives, such as political compromise or economic advantage. In different ways, each of the preceding points applies to the history of the several versions of the *DSM* (Colby and McGuire, 1981; Klein, 1995). DSM categories are neither as atheoretical nor as descriptively unbiased as has often been claimed, although for some conditions, *DSM-IV* appears to approximate the atheoretical ideal more closely than did *DSM-III*.

Taxonomies don't promise perfection or final truths, and the several versions of the *DSM* should not be judged as if they do. Further, relatively theory-free taxonomic systems have worked well for many nonbiological disciplines (e.g., chemistry). Thus, there are successful precedents. Nevertheless, it needs to be emphasized that nearly every chapter of every textbook of psychiatry questions features of the *DSM* system; that the number of signs and symptoms associated with *DSM*-classified disorders, which currently exceeds 250 (H. I. Kaplan and Sadock, 1989b), appears to be not only excessive, but also of questionable utility; that at least a dozen classification—explanation systems are in use for schizophrenia; and that certain conditions (e.g., schizoaffective disorder) continue to defy precise description. Further, as has already been noted, many signs and symptoms are associated with such a large number of conditions that their classificatory utility is questionable. For example, delusions are reported to occur in association with more than two dozen neurological disorders and two dozen mental disorders, as well as with a host of medical disorders (Manschreck, 1989).

When the rules that apply to reasoning from clusters of signs and symptoms to diagnostic categories are given a close look, still other issues arise. Summative reasoning, which is illustrated in the following equation, is the principal culprit:

$$a \pm b \pm c \pm d \pm e \pm f, \ldots, n = \text{Disorder X}$$

Summative reasoning works in the following way: Only a subset of attributes, such as a-b-d or d-e-f in the equation above, is required to make the diagnosis of Disorder X. Borderline personality disorder provides an example. In *DSM-IV* (APA, 1994), five (or more) of the following nine criteria are required for diagnosis:

- Frantic effort to avoid real or imagined abandonment;
- A pattern of unstable and intense interpersonal relationships characterized by alternating between extremes of idealization and devaluation;
- Identity disturbance: markedly and persistently unstable self-image or sense of self;
- Impulsivity in at least two areas that are potentially self-damaging;
- Recurrent suicidal behavior, gestures, or threats, or self-mutilating behavior;
- Affective instability due to a marked reactivity of mood;
- Chronic feelings of emptiness;
- Inappropriate, intense anger or difficulty controlling anger;
- Transient, stress-related paranoid ideation or severe dissociative symptoms. (p. 654)

Given that only five of the nine disorder-classifying criteria are required, it follows that two persons can be classified with borderline personality disorder yet share only one phenotypic attribute. This is far from precise classification. It also follows that estimates of comorbidity are likely to increase because individual attributes (e.g., anger, paranoid ideation, affective instability, chronic feelings of emptiness) are associated with more than one disorder-related summative reasoning equation. (According to one recent survey, 79% of lifetime disorders occur among persons with two or more disorders; Kessler et al., 1994). In short, developing a descriptive vocabulary that achieves a high degree of cross-clinician agreement may be served by summative reasoning, but its service to condition validity is another matter.

It can be argued that the development of valid, atheoretical diagnostic categories is critically important to psychiatry (Robins and Guze, 1970). It can also be argued that each iteration of the *DSM* more closely approximates the goal of developing reliable descriptive categories, and that establishing validity is a difficult and time-consuming task. One can agree with these points, yet to do so still does not solve the validity issue. In the few known instances in which there is unambiguous evidence of specific DNA profiles or tissue pathology associated with specific phenotypes, the view that *DSM* categories are valid is defensible. But as we have noted, the number of such instances is small. Further, it is unclear how either the concept of validity or the summative reasoning equation might apply to conditions without tissue pathology or how factors such as social context might be entered into summative reasoning equations. Given the preceding points, it is difficult to avoid the following question: Is there enough merit in the *DSM*-type approach to classification to justify the work it has already required as well as the work further iterations will require? In our view, the merit is questionable for a large percentage of conditions. Moreover, such efforts may obfuscate the search for causes and optimal treatments.

Explaining Conditions

Although the problems associated with psychiatry's efforts to identify and classify conditions are many, when the ways in which conditions are explained are brought under the microscope, identification and classification problems pale in comparison. Figure 2.1, which deals with prevailing model explanations of depression, provides an illustration.

In Figure 2.1, signs and symptoms (F) may be due to physiological or psychological events, or both, occurring at E, and events at E may be due to events occurring at B. This possibility represents perhaps the least complicated prevailing explanatory alternative. More frequently encountered explanations incorporate genetically influenced trait features (C), such as dysregulated neurochemical systems; possible contributions from D (adverse environments); and a variety of possible B-C-D-E-F interactions. A less frequently encountered explanation involves interactions between events at A and D leading to E and F.

The number of different combinations and possible explanations in Figure 2.1 is enormous, a point that is underscored by a sampling of findings from studies of de-

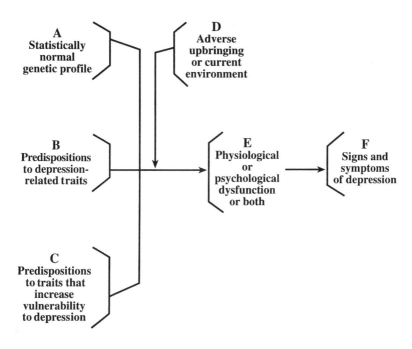

Figure 2.1 Possible explanations of depression.

pression. There are reported physiological correlates of risk for depression, such as degree of platelet monoamine oxidase activity (Murphy, 1990), evoked-potential augmentation and reduction (Haier et al., 1988), and reduced CNS serotonin concentrations (Meltzer, 1989). Among the prevailing models, such findings implicate events at B or C and E. There are also reports of personality features (e.g., tendencies toward anxiety, guilt, suspicion) that are frequently observed among persons who are depressed (Perris et al., 1984; Phillips et al., 1990), in which case, contributions from C are implicated. Pedigree studies point to contributions from B, and possibly C, with the caveat that genomic imprinting (gene expression that depends on the parental origin of a gene; Haig, 1993, 1995) may influence pedigree profiles. Further, there are some reports of high frequencies of disorders among relatives of normal controls (Zimmerman and Coryell, 1989). Such findings implicate events at B or C among relatives, but not necessarily among second-generation individuals. Numerous reports suggest A-D-E-F interactions, events at D (e.g., loss of parents during childhood) being viewed as a critical determinant of depression. There is also strong evidence that some persons with the same form of depression who receive the same treatment recover, while others do not (e.g., Shea et al., 1990; Burvill et al., 1991). These findings imply either classificatory ambiguity (Hudson and Pope, 1990), common final pathway phenomena, differential B or C effects, or differential responses to events at E (Andrews, Stewart, Allen, and Henderson, 1990; Andrews, Stewart, Morris-Yates, et al., 1990).

The potential for interpretive ambiguity and the difficulties associated with hypothesis disconfirmation have led some investigators to argue for a biochemical or neuroscientific approach to classification (e.g., Schatzberg et al., 1989; van Praag, 1989; Dubrovsky, 1995). These approaches also have their limitations. For example, the absence of norepinephrine autoreceptors, an abnormality in cell membrane lipids, and the hyperproduction of serotonin receptors (which interact with norepinephrine neurons) all might contribute to the same set of symptoms (Yuwiler, 1995). Moreover, those advocating the biochemical classification approach would be wise to accept that there are thousands of neuroproteins, and probably also thousands of receptor and cell membrane mechanisms, the functions and actions of only a few dozen of which are well known. Further, if the physiological data from studies of depression are closely analyzed, there is far less consistency than might be imagined; for example, measures of neurotransmitters and their metabolites range from statistically normal to statistically abnormal. Such findings raise yet other infrequently asked questions, such as: *Can statistically normal physiological states coexist with signs and symptoms?*

While the implications of Figure 2.1 could be elaborated, the discussion is sufficient to illustrate the two key points that need to be made here: Prevailing model explanations of conditions are far from satisfying, and alternative explanatory approaches deserve consideration. For example, a predisposition to depression might be thought of as the predisposition to emit depressive signals to initiate others' help and to reduce expletory behavior rather than as some type of genetic-physiological defect. This formulation would lead to numerous new inquiries, such as: When would a reduction of expletory behavior be useful, and when not? Do strategies or depressive signals reduce depression? Or do some signals make things worse?

Two Competing Explanatory Systems

Contrasting two explanatory systems will serve to pinpoint much of the preceding discussion, clarify some of the explanatory limitations of prevailing models, and further set the context for subsequent chapters. The first system deals with known or highly likely condition-contributing agents or events; the second, with possible evolutionary explanations. We will use the following definitions in this discussion and throughout the remainder of the book:

Etiology is the most general explanatory term that will be used. It is a biological concept and refers to the ingredients (e.g., dopamine receptor downregulation, adverse developmental environment) that are considered necessary for a condition to develop. Although ingredients may be present, a condition need not be (E. A. Murphy, 1987), as when a person has low baseline CNS serotonin concentrations but displays no condition-related features.

Predisposition refers to a postulated genetic loading for conditions and may be used in either a statistical or a biological sense. The statement that offspring of parents who both suffer from schizophrenia will have a greater than chance probability of developing schizophrenia is a statistical statement implicating probable genetic loading among the offspring. These kinds of statements are based on empirical findings showing positive correlations between parental and offspring condition type. The statement that Person X is predisposed to depression because depression runs in the family is a biological statement. How specific genes or their effects contribute to a condition may be implied, but predispositions alone do not explain conditions, a point that is consistent with the finding that there is less than a 50% concordance rate among monozygotic twins for many conditions that are thought to be genetically influenced.

Developmental disruption is a biological concept that refers to the interference of maturational programs. Disruptions can occur before the moment of conception and continue into adulthood. They may contribute to conditions apart from predispositions that may be present, or their effects may be additive.

Vulnerability is a biological concept that refers to an individual's susceptibility to developing a condition. Predispositions, developmental disruptions, or current environmental factors may be susceptibility-contributing factors.

Risk is a statistical estimate of vulnerability.

Pathogenesis or *pathogenic event* refers to proximate biological events that trigger the onset and continuation of conditions. The hypothesis that an alteration in CNS dopamine activity mediates schizophrenia is an example.

Explanatory System 1   Conditions Known or Highly Likely to Be Causal Agents

The first explanatory system considers known or highly likely condition-contributing agents, including trauma, living organisms (pathogens or other people, or both), environmental precipitants, and genetic disruptions. These agents are not mutually exclusive, yet for each, some type of tissue pathology or atypical DNA profile is either known or likely.

Trauma denotes a physical insult, such as cerebral concussion and CNS anatomical injuries. Insults can occur at any age, and all biological organisms are limited in their capacities to respond to trauma. In addition to physical injury, in utero events (e.g., maternal abuse of substances) can have traumatic effects on developing fetuses and can compromise normal maturation. While the consequences of trauma are not thought to be adaptive, selected responses, such as white cell invasion, increased blood flow to injured areas, and nerve regeneration, qualify as adaptive responses when they hasten recovery or offset further deleterious effects.

Competition between hosts and infectious or parasitic organisms, such as bacteria, viruses, or worms, can also contribute to conditions. Like trauma, infectious diseases can occur at any age. For most organism-mediated conditions, disorder-related signs and symptoms appear relatively late in the clinical course. Moreover, signs and symptoms often disappear when organisms are successfully treated, although there are notable exceptions, such as postmeningitic and postencephalitic syndromes and possibly certain cases of schizophrenia where viral infections have not been ruled out as a causal factor (Kendell and Kemp, 1989; Franzek and Beckmann, 1992; Pulver et al., 1992). Organismic responses to infections or parasites may be adaptive when they serve to counter the effects of destructive organisms or to confer immunity against subsequent infections (e.g., German measles, chicken pox, mumps).

Conditions can also result from environmental precipitants (e.g., metals such as lead) that disrupt genetic encoding, result in errors in genetic replication (e.g., chromosome dislocation), or compromise phenotypic expression. Stochastic accidents are another possibility. Conditions with any one of these causes may manifest themselves at any time in life, although the majority occur during childhood. They are not thought to be adaptive.

A fourth possibility deals with conditions for which the cross-generational transmission of condition-predisposing genetic information is implicated. Infantile autism (e.g., Ritvo et al., 1985), mood disorders (e.g., Kendler, Pedersen, et al., 1993), schizophrenia (e.g., D. Rosenthal, 1970; Suddath et al., 1990), selected personality disorders (Parnas et al., 1993), and alcoholism (Bohman et al., 1987; Pickens et al., 1991) are all examples of conditions that may be predisposed. These conditions are usually not thought to be adaptive. Again, however, features may represent attempts to adapt.

Putative dysfunctional physiological states are often discussed as if they belong to Explanatory System 1. But with the exception of a few conditions (e.g., vitamin deficiencies, Parkinson's disease), evidence meant to clarify the when and the how of their belonging to this system remains unconvincing. For example, a putative dysfunctional state, such as the downregulation of serotonin 2A receptors, may be a primary contributing factor to a condition, a secondary factor (e.g., a consequence of alterations in other physiological systems), or a physiological state associated with remission (e.g., Post and Weiss, 1992).

In summary, Explanatory System 1 is intuitively acceptable because different types of tissue pathology or putative atypical DNA profiles can be tied (or reasonably expected to be tied) to phenotypic features. However, at present, this system has two important limitations: The system satisfactorily explains only a small percentage of conditions, and possible adaptive functions of condition-related features may be ignored.

Explanatory System 2   Possible Evolutionary
Explanations of Conditions

Given the importance of natural selection and genes in neo-Darwinian thinking, it is reasonable to ask if some conditions have been selected or if they are inadvert by-products of selection. Here, we will take a quick first look at this question by discussing some conclusions from a recently published (and strongly recommended) book, and by reviewing a number of evolutionary concepts that have been regarded as possible explanations of disorders.

Why We Get Sick (1994) by Randolph Nesse and George Williams is about medical diseases viewed in evolutionary context. Key points from their analysis of disease-contributing factors include:

- The body is far from perfect, and diseases are unavoidable.
- Pathogens that attack humans evolve faster than humans. Thus, humans are involved in a neverending, nonwinnable "arms race" with pathogenic organisms.
- Current environments are in many ways out of step with evolved traits, and the lack of fit between traits and the current environment contributes to diseases such as athero-sclerosis, breast cancer, and obesity.
- Evolutionary trade-offs (e.g., selection favoring some traits over others) result in the lack of refinement or specialization of selected traits. No trait is perfect (e.g., one can process information only so fast, bones may be strong during adolescence but subject to deterioration in old age), although in different environments, some traits are more perfect than others (e.g., sickle cell trait in a malaria-infested area).
- Historical constraints (e.g., how a trait has evolved) result in evolved attributes or design features that may be especially susceptible to diseases. Diseases of the visual, cardiovascular, and nervous systems are possible examples. Back pain, appendicitis, and many forms of cancer may also qualify.

Nesse and Williams devoted only a single chapter to possible explanations of mental conditions. Nevertheless, their list of disease-contributing factors has clear implications for explaining conditions. For example, a poor fit between strongly evolved traits (e.g., dependency) and the current urban social environment may account for many instances of anxiety, depression, and chronic frustration. And evolutionary trade-offs may result in high degrees of cross-person variance for some traits, such as response to rejection.

Turning to specific evolutionary explanations, we look first at *pleiotropy*—the control of one or more phenotypic characteristics by one or a set of genes. Genetic information contributing to minimally adaptive traits may be carried along from generation to generation because such traits are controlled by the same genes that are responsible for an adaptive trait. Perhaps the clearest example of pleiotropy would be that of greater fecundity among females during reproductive years (the adaptive trait) coupled with late-life (postreproductive) vulnerability to conditions such as depression (the minimally adaptive trait; e.g., Williams, 1957; Albin, 1988; M. R. Rose, 1991). In this example, late-life conditions would have a reduced chance of being selected against, while increased fecundity would ensure that the pleiotropic gene(s) will be present in subsequent generations. Similar interpretations have been offered for unipolar depression and bipolar disorder, where a greater-than-chance occurrence of highly creative persons (the adaptive trait) with these disorders has been reported (e.g., Goodwin and Jamison, 1990), and for schizophrenia, where the genes thought to be responsible have been postulated to protect against a variety of illnesses (the adaptive trait; Nesse and Williams, 1994). In more complex examples, one monozygotic twin may have a disorder, while the other may be unusually talented.

Exaptation is a second possibility: A trait was adaptive in the past but is not adaptive in the present. This possibility is discussed in detail in chapter 3.

Genetic drift (change in gene frequencies by chance alone) is a third possibility (Beatty, 1992). The concept of genetic drift is often used to explain the apparently neutral (i.e., having no obvious phenotypic consequences) differences in DNA profiles across individuals, as well as neutral genetic variance across populations that have not inbred for extended periods (Cavalli-Sforza, 1991). In rare instances, isolation has nonneutral effects, as in the case of groups that lack certain enzymes and whose behavior is atypical (Brunner et al., 1993). Drift also may explain selection favoring the appearance of new adaptive traits; for example, chance may occasionally permit drift-influenced traits and environmental features to interact in ways that facilitate reproductive success and, therefore, a change in the species genome over subsequent generations.

Mutation is another possibility. Over the course of evolution, mutations are a major source of genetic change and genetic variation, and some traits that are currently viewed as adaptations are very likely products of mutations. Nevertheless, mutations remain a poor choice for explaining most conditions. Although some mutations are neutral and some adaptive, they generally have deleterious and often deadly effects. Hence, the genes affected by mutations usually do not long remain in populations. In severely debilitating disorders that are relatively uniformly distributed throughout the world, such as residual schizophrenia, and for which available evidence does not point to either an increasing or a declining prevalence, the possible part played by mutations is more difficult to infer. For example, if schizophrenia is due to a mutation(s), it would be necessary to reason somewhat as follows: The disorder-causing mutation(s) is very old (e.g., it occurred before ancestral migration out of Africa or Asia); the disorder has not negatively affected reproductive success in ways that can be easily demonstrated; and selection against the mutation (if it is occurring) is minimally influenced by such factors as the physical environment, cultural variables, and the social consequences of the disorder.

Homozygosity due to inbreeding is another possibility. There are a number of reports in which the degree and frequency of inbreeding correlate with greater asymmetry of physical features (Markow and Martin, in press), and these features covary with the frequency of conditions. However, it is doubtful that any of the widely distributed, severely debilitating conditions can be explained in this way. The near uniform cross-cultural prevalence of many conditions, when compared to cross-cultural differences in breeding practices (e.g., the acceptability versus the unacceptability of first-cousin marriages), suggests as much.

Minimal selection pressure is another possibility. Conditions such as dyslexia, which might go unnoticed in nonliterate societies, or conditions that appear after the critical reproductive years, such as postmenopausal depression, late-life depression in males, and Alzheimer's disease, might be explained in this way (Leckman et al., 1984). In principle, conditions resulting from minimal selection pressure can be distinguished from those to which a pleiotropy explanation might apply because of the absence of associated adaptive traits. Further, migration among members of populations in which selection pressures have been minimal could lead to the introduction of atypical genetic information into neighboring populations, but factors such as population size, the fitness characteristics of the genes in question, the rate of migration, and the genetic features of the population with which the migrants intersperse need to be considered.

Three related points, each developed further in subsequent chapters, are briefly mentioned here. First, there are often differences in the magnitude of genetic effects. An example of a strong genetic effect is the absence of a specific receptor that results in specific phenotypic consequences. An example of a moderate genetic effect is the reduced density of a receptor that may or may not have phenotypic consequences. Second, in many conditions, several moderate gene effects may be required for phenotypic expression. Third, for a variety of reasons, gene expression may skip one or more generations.

In summary, Explanatory System 2 raises a number of interesting possibilities, pleiotropy, exaptation (still to be discussed), genetic drift, and minimal selection pressure being the strongest candidates. Mutations and homozygosity due to inbreeding are less likely. But as noted, this analysis represents only a first pass at possible evolutionary explanations, and a narrow focus on gene-phenotype relationships rather than other products of evolution would severely limit the scope of our inquiry.

Explanatory Systems 1 and 2 and Prevailing Model Explanations

How do psychiatry's prevailing causal models fit within the two explanatory systems just discussed? The biomedical model is compatible with System 1, and it can accommodate explanations of conditions based on any of the four condition-causing factors (e.g., trauma, infectious diseases). In its current form, and with the possible exception of either deleterious mutations or homozygosity effects, it is less compatible with Explanatory System 2, primarily because it does not seriously entertain the possibility that selection may have resulted in conditions that are adaptive or have adaptive features. Answers are distinctly less clear for the psychoanalytic, behavioral, and sociocultural models. Because in their current forms they do not rely on genetic, tissue pathology, or evolutionary postulates, they are not compatible with System 2. As to System 1, at best they are compatible with the trauma category, but only if the trauma category is expanded so that experiences (e.g., adverse upbringing environments, dysfunctional learning, social stress) are viewed as forms of trauma. If this expansion is allowed, problems still persist, however. For example, since Freud, adverse environments have been a mainstay of many causal theories that do not include tissue pathology, and compelling evidence shows that adverse environments can influence ontogeny and contribute to conditions. Nevertheless, short of extremely adverse environments, what does and does not constitute an adverse environment is still far from clear; for example, seemingly adverse environments often don't have their expected effects. The opposite point holds as well: Many persons develop conditions in seemingly nonadverse environments.

Concluding Comments

The analysis of the prevailing practices of identifying, classifying, and explaining conditions leads to relatively straightforward conclusions. Methods for identifying conditions have questionable utility for clearly distinguishing types of conditions. Condition classification ends up much the same way. When prevailing models were assessed for their capacity to accommodate to the causes of conditions described for Explanatory System 1, the biomedical model fared the best, primarily because System 1 is based on tissue pathology or atypical DNA profiles. The other three prevailing models have a much narrower range of application and are limited primarily to explaining features of an expanded trauma category. However, had we examined other causal systems that more readily accommodate the psychoanalytic, behavioral, and sociocultural models, these models would have fared better and the biomedical model would have fared worse. None of the prevailing models fared well in Explanatory System 2, although the biomedical model could accommodate more explanatory alternatives than the other three prevailing models.

PART II

AN EVOLUTIONARY CONTEXT FOR DISORDERS

3

Evolutionary Concepts Important to Psychiatry

With the 1872 publication of *The Expression of Emotions in Man and Animals,* Darwin opened the door for evolutionary explanations of human behavior. The possibility remained largely unexplored until the 1930s, when R. A. Fisher, J. B. S. Haldane, and S. Wright began the integration of genetics and natural selection, an integration now referred to as *neo-Darwinism.* By the 1960s, possibility began to give way to reality. Three seminal publications, "The Genetical Evolution of Social Behavior," Parts 1 and 2 (Hamilton, 1964), *Adaptation and Natural Selection* (Williams, 1966), and "The Evolution of Reciprocal Altruism" (Trivers, 1971), introduced ideas that would serve as the building blocks for evolutionary explanations of human behavior from then until now. The book that attracted the most attention, however, was E. O. Wilson's *Sociobiology,* published in 1975.

Sociobiology was not the only book of the period addressing evolutionary issues (e.g., Dawkins, 1976), although it was arguably the most controversial. Its intellectual achievement was to weave together in a single text a mass of empirical data and evolutionary reasoning. Its message was unambiguous: A significant amount of the social behavior of nonhuman species, including nonhuman primates, could be explained by the use of evolutionary concepts. The message itself was not new, nor is it one with which the majority of biologists, then or now, would disagree. But there was more to *Sociobiology*. It was only a short step from explanations of nonhuman primate social behavior to speculations about the roots of human behavior. Wilson chose to discuss the possible evolutionary origins of human social organizations, barter and reciprocation, bonding, role playing, communication, culture, ritual, and religion. With this discussion, biology intruded itself into the intellectual territories of psychology, sociology, history, law, and moral philosophy, disciplines that had developed their views of human nature largely without recourse to the findings and reasoning of biology. A decade and a half of controversy followed.

While social scientists and philosophers argued with Wilson, biologists debated among themselves over such issues as group selection, punctuated equilibrium, sexual selection, and the importance of random drift. This is the place neither to recount the details of these interesting and often heated debates, nor to assess their influence on evolutionary thinking. A recent update is available (Keller and Lloyd, 1992). For the purposes of this book, the important points are that the basic structure and premises of evolutionary theory have withstood the many attacks aimed at discrediting them, and evolutionary thinking continues to extend its influence to disciplines outside biology.

Psychiatry's interest in evolutionary ideas parallels that of the social sciences. In the years between World War I and the early 1960s, evolutionary concepts interested only an occasional psychiatrist (e.g., Lewis, 1936; Meyer, 1948–1952; Price, 1967, 1969a, 1969b; Esser, 1968), and fewer than three dozen papers and books offering evolutionary interpretations of mental conditions were published. By the late 1960s, change was in the wind. Two years after Price (1967) published his evolutionary explanation of depression, Bowlby (1969) argued that one can best explain many of the features of infant-parent bonding, such as anxious attachment among infants, by taking their evolutionary origins into account. With the 1970s, the number of publications took a sharp upward turn (e.g., S. J. Hutt and Hutt, 1970; Jones, 1971; Davis, 1970; Ekman, 1971, 1976; E. A. Tinbergen and Tinbergen, 1972; Kellett, 1973; Jonas and Jonas, 1974, 1975; N. Tinbergen, 1974; White, 1974; McGuire, 1976, 1978, 1979a; McGuire and Fairbanks 1977; Daly and Wilson, 1978; Essock-Vitale and McGuire, 1979; D. A. Kraemer and McKinney, 1979). By 1980, it was reasonable to assume that the coming decade would find psychiatry eagerly embracing evolutionary findings and concepts (Leak and Cristopher, 1982).

Psychiatric studies from the 1960s through the 1980s were of two types. Either they emphasized the detailed, direct observations of behavior and its functions among psychiatric patients confined to hospitals (Scheflen, 1963, 1964; Fairbanks et al., 1977; McGuire et al., 1977; McGuire and Polsky, 1979; Polsky and McGuire, 1979; Bouhuys et al., 1987; Dienske, Sanders-Woudstra, and de Jonge, 1987; Pitman et al., 1987; Pitman, 1989), or they used evolutionary concepts to explain disorders such as schizophrenia (Kellett, 1973), depression (Price, 1967; Price and Sloman, 1987), anorexia nervosa (Surbey, 1987), sociopathy (Harpending and Sobus, 1987), anxiety and panic (Nesse, 1987a), and senescence (Nesse, 1987b). Years earlier, Nobel Prize winners Konrad Lorenz, Nikolaas Tinbergen, and Karl von Frisch had refined the techniques of direct observation (ethology) and demonstrated that novel insights into both the functions and the causes of behavior could be gained by the observation of behavior in natural settings. Where better than psychiatry might such a technique be applied? Most likely, the answer is: *Nowhere.* But answers don't necessarily predict outcomes. With the exception of the studies noted here, and perhaps a dozen more, the idea that detailed, direct observation of persons with disorders might be as informative as clinical interviews, paper-and-pencil tests, or laboratory measures of physiological variables never quite took hold. Psychiatry of the 1980s remained largely indifferent to the possibilities offered by ethological methods and evolutionary concepts.

There were many reasons for this indifference. Evolutionary theory was poorly understood outside biology (Charlesworth, 1986, 1992; Crawford, 1987, 1989). In the minds of most nonbiologists, topics such as speciation, territorial behavior, and mating practices among nonhuman species, not humans, were the worthy interests of evolutionary inquiry. The time required to collect detailed behavioral observations was another factor. Had ethologists been invited to participate in psychiatric investigations, things might have turned out differently. But such was not the case. Moreover, for their part, most ethologists steered clear of medicine—academic disciplines have their own territorial imperatives. More important, evolutionary models seemed to promise few new insights. Biomedical explanations of disorders (e.g., the norepinephrine hypothesis of depression and the dopamine hypothesis of schizophrenia), the search for biological markers, the hope for novel findings from new investigative techniques (e.g., brain scanning), and the therapeutic properties of drugs were more attractive prospects. In addition, large segments of psychiatry were facing intellectual and economic challenges: Sociocultural, behavioral, and psychoanalytic practitioners were struggling to maintain their identities in the face of evidence that many signs and symptoms could be rapidly and inexpensively ameliorated by drugs, and insurance companies and the government were progressively setting limits on the amount of money available for psychiatric care.

By 1990, the intellectual climate had changed once again. The controversy that for 15 years had surrounded *Sociobiology* and evolutionary explanations of human behavior had subsided. Evolutionary theory had gained footholds in fields as diverse as sociology, computer science, philosophy, and the law (e.g., Gruter, 1991; Dennett, 1995), and classical ethology had undergone a resurgence. Critics of the evolutionary interpretation of human social behavior continued to voice their views (e.g., Gould, 1992), but their critiques were much the same as they had been a decade earlier. In 1991, Williams and Nesse introduced the term *Darwinian medicine* into medicine's vocabulary, and few people objected, and by 1992, the suggestion that evolutionary biology should serve as the basic science for psychiatry found few opponents (McGuire et al., 1992). Yet, an in-depth exploration of the implications of evolutionary theory for psychiatry was a task still to be undertaken. It is to part of that task that we now turn.

Evolutionary Concepts

What evolutionary concepts are essential for developing a theory of behavior applicable to mental conditions? To begin to answer this question, the remainder of this chapter will review the concepts of natural selection, adaptation, function, ultimate causation, individual fitness, self-interest, inclusive fitness, reciprocal altruism, proximate mechanisms, development, traits and within-trait variation, learning, culture, and life history strategies. While they are occasionally discussed, clinical applications are left largely to later chapters. We begin with a brief discussion of evolutionary theory: what it is, what it is not, and what its relevance is to an understanding of both conditions and non-condition-related behavior.

Evolutionary Theory: An Overview

Evolutionary theory is in part about the replication of genes. It is in part about the modification and transmission of genetically influenced traits through time and across environments. It is in part about the internal rules that guide behavior and about the interactions between these rules. It is in part about learning, behavior, and environmental contingencies. And it is in part about function.

For the most part, it is a theory of gradual genetic and phenotypic change, although periods of rapid change are known. The time frame in which genetic and phenotypic changes have taken place extends back millions of years. If we pick up the story about 7 million years ago, chimpanzees, gorillas, and humans shared the same ancestor; 6.5 million years ago, the first primates appeared; 4.5 million years ago, bipedal hominoids first appeared; and 1.8 million years ago, *Homo erectus* appeared. One hundred and fifty thousand years ago is the estimated date for the appearance of *Homo sapiens*, and 50,000 years ago is the estimated date for the arrival of *Homo sapiens* in Europe. Across this time span, the physical environment underwent significant changes (e.g., ice ages), innumerable species became extinct, new species evolved, migrations occurred, and the genetic makeup of primates went through thousands of minor changes due largely to the pruning, honing, and facilitating processes of natural selection.

Because of the influence of genetic information on traits, humans, like other species, enter the world *predisposed* as well as *pre-prepared* to engage in certain behaviors more than others; to react in certain ways to specific stimuli and not to others; and to pursue certain goals more intensely than others. This array of characteristics is a product of our evolutionary past, a past that is in part carried in our genes. But genetic information only partially determines phenotypes. Other factors, such as experience, learning, and culture, are also relevant. Evolutionary theory is as much about these factors as it about the transmission of genetic material and its influence on traits.

When evolutionary biologists discuss predisposed traits, such as preferential investment in kin or differences in male and female mating strategies, nonbiologists often object to what they believe are its deterministic and negative social implications. But consider these concerns for a moment. In one sense or another, all sciences or would-be sciences are deterministic. The sociologist who postulates that criminal behavior is caused by peer-group influence, or the psychologist who predicts that a child will resent neglecting parents, is deterministic in the same way as the biologist who, for example, postulates that genetic information pre-prepares humans to rapidly learn certain things but not others. What is at issue is the *how* and the *why* of determinism. As to social implications, nonbiologists often worry that if behavior is viewed primarily as a product of genes, efforts at social reform will die out. Although this concern may be applicable to certain highly reductionist theories of behavior, it is not applicable to Darwinian psychiatry, which makes a strong case for the importance of both learning and social context as factors that influence development and its outcomes.

Much of our understanding of evolution comes from the work of population biologists and their analysis of phenotypic variation. Much also comes from evolutionary psychology, which combines evolutionary biology and psychology, and which focuses on evolved systems (algorithms) that process information and guide behavior. These are important distinctions. Findings from population biology permit inferences about interactions between environmental changes and the survivability of specific traits.

Findings from studies of evolved rules focus on short-term behavioral change, the external conditions that influence such change, and the internal systems that are thought to be responsible for change. Both bodies of research are integral to understanding and treating conditions.

Evolutionary theory does not assume that evolution has any plans or goals or that the products of evolution are ideal. Environments change, and a trait that is advantageous in one environment may not be advantageous in another (e.g., sickle cell trait, skin pigmentation). Because numerous factors (e.g., genetic information, learning, environmental options) influence phenotypic expression, individuals develop variant phenotypes. And because species evolve without advanced knowledge of future contingencies, only a portion of the variants survive and reproduce. In turn, both species' genomes and species' behavior change over time. What may appear to be an inconsequential genetic change in one generation—say, a 0.1% increase in the frequency of a specific gene in a population—may have significant effects several generations later. Evolutionary theory can accommodate these outcomes. The theory is as much about normal phenotypes as it is about variant phenotypes, and both phenotypes can be understood within the same theoretical framework.

Natural Selection

At any moment in time, the prevalence of genetically influenced traits is a consequence of prior interactions between the trait and reproduction. Not all traits replicate equally well. Natural selection is about the conditions that influence the differential survival of traits (Darwin, 1859). Proper use of the concept requires that three conditions be met: (1) Traits vary among individuals (phenotypic variation); (2) consistent interactions exist between specific traits and survival (fitness variation); and (3), for certain traits, there is a consistent relationship between parents and offspring (inheritance; Endler, 1992). Differential cross-generation trait survival may be due to numerous factors, such as a high degree of phenotypic mortality in particular environments, intraspecific and interspecfic competition, or differential mating success in which individuals with specific traits mate and have offspring more successfully than individuals with other traits (Endler, 1992). Because environments change (e.g., the arrival of new parasites), trait survival is neither predictable nor linear.

Natural selection enters ensuing discussions as explanations of cross-person phenotypic differences and how condition-contributing genetic information can remain in the human genome. Further, different types of selection (gene, individual, family, group) have different implications for explaining conditions. If selection occurs at the level of the individual, and disorders are minimally adaptive, then a decline in the frequency of disorders would be predicted. On the other hand, if selection occurs at the level of the gene, the frequency of disorders will fluctuate because of possible intergenetic cooperation or competition; for example, if one function of DNA is to replicate itself, replication priorities may override condition-related effects of DNA (Dawkins, 1976).

Adaptation

In biology, any anatomical structure, physiological or psychological process, or behavior that makes an organism more fit to survive and reproduce in comparison to other members of the same species is an adaptive trait (Sober, 1987; West-Eberhard, 1992). It is the *in comparison to* part of the definition that requires emphasis. One's degree of adaptation is relative. There is no absolute measure.

For evolutionary biologists, the history of species is not indiscriminate with respect to which adaptations have been preserved and which have been lost (Plotkin, 1994). Studies of adaptation overlap with studies of natural selection and focus on such issues as the environmental conditions that favor certain traits over others; interactions between traits, reproductive success, and survival; and the changing characteristics of traits over time. Because reproduction is related to adaptation, adaptations during one generation can extend to subsequent generations.

Earlier, we noted that the influence of genetic information on traits ranges from strong to weak. The majority of traits that are of interest to psychiatry are in the strong to moderate part of the range. Examples include capacities to bond, to interpret information, to assess the costs and benefits of social interactions, to build mental scenarios (e.g., develop contingent strategies), to learn from one's mistakes, and to efficiently navigate the social environment (e.g., Alexander, 1990a, 1990b). For each of the these capacities, genetic information, experience, and the environment combine to influence the final products and their functions. Language provides a convenient example. Literally all humans use language, and the capacities essential for acquiring and utilizing language are part of *Homo sapiens* genetic makeup (Chomsky, 1957, 1980; Lieberman, 1984; Pinker, 1994). However, which language one speaks is determined by one's experience, not one's genes, and the manner in which one speaks and what one says can have survival and reproductive consequences.

Although the words are the same, the meaning of adaptation differs among biologists and psychiatrists. In psychiatry, adaptation is usually synonymous with adjustment, a point illustrated by the definition of adaptive functioning used in *DSM-IV* (APA, 1994):

> Adaptive functioning refers to how effectively individuals cope with common life demands and how well they meet the standards of personal independence expected for someone in their particular age group, sociocultural background, and community setting. (p. 40)

Adjustment can be subdivided into autoplastic or alloplastic adjustment. *Autoplastic adjustment* refers to short-term behavior changes, such as self-improvement or conforming to the demands of one's social environment (Futuyma, 1986). Efforts at self-change through insight, behavior modification, and satisfaction of the requirements of a new job are examples. *Alloplastic adjustment* refers to changing the environment to one's advantage (Linn, 1985). Examples include changing friends, seeking a new job, or building a fence around one's property. Autoplastic and alloplastic adjustments apply within an individual's lifetime. They may or may not influence the genetic makeup of subsequent generations.

A number of evolutionary psychologists have argued that the last period of intense selection for many of the present-day traits of *Homo sapiens* occurred during a period referred to as the *environment of evolutionary adaptation* (EEA; Tooby and Cosmides, 1990b). Scholars differ on the exact dates of the EEA, but most agree that it was sometime between 100,000 and 10,000 years ago. This view holds that *Homo sapiens* largely ceased to evolve genetically, morphologically, or psychologically following the EEA; that psychological capacities for mediating behavior, rather than behaviors per se, were the traits favored by selection during the EEA; and that selection favoring psychological capacities most parsimonously accounts for human behavioral plasticity (behavioral accommodation to multiple environments and contingencies). The view is consistent with several facts: Humans have a far greater array of behavioral strategies and exhibit far greater behavioral plasticity than any other known species; they have survived and reproduced through periods of significant change in the physical environment (e.g., ice ages, periods of drought, plagues); and they occupy a significantly greater number of niches than any other known species.

In the time between the EEA and the present, features of the social and physical environments have undergone significant change (e.g., urbanization, agriculture, industrialization, communication technology, medical therapeutics). These changes are thought to have exceeded the rate of within-species genomic change. If we apply these points to mental conditions, this lack of parity, which amounts to a form of postulated *genome lag*, has led some theorists to suggest that the lag precludes optimal fits (mismatches) between current environments and many of the traits selected during the EEA. In turn, the probability of psychological and physiological dysfunction is thought to increase (e.g., Jonas and Jonas, 1974, 1975; Glantz and Pearce, 1989; Tooby and Cosmides, 1990a; Nesse and Williams, 1994; Crawford, 1995; Bailey, 1996; Stevens and Price, 1996).

Despite the attractiveness of the genome-lag hypothesis for explaining many features of mental conditions (Stevens and Price, 1996), important questions remain unanswered. For example, given the known wide distribution of *Homo sapiens* during the latter part of the EEA (Africa, Europe, Asia, Australia, North America), the physical and social environments of the EEA are highly likely to have been far more varied than is usually assumed. And while it is no doubt true that many features of the environment have changed more rapidly than the species genome, advocates of the genome-lag hypothesis have yet to explain why so many traits presumably selected before and during the EEA remain so well adapted in the present, for example, reproduction, developing novel strategies, and so on. A possible explanation is that many features of the past and present social environments that might have influenced selection are more similar than not; for example, the number of persons with whom one regularly interacts may now differ minimally from thousands of years ago (Dunbar, 1993). There is also the possibility that *Homo sapiens* is still undergoing genetic change. According to most investigators, *Homo sapiens* is recent in origin. If so, speciation may be continuing, and the current version of *Homo sapiens* may have significantly different adaptive capacities from those of our ancestors even 20,000 years ago. Similar points may apply to mental conditions.

A further assessment of the genome-lag hypothesis and the views of evolutionary psychologists is deferred to later chapters. For the moment, it is best to keep an open mind about selection favoring psychological capacities rather than behavior and about their explanatory relevance to explaining conditions.

Efforts to assess the adaptiveness or nonadaptiveness of traits—*adaptationism* is the label sometimes applied to such efforts—are not without critics (Gould and Lewontin, 1979; Lewontin, 1979; Symons, 1990). Gould and Lewontin (1979) made the point in this way: "It [adaptationism] proceeds by breaking an organism into unitary 'traits' and proposing an adaptive story for each considered separately" (p. 581). The criticism is sometimes valid. The concept of adaptation has been misused, for example, when a specific trait, such as a psychic defense, is discussed separately from related traits, environmental contingencies, and its functional consequences.

Nevertheless, concerns about misuse are manageable provided the following points are kept in mind:

1. Because of past evolutionary compromises and trade-offs, organisms are not optimally designed (Mayr, 1983). Even the best adapted organisms possess many features that either have no apparent adaptive value (e.g., the chin, color of the blood) or for which more efficient designs might have evolved (e.g., muscles of the back, strength of bones). Less than optimal designs occur in part because in most instances the target of selection is the whole individual, not individual traits. While some traits, such as perfect pitch, appear to be distinct (discontinuous variation), and while other traits, such as poor visual acuity, are more likely targets of selection than other traits, such as ear lobe form, it is more accurate to view each trait as only one part of an interconnected anatomical-physiological-psychological-behavioral system on which selection works. In biology, speaking of individual traits is simply a shorthand convention, one acknowledging that selection occurs at the level of the individual but focusing on the history of individual traits (Mayr, 1983; Alexander, 1990a, 1990b). Dissection of phenotypes into individual features is necessary because it is the only operational means of implementing the study of the function of a given feature (Mayr, 1983). Thus, one moves back and forth from the individual to traits: "The student of adaptation has to sail a perilous course between a pseudoexplanatory reductionist atomism and stultifying nonexplanatory holism" (Mayr, 1983, p. 329). Psychiatry, no less than evolutionary biology, continually struggles with similar interpretive problems (Mandell and Selz, 1992).
2. Selection does not optimize adaptive traits or strategies as much as it gradually eliminates unfit traits or strategies (Tuomi, Hakala, and Haukioja, 1983).
3. Traits can differ in their degree of adaptiveness for reasons other than selection. Fetal poisoning, maternal viruses, and physical accidents can compromise maturational programs and, in turn, trait expression and refinement.
4. There are important distinctions between the beneficial effects of traits that have been selected and the possible, nonselected beneficial effects of traits. For example, immunological responses to viral infections were probably selected not for their subsequent immunity, but to counter the short-term effects of diseases (Futuyma, 1986). A similar point may apply to emotions. Initially, emotions may have been selected because they led to rapid behavioral responses. Subsequently, they may also have come to provide information about the effectiveness of behavioral strategies.
5. While there is a positive correlation between behavioral plasticity and the capacity of individuals to adjust, as noted, adjustment is not synonymous with adaptation. Adjustment to a pathological family environment may reduce intrafamily conflict, yet it may also diminish the chances of reproductive success in the person who adjusts.
6. A trait that is adaptive in one social environment (e.g., verbal intelligence in the United States) does not predict its adaptiveness in other environments, for example, in an environment in which social intelligence has greater survival value than verbal intelligence (Borgerhoff Mulder, 1987b).
7. Adaptive behavior does not imply that the actors are aware of all the factors contributing to their behavior.

In our view, the question is not whether the biological concept of adaptation should be incorporated into psychiatric thinking, but how the concept can inform our understanding of conditions and can improve treatment. Yet, any attempt at incorporation raises a number of questions about how to measure adaptation.

Several measurement approaches have been proposed, including the influence of a trait on reproductive success; the reproductive advantage conferred on the bearers of a trait (Caro and Borgerhoff Mulder, 1987); the expenditure of energy required to sustain existence (Bock, 1980); and the relative frequency with which an adaptive trait appears in subsequent generations. Unfortunately, none of these measures has much clinical utility. Thus, an alternative way of assessing adaptation needs to be developed. The one we will use is: A *trait is adaptive if it contributes to achieving biological goals.*

We will have to wait until chapter 4 and the discussion of biological goals before nailing down the details of how traits and goal achievement interact, as well as how these interactions can be measured. Nevertheless, an example and a point of clarification can be offered here. First, the example. If reading others' behavior rules—that is, accurately predicting how another will respond to different social contingencies—can be shown to increase the probability of acquiring a mate (a biological goal), then two persons can be compared for their rule-reading capacities, and the person whose capacities lead to a greater percentage of accurate readings at a reduced effort has an adaptive advantage. Second, the point of clarification. In evolutionary theory, the ultimate function of any adaptation is to increase the chances of gene survival. In most instances, gene survival is achieved through a chain of events or short-term goals, for example, acquire a mate → have offspring → raise offspring → offspring reproduce. We can measure the adaptiveness of a given behavior by assessing the effort (costs) required to achieve each step in the chain. This approach not only allows for manageable measures of adaptation but also avoids the problem of trying to link assessments of adaptiveness to the survival of genes in future generations. Further, it retains an essential feature of evolutionary theory, namely, that individuals compete with one another to achieve biological goals (e.g., competition among males for mates). It permits the comparison of goal-related traits and strategies across individuals without introducing endless qualifying statements. And it invites inquiries into whether some conditions and features of conditions are selectively neutral (i.e., have minimal impact on goal achievement) or selectively nonneutral, in which case traits and condition probability should interact.

Turning to terminology, the adaptiveness of a trait can be discussed in at least two ways: on a scale ranging from negative to positive, or on a positive-only (0–1) scale. We will use the latter convention because traits with different degrees of adaptiveness, not the presence or absence of traits, most often distinguish persons with conditions from those without.

The biological usage of adaptation enters discussions in numerous ways, for example: Do minimally adaptive traits increase the likelihood of conditions? In what ways do conditions compromise adaptations? And are some conditions or features of conditions adaptations?

Function

The function of a behavior is its purpose. Less teleologically, the function of a behavior is the beneficial consequence through which natural selection acts to maintain the trait in question (Hinde, 1982). The function of foraging is obtaining food. The functions of social interactions include recognizing possible mates and good reciprocators, developing social support networks, and obtaining information about resources. Some functions, such as foraging, are closely tied to specific behaviors, while others, such as acquiring financial resources or making friends, are associated with a large, but far from infinite, set of behaviors.

As with *adaptation,* the proper evolutionary use of the term *function* must meet certain requirements. The statement *The function of Trait T is F* requires that Trait T has been shaped by selection; that it serves Function F; and that Function F increases individual fitness (the replication of one's genes in the next generation; Nesse, 1988a). We will use the term *function* only for those traits for which there is evidence that meets the three requirements or for which there are good reasons to assume that research would provide such evidence. Two examples illustrate this usage. When a person washes his or her hands 230 times a day, the behavior may result in above-average cleanliness, but it is doubtful that this degree of hand washing positively correlates with increased fitness, if only because excessive hand washing severely interferes with carrying out other potentially adaptive activities. Thus, it is unlikely that this behavior has been favored by selection. The opposite interpretation may apply to time-limited depression. If it can be demonstrated that depression warns a person that she or he is failing competitively, that it initiates physiological slowing and social withdrawal, both of which reduce ongoing social interaction costs, and that the signaling of symptoms to others increases the probability that others will provide assistance, then a case can be made that these features of depression have been favored by selection, and that they have specific functions.

Two related points should be mentioned. First, functions and the capacities to carry them out (functional capacities) need to be distinguished. Two persons may have the same short-term goal—say, attracting a mate—yet differ in their capacities to achieve the goal. Second, many of the functions that are of interest to psychiatry (e.g., optimally navigating the social environment, reciprocating favors, investing in kin, acquiring mates and resources) are carried out in the social arena. Because features of the social environment change, carrying out the same function often requires the use of different strategies and capacities. Thus, if functions, functional capacities, and outcomes are to be accurately evaluated, behavior and its outcomes need to be assessed on a moment to moment basis; for example, the success of a social signal is defined in part by the response of the person receiving the signal.

As noted earlier, the moment-to-moment evaluation of behavior is not psychiatry's strong suit. Global assessments of capacities, or inferences about such capacities, developed from historical data are the usual bases for functional evaluations. These preferences are not without consequences. Not only is behavior incompletely understood, but numerous opportunities are lost to develop testable hypotheses dealing with condition-contributing variables: *Whatever else they are, most mental conditions are conditions of failed functions.* In what follows, the assessment of function and functional capacities in social context play central roles in the characterization and explanation of conditions.

Ultimate Causation

Ultimate causation explains why in the remote past some traits were selected over others. For example, selection favoring the capacity to detect specific sounds is likely to have increased chances for identifying predators and thus for survival. Because the past cannot be re-created, explanations of ultimate causes are unavoidably speculative. Nevertheless, strong inferences about past events can be developed and tested. For instance, if selection has favored helping nonkin when the chances of reciprocation are high, selection is also likely to have favored strategies of retaliation if reciprocation fails to occur, and if males are uncertain if they are the biological fathers of offspring, selection is likely to have favored tendencies to possess females. Evidence supports both of these predictions (Trivers, 1971; Daly, Wilson, and Weghorst, 1982). When investigators using different research methods reach similar answers to closely related questions, so much the better (e.g., Blurton Jones, 1984).

While the list of ultimately caused traits is long, its length is not surprising: Evolution has a long history. Examples include parent-offspring bonding (Bowlby, 1958, 1977); male and female possessiveness and jealousy (Daly et al., 1982); cooperative and reciprocal behavior among nonkin (Trivers, 1971); parent-offspring conflict (Trivers, 1974; Haig, 1993); sibling rivalry; preferential investment in kin (W. D. Hamilton, 1964); menopause (Peccei, in press); and deception and self-deception (Trivers, 1985; Whiten and Byrne, 1988). Although their strengths may differ, predispositions to engage in these behaviors are assumed to be present in literally all individuals. Predispositions can be refined and directed (Bohman et al., 1982; D. Reiss, Plomin, and Hetherington, 1991). Yet, there are limits; for example, offspring-offspring conflict occurs despite the efforts of millions of parents to prevent it.

Evolved psychological capacities, often referred to as *algorithms,* again enter the discussion. Algorithms are postulated psychological systems that are the result of ultimate causes (Cosmides, 1989; Cosmides and Tooby, 1989; Barkow, Cosmides, and Tooby, 1992) and that partly contribute to the mediation of behavior and function, such as signal detection, contingency evaluation, cost-benefit calculations, and novel strategy development. In effect, they are the systems responsible for information interpretation, decision making, and behavior. Evolutionary reasoning rejects Locke's *tabula rasa* view of the brain and replaces it with the view that many special-purpose systems have evolved. Some have highly specific functions, as studies suggest is the case for identifying cheaters (Mealey, Daood, and Krage, 1996). Others have more general functions. Social comparison provides an illustration. Social comparisons involve calculations that balance opportunities with threats and potential danger. They are made across a variety of contexts and in association with a variety of motives (e.g., mate choice, job success, competitive sports). And the information used for comparisons, as well as the criteria for comparison interpretation, differ significantly across contexts (Gilbert and Allan, 1994). It is the use of social comparisons across different situations, along with the diversity of information, that requires interpretation, and that suggests that algorithms also have general functions. Moreover, algorithms are influenced by learning, a point that is covered later in the chapter and that is important in explaining how algorithms that are usually highly adaptive are sometimes minimally adaptive. And algorithm function may change dramatically, as is often the case during extreme mood changes; for example, a nondepressed individual may enjoy meeting people and may view them as opportunities to make new relationships; when depressed, the same person may dread such events and avoid social contact whenever possible.

The algorithm concept is consistent with, but not necessarily isomorphic with, evolutionary propositions that the brain functions modularly (Gazzaniga, 1989, 1992), and that modules have evolved as specialized systems for responding to specific ecological conditions and options (Sawaguchi, 1988; Cosmides, 1989; Pinker, 1994). It follows that many behaviors often attributed to learning are more parsimoniously explained in terms of algorithm function; for example, humans enter the world with much of the learning apparatus and its apparatus-relevant content already in place.

An appreciation of ultimately caused traits is critical to the evolutionary interpretation and treatment of conditions. To a large degree, effective treatment hinges on identifying how these traits contribute to conditions. But ultimately caused traits can also enhance therapeutic options, as often occurs when individuals are capable of assimilating and using models provided by therapists.

Subsequent discussions of ultimately caused traits will focus on their identification, their functions, their contributions to and interactions with conditions, and the degree to which they can be modified by interventions.

Individual Fitness

Individual fitness refers to the within-population contribution to the next generation of one genotype relative to other genotypes (E. O. Wilson, 1975). Like *adaptation,* the term requires a referent: Person A has greater individual fitness than Person B if more of Person A's genes appear in the next generation. The genetic arithmetic of individual fitness is straightforward. Parents share approximately 50% of their genes with each offspring; 25% with each grandoffspring, niece, and nephew; and so forth.

The concept of individual fitness does not imply that selection has favored species whose members reproduce as rapidly as possible, or that reproduction is always the highest priority goal. Some species (e.g., insects) reproduce frequently. Others, such as elephants, have a more paced reproduction. When all species are considered, a variety of reproductive strategies have evolved, and literally all strategies that have been well studied turn out to be responsive to environmental contingencies. For example, among nonhuman primates, reproduction is influenced by such factors as available nutrients and social status (Fairbanks, 1988a; Fairbanks and McGuire, 1988). Among large samples of humans, it often positively, and rarely negatively, correlates with wealth and social status (Essock-Vitale, 1984; Low, 1991). And for as yet unexplained reasons, it is known to fluctuate in societies in which there is no evidence of conscious birth control (Low, Clarke, and Lockridge, 1992).

Are mental conditions associated with reduced individual fitness? This question has been a topic of research for at least five decades. Answers remain far from clear, however, at least for the majority of disorders. There are some exceptions, such as infantile autism, where findings suggest that reproductive success is well below average (Ritvo et al., 1985). On the other hand, for schizophrenia, reproductive rates do not appear to be reduced (e.g., McSorley, 1964; Slater, Hare, and Price, 1971; E. H. Hare, Price, and Slater, 1972; Erlenmeyer-Kimling and Paradowski, 1977; Erlenmeyer-Kimling, Wunsch-Hitzig, and Deutsch, 1980; Hilger, Propping, and Haver-kamp, 1983; Der, Gupta, and Murray, 1990; Jönsson and Jönsson, 1992; Lane et al., 1995). The most comprehensive long-term findings come from Norway, where studies indicate that there is a slightly higher fertility rate among persons suffering from manic-depressive disorder than among those suffering from schizophrenia, but both groups approximate the fertility rates for control populations not suffering from disorders (Ødegard, 1960, 1980). Many persons who were subjects in these studies were diagnosed and had offspring before the introduction of modern psychotropic medications,and therefore the possibility that salutary drug effects mask disorders was reduced.

The absence of clear differences in birthrates between persons with and without conditions has a number of possible explanations:

1. A percentage of persons with conditions who carry condition-influencing genes may go undetected. Possible examples include persons with mild phobias; mild forms of antisocial, histrionic, or obsessive-compulsive personality disorders; and the nonafflicted monozygotic twins of individuals with conditions such as schizophrenia.
2. Conditions may not be associated with a reduction in an individual's desire or willingness to engage in sexual activity. Moreover, some persons with disorders, such as borderline personality disorder and hypomania, may have increased sexual appetites.
3. A certain percentage of conditions (e.g., late-life depression, dementia) exhibit clinically relevant characteristics after the prime reproductive years and thus have no direct impact on reproduction. Pleiotropy may be applicable here.
4. Females with certain conditions may be especially vulnerable to sexual advances and sexual coercion by males.
5. Clusters of persons with conditions who live in social enclaves may influence reproductive outcomes.
6. Reproductive rates for persons with conditions may approach a maximum, while those among persons without conditions may decline for other reasons.
7. Behavior associated with many conditions may not be selected against.

While the relevance of each of these possibilities remains to be determined, it is Number 7 that has the most interesting implications, and that will return to the discussion both below and in subsequent chapters.

Individual fitness enters discussions about conditions in the following way: Does actual or potential reduced fitness increase condition risk?

Self-Interest

A basic premise of evolutionary theory is that individuals have evolved to act in their own interest, not in the interests of the group or the species (Williams, 1966). In evolutionary usage, self-interest does not equate with *greed, selfishness*, or *narcissism* as these terms are used in psychiatry and everyday language. Moreover, to act in one's self-interest does not mean that others will not benefit. In evolutionary reasoning, a mother's care for her offspring is not only self-interested behavior, but also fitness-enhancing behavior for both mother and offspring. Because maternal care increases the chances of offspring maturation and reproduction, both mother and offspring benefit. From another perspective, achieving biological goals normally requires the participation of others, and because it does, one often acts in ways that benefit others as well as oneself.

Both evolutionary biologists and philosophers have asked: How is it possible that individuals can be both self-interested and altruistic? Evolutionary answers to this question are found in the theories of kin selection and reciprocal altruism discussed in the next two sections.

Kin Selection and Inclusive Fitness

Increasing the number of one's genes in subsequent generations is not restricted to having offspring. Nonoffspring kin share a percentage of one's genes by direct descent from a common ancestor. Inclusive fitness is a measure of one's total genetic replication, including genetic kin other than direct descendants (W. D. Hamilton, 1964). Kin selection theory is an ultimate-cause explanation for the selection of altruistic behavior toward kin when immediate benefits may not be forthcoming (W. D. Hamilton, 1964). Investment in kin will occur when, on average, the loss in the investor's individual fitness is more than offset by an increase in the investor's inclusive fitness. While investment in kin can be costly (e.g., the expenditure of time, energy, resources), the costs may be offset if the recipients of the investment reproduce, in which case, one's genetic replication is the benefit. In evolutionary logic, increasing the possibility of genetic replication is one reason that parents invest in both their offspring and their kin. In a more complex instance, an altruist may be childless yet invest in his or her siblings, each of which share 50% of his or her genes. If the investment results in each of his or her siblings' having two additional offspring (each of the siblings' offspring carry one fourth of the altruist's genes), the altruist will have the same number of genes in the next generation as if the altruist had two offspring of his or her own, each of which shared half of his or her genes ($1/4 \times 4 = 1.0$; $1/2 \times 2 = 1.0$).

Given that there are limits to the amount individuals can invest, kin investment is likely to flow in ways that both minimize reductions in altruists' individual fitness and maximize their inclusive fitness. Because parents share more genes with their offspring than with the offspring of collateral kin, an obvious prediction from kin selection theory is that parents will invest more in their own offspring than in more distant kin. Studies of investment confirm this prediction (Essock-Vitale and McGuire, 1980, 1985b; Burnstein, Crandall, and Kitayama, 1994). Evolved cognitive and perceptual biases (e.g., viewing offspring in the best possible light) may also contribute to this outcome (Janicki and Crawford, 1992). Paternity certainty is yet another investment-influencing factor. A maternal grandmother is certain of her genetic relatedness to her grandoffspring via her daughter, but less so via her son. A paternal grandfather is less certain. Thus, grandmothers would be expected to invest more in the offspring of their daughters than in those of their sons, and grandfathers would be expected to invest proportionally less in both instances.

Kin selection theory enters the ensuing discussions in several ways: Do persons with conditions receive less than average investment from kin? Do persons with conditions invest in kin less than average? And if the investments received from kin are less than average, does the likelihood of conditions increase?

Reciprocal Altruism

Reciprocal altruism theory explains helping behavior among nonkin and, in certain instances, among kin. Person A will help Person B (a cost to Person A and a benefit to Person B) if there is a high probability that Person B will reciprocate (a cost to Person B and a benefit to Person A). As long as the benefit received by Person A exceeds the cost to Person A, selection should favor such behavior (Trivers, 1971; Blurton Jones, 1984). Because there is a delay between helping and repayment, helping others requires that a potential altruist take into account the future and its uncertainties (Trivers, 1985).

Reciprocal behavior is observed in a variety of nonhuman species, including bats (Wilkinson, 1988), lions (Packer, 1988), dolphins (Norris and Schilt, 1988), and nonhuman primates (de Waal, 1989; de Waal and Luttrell, 1988). A detailed review of studies and the uses of reciprocal and cooperative behavior among humans can be found elsewhere (Argyle, 1991). And as would be expected on the basis of kin selection theory, reciprocal behavior should occur with greater frequency among monozygotic twins than among either dizygotic twins or nontwin siblings. Findings are consistent with this expectation (Segal, 1984). *True* altruism would occur when a person adopts a nonrelated child off the street or from another country. Among some samples, 15% of homeless children are adopted by nonrelated adults (J. Lancaster, personal communication, 1996). This finding suggests the possibility of strong predispositions for helping others and for parenting. Adoption would be expected to occur in instances in which the thought of, or the presence of, a child triggers maternal and paternal emotions in potential adopting parents.

Reciprocal altruism theory is an ultimate-cause explanation of behavior, and it is consistent with the principle of self-interest. The theory presupposes the presence of a stable social environment in which there are consequences (e.g., social ostracism) for failing to reciprocate help. Yet, even in such environments, those who receive help may socially defect (accept help but not reciprocate). It follows that a complex set of assessments is essential to determine whether another person should be helped, how much he or she should be helped, when repayment is due, whether the repayment is sufficient for the help provided, and how to respond if repayment is not forthcoming (e.g., Cosmides and Tooby, 1987, 1989; Cosmides, 1989). Of all the ultimately caused traits we will discuss, none is more important for successful long-term social navigation than reciprocation.

A number of authors have called for revisions of the theory by suggesting that reciprocal behavior is a special form of kin selection behavior (Rothstein and Pierotti, 1988) or that it is a secondary consequence of evolved capacities to communicate (Caporael et al., 1989; Buck and Ginsburg, 1991). Resolving these differences is not critical. For our purposes, the key points are as follows:

1. Reciprocal behavior is observed frequently.
2. It occurs in all known cultures, and all known cultures have rules dealing with reciprocation.
3. The degree to which individuals do or do not engage in reciprocal behavior has measurable social consequences; for example, reciprocal relationships are reduced or discontinued if helping is not reciprocated (Essock-Vitale and McGuire, 1990).
4. Nonkin reciprocal relationships complement kin relationships by offering helping and repayment options that are often unavailable among kin (Essock-Vitale and McGuire, 1985a, 1985b).
5. Different attributes influence the likelihood of others' offering help; for example, persons who are socially responsible, who have religious affiliations, and who enjoy

life are more likely to be helped than persons lacking these attributes (Benson et al., 1980).

Reciprocal altruism theory and reciprocal behavior among persons with conditions have an important place in what follows: Do persons with conditions differ in their capacities to identify good and bad reciprocators? Do conditions compromise reciprocation capacities? Do conditions or their features alter the probability that others will provide help?

Proximate Mechanisms

Short-term changes in behavior are mediated by nervous system structures that have physiological, psychological, and anatomical properties. *Proximate mechanisms*—or *proximate causes,* another term often used—is the term applied to systems that are responsible for short-term behavioral changes. Ultimate causation explains why proximate mechanisms have been selected. Proximate mechanisms explain the workings of mechanisms within specific time frames. Ultimate and proximate causes are not alternative explanations of behavior. Rather, they are complementary. Behavior has both ultimate and proximate contributions.

Proximate mechanisms are often influenced by external information. For example, if one unexpectedly has the opportunity to acquire a mate or a valued resource, a cascade of psychological, physiological, and behavioral events may follow. However, except for reflex behavior (e.g., withdrawal from a painful stimulus), one-to-one relationships between external stimuli and responses are not uniformly observed. There are several reasons. Stimulus-response relationships are influenced by motivational priorities, and stimuli are seldom unitary. Other persons signal multiple, often conflicting, messages, and persons respond to different features of messages. Moreover, proximate systems compete among themselves, as is the case when one desires to insult someone but does not, or when one is tired but struggles to stay awake.

Because proximate systems can be manipulated, they have been attractive targets of psychiatric research, and over the past three decades, most of the research in psychiatry has focused on putative condition-causing and condition-ameliorating proximate systems, for example, genetic encoding, receptor density, and drug effects. Although we have suggested that this research focus is too narrow, we do not mean that it is unimportant. Proximate systems may be the principal cause of a disorder (e.g., chromosome breakage, missing enzyme), may influence the course of conditions (e.g., physiological states), or may alleviate conditions (e.g., medications, reassurance, or removing persons from frightening environments; Crawford, 1989; Reiss et al., 1991).

Proximate mechanisms enter the ensuing discussion in a number of ways, including their contributions to conditions; their interactions; the influence of social signals and environmental change on system function; and their response to medications.

Development

In evolutionary biology, the term *development* refers to the unfolding of ultimately caused maturational programs; interactions between these programs and external events (Blurton Jones, 1972; Ebbesson, 1984; Fairbanks and McGuire, 1988) and epigenesis (Alexander, 1990a). Maturational programs are thought to be similar among closely related species, a point suggested by studies demonstrating close analogies between human and nonhuman primate anatomical, hormonal, psychological, and behavioral development (e.g., Nyborg, 1994). These and related findings lend weight to such views as that tendencies to bond are anchored deep in our evolutionary past; that certain types of interpersonal interactions are essential for normal development (e.g., Harlow and Harlow, 1962; van de Rijt-Plooij and Plooij, 1987; Reite et al., 1981, 1989; G. W. Kraemer, 1992); and that environmental influences on phenotypes are distinct from the copying features of genes (Dawkins, 1982).

Psychiatry has a century-long interest in events influencing maturation. Freud devised his models of neuroses on two basic premises: that parent-offspring bonding is an essential prerequisite for the development of a healthy psyche, and that early psychological trauma can lead to intrapsychic conflicts and distortions that are the bases of an unhealthy psyche and neuroses. To this day, his views have remained topics of research and theoretical interest (Bowlby, 1969, 1973; Ainsworth et al., 1978; N. G. Hamilton, 1989; Stevens and Price, 1996).

Few investigators would dispute the importance Freud and Bowlby placed on mother-infant bonding, arguably the most important bond in one's life. However, when one turns to the details of bonding, things are less clear. The literature of development, which includes thousands of bonding-related findings, attests to this point (e.g., Savin-Williams, 1987; MacDonald, 1988a, 1988b). Only a few of the findings will be discussed here. Parents interact differently with their offspring (Weintraub and Frankel, 1977; Daly and Wilson, 1980; Eibl-Eibesfeldt, 1983, 1989; Lancaster et al., 1987), and parental temperament influences parent-offspring interactions (Kagan, Reznick, and Snidman, 1987, 1988). Different parental behavioral styles and family and social environments are associated with different outcomes in cognitive development (Sigman et al., 1990, 1991); representational tactics and the frequency and type of play (MacDonald, 1988b); personality features and the degrees and types of psychopathology (Plomin and Daniels, 1987); and resource acquisition strategies (Charlesworth and LaFreniere, 1983). Ethnic differences in the achievement of maturational milestones are known (Freedman, 1974), and relationships between parental investment (Davis-Walton, 1995), birth order, and personality type have been reported (Sulloway, 1995). Research among nonhuman primates confirms that a wide range of mothering styles is associated with the development of competent offspring (Fairbanks 1988a, 1988b, 1989). A reasonable expectation is that the same finding applies to humans.

Effects of adverse upbringing environments are also well documented. Among both human and nonhuman primates, adverse environments can have psychological, physiological, cognitive, and immune system consequences, some of which may continue throughout life (Spitz, 1945; Hinde and Spencer-Booth, 1971; Blomberg, 1980; Reite et al., 1981; Kraemer et al., 1984, 1989, 1991; Fairbanks, 1989; M. H. Lewis et al., 1990; Beauchamp et al., 1991; Kraemer, 1992; Schneider and Coe, 1993). Among humans, early and secure bonding has been shown to lead to both better physical and better mental health during the adult years (Vaillant and Vaillant, 1990), while disrupted bonding increases the chances of the opposite outcomes (Cloninger et al., 1982; Erickson, 1993).

The social environment is only part of the development story, however. Genetic information serves to channel and constrain maturational programs. Evidence suggests that many features of personality, such as fearfulness, shyness, extroversion, and neuroticism, reflect channeling and constraining influences (Kagan et al., 1987, 1988; Plomin, 1990). Compared to dizygotic twins raised apart, monozygotic twins raised apart show greater similarities in a variety of behavioral and personality measures (Bouchard et al., 1990), and hyperactive, attention-deficit adolescent males are at risk for later antisocial behavior and substance abuse disorders (Mannuzza et al., 1989, 1991). Yet, it is also true that behaviors that are marginally functional during the early years of development (e.g., impulsivity, quickness to anger) sometimes disappear during adolescence (Caspi et al., 1996). Given that age, biological goal priorities, and physiological changes interact, such changes are to be expected.

Evolutionary interpretations of development often differ from the prevailing model interpretations. Childhood phobias, parent-offspring conflict, and sibling rivalry will serve as illustrations. If children fear the dark when there is nothing to fear, their behavior is often interpreted as irrational and given a psychiatric diagnosis, such as childhood phobia. However, if it is allowed that fear of the dark and strange places may be predisposed, that in the past such fears were adaptive, and that a child's communication of his or her fears often results in protective behavior by caretakers, then it is worth considering the possibility that selection has favored both fear of the dark and parental responses to children's fears (Troisi and McGuire, 1992). Parent-offspring conflict is expected because self-interest leads to attempts by offspring to maximize parental investment (Trivers, 1974). In turn, parents will limit their investment in already-born offspring so that they can invest in subsequent offspring, assist collateral kin, or attend to other goals, such as acquiring resources. Thus, offspring seldom receive the amount of parental investment they seek. Haig (1993, 1995) has convincingly demonstrated that such conflicts begin in utero, where mother-fetus competition frequently escalates over the availability of nutrients: Fetal hormones increase blood flow to infants while reducing the available blood to mothers, and maternal uterine cells respond by opposing the invasion of fetal cells. Such conflicts can extend into later life, and extension may explain in part the high frequency of conflicts among adolescents with their parents, peers, and teachers. Similar points apply to sibling rivalry. Because of self-interest, siblings compete with one another over the available resources, including parental attention. When parents preferentially invest in offspring of the same sex—and studies suggest that they do (e.g., Weintraub and Frankel, 1977)—the degree of sibling rivalry should be greater in families in which the offspring are all either male or female.

What is to be made of the mass of developmental data, much of which remains to be satisfactorily explained and integrated (Ainsworth et al., 1978; MacDonald, 1988a, 1988b)? Despite the reluctance of many developmental psychiatrists and psychologists to embrace evolutionary ideas seriously (Charlesworth, 1986, 1992), it is likely that evolutionary concepts will serve to organize and explain a large percentage of the findings of developmental studies: "The evolutionary perspective can unite the study of both species-typical development and individual variation" (Scarr, 1992, p. 1). To take an obvious example, monozygotic twins are more similar in their mental development than either siblings or dizygotic twins (R. S. Wilson, 1978). Or consider parental warmth. In evolutionary context, warmth is not simply an inconsequential behavioral variant but an indication of parental investment, as well as a way of reducing the physiological consequences of stress in both infants and caretakers (MacDonald,

1992). In effect, the evolutionary proposition "that human social behavior, and the mechanisms of perception, cognition, and emotion, . . . is the product of . . . individual selection in the context of dealing with change and unpredictability in the social environment" (Thornhill, 1990, p. 13) opens a variety of investigative and theoretical doors for new and synthesizing looks at the effects of different upbringing environments and their interaction with maturational programs (van den Berghe, 1988). If findings from our closest nonhuman relatives and behavioral genetics are taken as guides, studies of development will seek a balance among the three factors that appear to most influence maturational outcomes: genetic influences on traits; interactions between features of the social environment and predisposed preferences for specific environmental features (Plomin et al., 1994); and the psychological and physiological effects of different types of caretaking. Each of these points enters subsequent discussions.

Traits, Traits and Genetic Information, and Trait Variation

Traits

Traits are measurable (e.g., height, anatomical symmetry, information processing speed, baseline CNS serotonin concentrations) or inferable phenotypes (algorithm function) that are influenced by genetic information and that have specific functions. For example, allergic responses may serve as immunological defenses against toxins (Profet, 1991); menstruation may function to remove pathogens transported by sperm (Profet, 1993); and language serves such functions as thinking and communicating.

Minimally adaptive traits can be characterized as either suboptimal or dysfunctional. The term *suboptimal* refers to a trait's degree of refinement relative to a defined standard for a person's age and sex, which will be defined as the modal measure for a specific trait in a culture. Persons differ in degrees of suboptimality, and these differences have a central place in explaining both social options and conditions. For example, some individuals lack the capacity to empathize and thus to accurately read others' behavior rules. And some individuals are physically attractive or intelligent while others are not. Social consequences follow (Cairns et al., 1988; Asher, 1990; Coie et al., 1990; Gilbert and Allan, 1994). *Dysfunctionality* refers to a temporary or state change in a trait, such as the inability to concentrate when one is extremely anxious. Suboptimal and dysfunctional traits and their consequences are topics throughout the remainder of the book.

Traits and Genetic Information

Chapter 9 addresses this topic in detail. Here, only two points will be noted. First, there is compelling evidence that genetic information is an important contributing factor to a host of traits, such as neuroticism, intelligence, and memory, that are of special interest to psychiatry (e.g., N. L. Pedersen et al., 1988; Brunner et al., 1993; Plomin et al., 1994; Benjamin and Gershon, 1996). Second, the degree to which many genetically influenced traits are independent is still a topic of research (Braungart et al., 1992; Bouchard, 1993, 1994). Determining the degree of independence is important in the assessment and understanding of phenotypes, in the identification and classification of conditions, and in the design of interventions. If traits are largely independent, the likelihood is great that conditions represent clusters of suboptimal and dysfunctional traits and that multiple causes need to be considered. Conversely, if seemingly separate traits actually represent different features of a single trait, the possibility increases that disorders are due to a few factors.

Within-Trait Variation

Within-trait variation refers to the degree to which a given trait varies within a population. Some people are tall, some short. Some have high baseline MAO levels, others have low levels. Some are shy, some outgoing. And some traits run in families; for example, among rhesus monkeys, families with high and low levels of the CSF biogenic amines norepinephrine, homovanillic acid, and 5-HIAA are known (Clarke et al., 1995). As with development, the factors contributing to trait variation are as important for explaining conditions as the variation itself. For example, a trait that is strongly predisposed may fail to undergo refinement because of a depriving upbringing environment, or changes in physiological states may reflect environmental information and strategy changes, as appears to be the case for CNS serotonin sensitivity in vervet monkeys (Raleigh et al., 1984).

A large percentage of within-trait variation is due to the genetic mixing that occurs with every conception. While such mixing is usually thought of in terms of its effects on offspring attributes, its primary evolutionary function may be to facilitate immunological defense. Genetic mixing is the postulated basis of *moving immunological targets* that counter the invasiveness of hostile, disease-causing parasites (W. D. Hamilton and Zuk, 1982). Because of faster reproductive rates and genotype-phenotype change, parasites have the potential of becoming increasingly effective in their attacks on slowly reproducing host species if the host species fail to alter their immunological makeup. Sexual reproduction among species with two sexes provides a basis for distributing genes that influence immunological capacities in ways that are unpredictable to parasites. The evolution of two sexes (sexual selection) rather than one is thought to have occurred in part as an evolutionary counterstrategy to offset the destructive effects of parasites and improve survival probabilities (W. D. Hamilton and Zuk, 1982; see also Ewald, 1988, 1991a, 1991b). Further, to the degree that sexual selection fosters genetic heterozygosity, the survival probability of heterozygotic individuals may be increased because of a greater number of immunologically active genes (Nesse and Williams, 1994). Put differently, altering immunological defenses while preserving immunological processes that distinguish between the self and the nonself is an adaptation to future environments that, because of new parasites, will be more difficult than the current environment (Trivers, 1985). Not surprisingly, parasite infection and social attractiveness interact; for example, studies show a negative correlation between degree of infection and attractiveness in a potential mate (Gangestad and Buss, 1993).

There is another important implication in this line of reasoning: If selection against parasites has taken precedence over selection for traits that are responsible for, say, cognitive capacities, cognitive capacities may be less well honed than if they had been more primary targets of selection. Studies that repeatedly demonstrate significant cross-person differences in cognitive capacities (e.g., reading comprehension) are consistent with this possibility. There are, however, alternative explanations. Both parasite defenses and cognitive abilities might be poorly honed because selection has worked primarily on other traits, such as the capacity to reproduce. Or it is possible that parasite defenses and cognitive abilities interact, rather than one's being primary and the other's being secondary; for example, effectively dealing with parasites may depend on cognitive abilities (e.g., gaining knowledge about parasites' habits and the conditions under which they invade organisms).

As the preceding points are developed further in later chapters, it is important to keep two points in mind: *Evolution has thrived on trait variation, and species evolve, in part, because of the reproductive and survival advantages associated with different traits.*

Traits, traits and genetic information, and within-trait variation enter the ensuing discussions that deal with trait optimality, trait suboptimality and dysfunctionality, condition likelihood and risk, and arguments favoring the use of an alternative classification system for conditions.

Learning

Learning gives the benefits of flexibility and efficiency in dealing with environmental contingencies while incurring the cost of sometimes learning the wrong thing. In evolutionary writings, discussions of learning deal primarily with three topics: selection favoring specific types of learning; the refinement of predisposed capacities to learn; and relationships between learning and adaptation (Staddon, 1983). From an evolutionary perspective, it makes little sense to assume that animals learn everything anew (Hinde and Stevenson-Hinde, 1973; Marks, 1987; Turke, 1990). A more likely scenario is that blueprints for learning are a part of the *Homo sapiens* genome, and that learning is guided by a set of internal instructions in combination with highly probable environmental events (e.g., imprinting in birds, adequate parental caretaking among primates) that shape both learning capacities and what is learned (Keil, 1981; Garcia y Robertson and Garcia, 1987; Tooby and Cosmides, 1990a; Avital and Jablonka, 1994; Plotkin, 1994; Maestripieri, 1995). Humans, like most animal species, are born into environments that foster learning what is essential for them to survive and reproduce, and learning and intelligence may have evolved to establish adaptive matching between the learner and certain short-term stabilities and instabilities of the world (Plotkin, 1994). The implications of this proposition for efficiency are hard to ignore.

The literature on learning is vast, and because so much is learned and so many different types of learning have been proposed, learning is not easily defined. In respondent or classical conditioning, learning takes place because of the continuity of environmental events. When events occur closely together in time, individuals are likely to associate them. In operant conditioning, learning is a consequence of one's action and its effects on the environment. Social learning theory combines classical conditioning and operant learning and postulates that there are reciprocal interactions between individuals and the environment: The environment determines aspects of behavior, and individuals can change the environment. For each of these learning types, cognitive processes have a central place in the mediation of responses to environmental events (Agras, 1978, 1989; Rescorla, 1988), and for each type, selection is assumed to have different effects (e.g., Avital and Jublonka, 1994). Specific types of learning appear to be of greater importance at different stages of development. For example, trial-and-error learning (operant learning) is probably more critical during childhood than during later life. And the situations in which learning occurs can affect both what is learned and subsequent behavior, a point illustrated in a study of snake phobias in rhesus monkeys (Mineka and Cook, 1986). In this study, the investigators allowed naive monkeys, which on initial exposure are fearful of snakes, to observe a nonfearful monkey in the presence of snakes. The naive monkeys were then introduced to snakes, and the majority showed none of the usual signs of fear. Prior exposure to a nonfearful conspecific thus appears to have *psychologically immunized* the observing monkeys from behaving fearfully. Such findings imply that parents' responses to events may influence responses to similar events by their offspring. These authors also call attention to differences between fearful affects and fearful avoidant behavior: some persons take avoidant action without much affective fear, whereas others may be fearful but do not avoid.

Not unexpectedly, learning and genetic information interact. For example, studies suggest that a significant percentage of the variance in information-processing speed and memory is influenced genetically (Plomin, 1990). Other studies suggest that the rate of learning is influenced by a variety of external contingencies (Agras, 1989). Given these points, it is not surprising that people also differ in the degree to which their learning systems are open-ended and subject to modification in response to new information. Indeed, most of us probably learn many things incorrectly and spend a good part of our lives trying to correct our learning mistakes; for example, inadequate social learning may result in a failure to calculate accurately the social costs of specific acts such as deceit, although the opposite point has also been made, namely, that social learning and memes may make individual learning more accurate (Dawkins, 1982; Boyd and Richerson, 1995). All this is not to say that the evolutionary explanations of learning are entirely satisfactory. For example, behavioral flexibility, a trait that is usually assumed to develop only after complex cognitive systems are in place, may be more phylogenetically primitive than is usually supposed: Learned behaviors that have adaptive consequences may precede genetic influences on such behavior (Smith-Gill, 1983; Tierney, 1986).

Much has been made of the idea that some conditions result from inadequate learning or dysfunctional learning. Two types of findings bear on this point. On one hand, research shows that persons learn at different rates: Intelligence, learning, and CNS glucose metabolic rates all correlate positively (Haier et al., 1992). These findings could be compatible with the inadequate-learning hypothesis. On the other hand, significant social benefits are associated with learning things correctly (e.g., one's manners often determine if one is accepted in a social group). Thus, there is an expectation that experience will gradually correct wrong learning. How these possibilities fit with trait explanations of learning differences must wait until Chapter 9, where a strong case is made for the constraining influence of suboptimal traits on learning.

Culture

Few parts of evolutionary theory have caused more debate than its explanations of culture. Views range from close ties between genes and culture (Lumsden and Wilson,

1981); to moderate but identifiable associations between migration patterns, genes, language, and culture (Cavalli-Sforza, 1991); to culture's being decoupled from genes (Bock, 1980). Here, we will limit the discussion to two points: our view of the capacity of evolutionary theory to explain key features of culture and the implications of this view for understanding disorders.

We agree with Richerson and Boyd (1978), who argue:

> Both genes and culture evolved by natural selection; the reproductive fitness optimum as a function of phenotype is different for genes and culture because the rules of inheritance of the two systems are different; and, a genetic capacity for culture is assumed to be optimized by selection with respect to genetic fitness. (p. 127)

This quote builds on the assumption that the capacity for culture is a trait or set of traits that have coevolved with predispositions for specific behavioral traits, and that these traits permit the development and perpetuation of certain cultural features while reducing the probability of others. Some features of culture are under genetic control, a relationship sometimes characterized by the term *distance,* and the distance between specific cultural traits and genes varies (Barkow, 1984). Other features of culture are outside genetic control, for example, multiple parents due to divorce and remarriage. Because environments differ, different cultures adopt and discard different explanations, social rules, and values. Those that endure, such as socially articulated reciprocation rules, are likely to be amplifications of predisposed traits. Said another way, there is good reason to believe that culture adjusts to the human genome, although often rather slowly, and often with significant digressions. Further, different cultures emphasize different attributes. For example, in tropical high-parasite climates, individuals involved in selecting a mate are likely to assign a significant weight to signs of immunological competence as revealed through the absence of disease. Conversely, in northern low-parasite climates, immunological competence may have a lower priority (Gangestad and Buss, 1993). It follows that there should be a high degree of agreement within specific cultures concerning the identification and labeling of behaviors that enhance fitness as well as those that reduce it: Culture tends to normalize behavior by reducing trait variation through positive responses to socially condoned behavior and negative responses to socially deviant behavior. It also follows that psychiatric labeling practices which focus on behaviors not easily tied to fitness may reflect transient social values rather than scientific criteria for condition identification and classification.

From another perspective, a key feature of culture is its establishment and perpetuation of the conditions for influencing the minds and behavior of others. Social attractiveness provides a convenient example. Gilbert (in press) suggested that the motivation for attractiveness is a major contributing factor to culture, as well as one that allows cultures to endure. In this model, social attractiveness makes one more desirable to others, and one's degree of desirability correlates positively with one's ability to influence others' minds and, in turn, one's social success (Gilbert, in press). In addition, moral signals from others, such as shaming, devaluing, or ostracizing, have powerful effects on controlling and directing the behavior of those receiving such signals (Alexander, 1987). The greater the frequency with which one receives such signals, the less one's social attractiveness. This view provides an explanation of why individuals go to considerable lengths to avoid being shamed or ostracized, and why numerous trade-off strategies are associated with such behavior, including deception, self-deception, altruism, self-sacrifice, and self-negation.

One other important feature of culture deserves mention. Culture takes into account sex differences (Fox, 1989; Boyd and Richerson, 1995). Evolutionary theory predicts that the ways in which males and females are brought up and educated will be tied to prevailing male and female reproductive strategies and the interpretations cultures give to these strategies. The available data support this prediction with respect to both child rearing (Chagnon and Irons, 1979; Low, 1989; Low et al., 1991) and strivings for cultural goals (Borgerhoff Mulder, 1987b).

Life History Strategies

Life history strategies are ultimately caused, genetically influenced programs for allocating resources to achieve biological goals. The life history strategy concept has its origins in studies of interactions between such demographic traits as size at birth; growth pattern; age at maturity; size at maturity; number, size, and sex ratio of offspring; age- and size-specific reproductive investments; and age- and size-specific mortality schedules (Maynard Smith, 1982; Harvey and Clutton-Brock, 1985; Charnov, 1991). Life history theory makes the simplifying claim that

> the phenotype consists of demographic traits—birth, age, and size at maturity, number and size of offspring, growth and reproductive investment, length of life, death—connected by constraining relationships, trade-offs. These traits interact to determine individual fitness. (Stearns, 1992, p. 10)

Life history theorists have taken the central organizing principles of evolutionary theory and developed models that explore the consequences of different patterns of energy allocation for achieving biological goals. For each species, selection favors the pattern of energy or resource dispersal that is associated with the greatest reproductive success. Some resources go toward growth and development, some go toward the basic body functions (e.g., acquiring nutrients), and some go toward reproduction (Fisher, 1930). How resources are allocated differs between the sexes and across age groups. One can observe other species only so long without concluding that they allocate their time and energy in age- and sex-characteristic ways, and that these allocations have reproductive and life expectancy consequences. Indeed, species are often referred to by terms that reflect their time-energy budgets: Beavers are industrious, sloths are lazy, and so forth. Similar points hold for nonhuman primates (Harvey and Clutton-Brock, 1985), which have species-characteristic patterns of foraging and socializing (Cords, 1988; de Waal and Luttrell, 1988; Partridge and Harvey, 1988), competing for mates, contesting status, and raising infants (McGuire, 1974). From an evolutionary perspective, these allocations are not a matter of chance or individual choice.

Resource allocations are bound together by numerous trade-offs (e.g., cognitive, physiological, life expectancy), and the outcome of trade-offs influences individual fitness; for example, having one's first offspring at age 20 or age 40 has different fitness consequences. Life history traits are also influenced by environmental contingencies. For example, some well-fed females achieve menarche at age 10, while some poorly fed females do not achieve menarche until age 20. Upbringing environments can have similar effects (Kim et al., in press). Such differences reflect interactions between the amount and quality of nutrients and hormone-mediated feedback systems that determine the onset of reproduction. Sexual selection is also a factor. Compared to males, females allocate more time and energy to the production and care of offspring, and to developing and maintaining social support systems. Males invest more time and energy in competing over mates, in acquiring external resources, and in obtaining and defending status and reputation. These differences lead to predictions of sex-related differences in condition frequency (chapter 7).

In the chapters that follow, life-history-strategy concepts are used to explain the amount of time and energy individuals devote to achieving biological goals. Motivations are viewed as synonymous with energy allocations, and allocation programs influence behavior through the prioritization of goals and through directing and setting constraints on the amount of energy and time devoted to achieving goals. While there are detectable differences in life history trait features (e.g., early versus late menarche), we have assumed that the basic allocation of resources for achieving goals, such as acquiring external resources and reproducing, does not significantly differ among persons with and without conditions. For example, women with schizophrenia and personality disorders often have normal menstrual periods, reproduce, and are possessive of their offspring and mates. Such findings suggest that life-history-strategy differences are not a major cause of conditions. Rather, cross-person similarities in motivations and energy allocations explain how persons with conditions are *locked in* to trying to achieve specific goals in their social environment. When their efforts are mediated through suboptimal or dysfunctional capacities, conditions become far more understandable.

Concluding Comments

There are nonevolutionary explanations for many of the points discussed in this chapter (e.g., Leak and Christopher, 1982; Crawford, 1987). For example, all human cultures of which we are aware recognize the importance of and have special names and explanations for mother-infant bonding, reciprocation, and cheating. Often, these behaviors are explained as traditions, as the wish of deities, or as common sense. These explanations are not surprising. To the degree that our behavior is strongly influenced by our evolutionary past, the products of evolution are likely to be recognized no matter what vocabulary is used or how the products are explained.

This brings us to a theory of behavior.

4

A Theory of Behavior

For an evolutionary theory of behavior, the following concepts are essential: ultimate causation, development, proximate mechanisms, function, sex differences, and the social, physical, and nonhuman animal environments. These concepts are discussed in chapter 3. Ultimate causation explains past conditions that rendered some traits better suited than others to solving adaptive problems. Development explains the refinement of traits and the effects of learning and experience. Proximate mechanisms explain moment-to-moment changes in behavior. Function is the purpose of behavior. The sexes differ in their anatomy, physiology, behavioral strategies, and capacities. Competing organisms influence health, behavior, and function. The environment is the context in which behavior takes place and most biological goals are achieved.

This chapter presents an overview of a theory of behavior. Systems of behavior are reviewed first; then, we focus on the four infrastructures responsible for behavior: motivations-goals, automatic systems, algorithms, and functional capacities. The social environment and environment-condition interactions are the topics of the third section, and the last section discusses the functional analysis of behavior—psychiatry's *missing link*. Clinical cases illustrate key points. And topics introduced in this chapter are developed further in subsequent chapters.

Behavior Systems

The term *behavior system* refers to *functionally and causally related behavior patterns and the systems responsible for them* (Hinde, 1982, p. 8). Four ultimately caused behavior patterns or systems are thought to apply to *Homo sapiens*: the survival, reproductive, kin assistance, and reciprocation systems. These are the systems that now command our attention. They are systems that were selected to meet past necessities: Organisms needed to engage their environments in ways that allowed them to efficiently obtain nutrients and mates and to avoid death prior to reproduction. Up to a point, if the systems responsible for carrying out these functions are already in place, so much the better (e.g., N. Tinbergen, 1951). In the theory of behavior outlined here, behavior systems serve three critical functions: (1) They are the basic frameworks for interpreting behavior patterns among individuals both with and without conditions; (2) they are the basis for grouping disorders for causal analysis (chapter 7); and (3) they serve as guides for the design of interventions (chapter 15).

The behavior system concept is not new to psychiatry, and ideas similar to those developed here can be found in the works of Bowlby (1969, 1973), Freedman (1979), McGuire et al. (1981), Gardner (1982), Bailey (1987), Chance (1988), Gilbert (1989, 1992), and Badcock (1992). The concept is consonant with the view that selection has favored systems which produce integrated patterns of behavior; that during evolution the human brain has progressively added capacities which permit increasingly complex and sophisticated information processing, particularly symbol processing (MacLean, 1985, 1990; Tooby and Cosmides, 1990a, 1990b); and that brain function can be understood by means of modular models (Gazzaniga, 1989, 1992). The four systems are inferable from observations of our closest relatives: Chimpanzees forage; they defend themselves; they select mates and reproduce; they invest in their offspring and collateral kin; and they develop reciprocal relationships (e.g., de Waal, 1989). The majority of biologically important human behavior patterns can be understood within the same systems. Systems interact, as when one discontinues reproduction-related behavior if one's survival is threatened. And behavior associated with one system influences the functionality of other systems; for example, developing reciprocal relationships may improve one's chances of survival (Hinde, 1982). Each behavior system can be further dissected into four key components or infrastructures: *motivations-goals, automatic systems, algorithms,* and *functional capacities.* A schematic representation of these components and their interactions is shown in Figure 4.1.

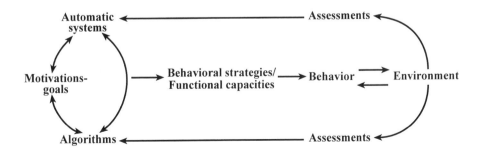

Figure 4.1 A schematic of behavior systems and infrastructures and their interactions.

Figure 4.1 depicts interactions among infrastructures and the social environment. Motivations-goals are essential conditions for behavior. Automatic systems and algorithms interact, process information, translate motivations into behavioral strategies, and mediate behavior via functional capacities. Depending on one's functional capacities and environmental contingencies (e.g., others' cooperativeness), one's behavior is more or less successful in achieving one's biological goals. Assessments of one's efforts to achieve goals can result in automatic system, algorithm, and strategy alterations.

To be sure, there are more complex representations of the points discussed in Figure 4.1. For example, in analyzing behavior-influencing factors, Hinde (1992) differentiated between society, group, relationship, interaction, individual behavior, and physiological variables. These factors interact among themselves, and they influence and are influenced by sociocultural structures and the physical environment. We will have more to say about these complex representations shortly. Here, the task is to elucidate some of the basic features of infrastructures.

Four Infrastructural Systems

Biological Motivations-Goals

Each of the behavior systems is associated with a characteristic set of ultimately caused biological motivations and goals. Motivations-goals are essential for behavior; they are influenced by both internal or external information (e.g., hunger, a threat); there are sex-characteristic differences in motivational priorities and intensities; and intensities and priorities change with age.

Table 4.1 presents the principal motivations-goals for the reproductive system. (The appendix presents motivations-goals for the survival, kin investment, and reciprocation systems.) Listed beneath each of the motivations-goals are functional-psychological and physiological events that are integral to goal-related behavior. Responses to goal-related failures are listed last.

Although Table 4.1 is consistent with the premises of evolutionary theory, other investigators would no doubt introduce modifications. For example, Scott (1950) listed the following basic motivation and behavior categories for carnivores: *ingestive* (eating and drinking); *investigative* (exploring social, biological, and physical environments); *shelter seeking* (seeking out and coming to rest in the most favorable part of the environment); *sexual* (courtship and mating behavior); *epimeletic* (giving care and attention); *agonistic* (any behavior associated with conflict, including fighting and escaping); *et-epimeletic* (soliciting care and attention); and *allelomimetic* (doing the same thing as another with some degree of mutual stimulation). For humans, Maslow (1971) argued that needs (e.g., self-esteem, individuality, perfection, goodness) and their underlying motivations are hierarchically arranged, and that fulfilling needs is an essential requisite for higher order activities (e.g., cognitive and aesthetic functioning). Buck (1988) developed a hierarchical system of motivations that includes reflexes, instincts, drives, acquired drives, affects, effectance systems (systems affecting the environment), and language. And Reiss and Havercamp (1996) postulated that

Table 4.1 Motivations-goals and associated features for the reproductive behavior system.

Motivation-goal: IDENTIFY AND SELECT MATE

Functional-psychological events: ↑ psychological regulation, joy satisfaction

Physiological events: ↑ physiological regulation, (*) opioid activity, serotonin activity; ↓ stress-induced hormone activity

Responses associated with goal-related failure: ↑ dysregulation, irritation, loneliness, depression, anxiety, self-deception, (*) dopamine activity; ↓ serotonin activity; development of alternative strategies to identify and select mate

Motivation-goal: RETAIN MATE

Functional-psychological events: ↑ psychological regulation, joy, satisfaction, power and control, self-esteem

Physiological events: ↑ physiological regulation, (*) opioid activity, serotonin activity; ↓ stress-induced hormone activity

Responses associated with goal-related failure: ↑ dysregulation, anger, possibility of retaliation, depression, social withdrawal, self-deception, (*) dopamine activity; ↓ serotonin activity; development of alternative strategies to retain mate

Motivation-goal: HAVE AND RAISE OFFSPRING

Functional-psychological events: ↑ psychological regulation, joy, satisfaction; focus of energy allocation

Physiological events: ↑ physiological regulation, (*) opioid activity, serotonin activity

Responses associated with goal-related failure: ↑ dysregulation, anxiety, depression, social withdrawal, self-deception, (*) dopamine activity; ↓ serotonin activity; alternative strategies devised to have or adopt offspring

Motivation-goal: PROTECT MATE AND OFFSPRING FROM ATTACK

Functional-psychological events: ↓ psychological tension and fear; ↑ the chances that mate and offspring will experience psychological regulation and pleasure

Physiological events: ↑ physiological regulation, (*) opioid activity, serotonin activity

Responses associated with goal-related failure: ↑ dysregulation, anxiety, depression, social withdrawal, self-deception, (*) dopamine activity; ↓ serotonin activity

Note. Because the details of very few physiological behavior relationships are fully understood, an asterisk (*) is placed in front of those physiological entries for which more research is required (↑ = increases, ↓ = decreases).

individuals differ in the amount of reinforcement they need to satiate their motivations.

Scott's categories can be integrated without difficulty within the survival, reproductive, kin investment, and reciprocation systems. In principle, integrating Maslow's, Buck's, and Reiss and Havercamp's categories is also possible (e.g., motivations are likely to have hierarchical features), although Maslow's emphasis on needs that are only remotely tied to biological systems (e.g., justice, beauty, truth) would increase the difficulty of integration. Integration issues aside, in our view the most important points to emphasize are that (1) motivation-goal systems develop in characteristic ways; (2) once developed, they act independently or semi-independently just as often as they do hierarchically; and (3) the motivation system that has the highest priority is largely a consequence of interactions between age- and sex-characteristic motivations-goals and environmental contingencies. Thus, motivational states are products of both external and internal variables. This characterization differs from earlier motivation models based primarily on internal systems, for example, "hydraulic" concepts of instinct and motivation (Freud, 1922; Lorenz, 1965).

When we look closely at Table 4.1 and the appendix, one of the first things we notice is that none of the items is obscure. Persons of all known cultures act to reproduce, to survive, to assist kin, and to trade favors with nonkin. Moreover, they do so in similar and predictable ways (Essock-Vitale and McGuire, 1980, 1985a, 1985b; Eibl-Eibesfelt, 1983, 1989). A second point concerns motivation-age interactions. Attempts to survive are observed from the moment of conception (Haig, 1993). Kin investment in the form of seeking proximity and bonding with caretakers occurs during the weeks following birth and continues until late life. Nonkin relationships develop toward the end of the first year of life and remain important until death. Concerns about acquiring a mate and having offspring intensify during the second decade and may remain important for several decades. And protecting a mate and offspring from harm begins after one has acquired a mate and offspring. (Differences in male and female motivational intensity are discussed in chapter 7.) The third point to note is that *the participation of others is essential for achieving goals* (e.g., acquiring a mate). It is this requirement that *locks* individuals into their social environment (the *locked-in* principle).

There are exceptions to the preceding points. Within limits, cultural values, social roles, and physical features of the environment influence both the time at which and the intensity with which individuals strive to achieve goals (Kim et al., in press), although this point probably applies more to adults than to infants or adolescents. There are also cultural practices (e.g., religious orders) in which age- and sex-characteristic goals are constrained (Steadman and Palmer, 1995). Such practices do not negate the points developed here; rather, they call attention to both human variance and the often strong impact of cultural features.

Motivations are subject to genetic influences and thus to trait variation. However, both evidence and theory favor the view that the degree of cross-person variation in motivations-goals is narrow, not wide. On the evidence side, studies show that the patterns of kin investment and nonkin reciprocation vary minimally for females between ages 20 and 40 (Essock-Vitale and McGuire, 1985a, 1985b), and despite the often striking differences between cultures, the opposite sex takes on a new meaning for nearly all individuals when they reach adolescence. Further, the vast majority of women have the vast majority of their offspring during a 10- to 15-year period beginning at age 18. On the theory side, motivations-goals, including motivational intensity, are likely to be similar across persons; otherwise, those with diminished intensity would be competitively disadvantaged and obvious targets of selection (e.g., reproduce less often, engage in less intense self-defense; cf. Marin, 1997).

The preceding points are relevant to explaining mental conditions: Although low-motivation hypotheses have often been entertained as a causal factor, with the possible exception of a handful of disorders, such as residual schizophrenia, hypoactive sexual desire, and postpsychotic depression, such explanations are neither compelling nor easily reconciled with much of the evidence discussed here. Motivations-goals appear to be buffered from many of the effects of conditions just as reproduction appears to be buffered.

Mrs. P

Mrs. P was 28 years old, married, with no children. She sought psychiatric help because of symptoms of depression associated with weight loss, an inability to perform routine tasks efficiently, frequent periods of crying, and feelings of inadequacy.

Until age 26, Mrs. P's history was a model of normal development. She was healthy, happy with her family and social life, pleased with her part-time job, and an active member of her church. Events leading to her depression began when she failed to become pregnant. A medical evaluation revealed fallopian tube obstruction. Surgery to correct her condition proved unsuccessful, and the doctors who performed the surgery recommended against further operations. Her depression worsened thereafter, and within months, there was a notable decline in her interest in her family and friends and an associated weight loss. Mrs. P had supportive parents, siblings, and friends, as well as a loving and caring husband. Nevertheless, she continually rejected their efforts to persuade her to obtain a second medical opinion or to adopt a child. At her husband's insistence, Mrs. P consulted a psychiatrist.

The psychiatric evaluation revealed that Mrs. P was suffering moderate to severe depression. Antidepressant medications were initiated and continued for six months. Her symptoms persisted. Treatment then switched to psychotherapy, which focused on her feelings about being childless and her resistance to considering either an alternative medical evaluation or adoption. After 20 treatment sessions, she agreed to obtain a second medical opinion. Several months later, an operation was performed to correct her condition. Six months after the operation, she became pregnant. A month later, the signs and symptoms of depression began to decline, and by the fifth postoperative month, there were no clinical indications of depression. Three years later, Mrs. P and her husband had a second child. At a five-year follow-up, Mrs. P had not suffered further from depression.

The case of Mrs. P suggests a number of important points: (1) Disorder assessments should consider the degree to which persons are achieving biological goals; (2) signs and symptoms interact with the degree of goal achievement; and (3) interventions designed to facilitate goal achievement are essential if some conditions are to be successfully treated. Sex differences also need to be considered. For example, studies of infertility show that 37% of the women but only 1% of the men in infertile marriages experience psychological disturbances (McEwan, Costello, and Taylor, 1987). From an evolutionary perspective, such differences are expected: Having offspring is a more central component of female than of male identity.

The case of Mrs. P also raises a number of critical questions, such as: Why do some individuals become depressed? Why does depression have such an impact on social cognition and behavior? And why is depression associated with outward signs of giving up rather than trying harder? Satisfactory answers to these questions require more discussion, but partial answers can be offered here. Individuals become depressed when their efforts to achieve biological goals continually fail while their motivations to achieve specific goals remain intense. Depression has its impact on social cognitions and behavior in part as a *warning system* that one's goal-seeking efforts are failing and in part as a social signal to others to elicit their assistance in achieving goals. And depression frequently leads to individuals' giving up trying because they believe they lack viable strategies to achieve their goals.

In summary, biological motivations-goals are ultimately caused systems that exert major influences on the direction, persistence, and intensity of behavior. They are assumed to vary minimally across persons of the same age and sex. Individuals bring their motivations to their environments, and environments influence the priority and intensity of motivations. Behavior is the outcome of these interactions. Persons are *locked in* to their social environment to achieve goals. And their degree of goal achievement inversely correlates with their signs and symptoms.

Automatic Systems

Automatic systems are ultimately caused, physiological-psychological-anatomical systems that *initially filter, select, and prioritize internal and external information. They are responsible for such information-processing activities as organizing visual information in three-dimensional space; associating sounds with location and distance; approximating the location of pain; recognizing familiar persons, environments, sounds, and smells; detecting others' emotional states; and emotions—in effect, organizing and prioritizing information that has adaptive value.* We have chosen the term *automatic system* to underscore the fact that neither are these systems volitional nor are their workings fully available to awareness.

The concept of an automatic system is not new to psychiatry. Writings suggesting or implying a high degree of pre-preparedness to process, prioritize, and selectively respond to information can be found in Bowlby's (1958, 1969, 1973) discussions of mother-infant attachment, Plutchik's (1980, 1991) and Nesse's (1990a, 1991b) formulations of emotions, Chance's (1988) models of agonistic and hedonistic systems, the writings of Price and his colleagues (Price et al., 1994) on depression, and Gardner's (1982) formulation of mania.

Clinical data are consistent with the idea that each of the four behavior systems is associated with a functionally distinct automatic system. Automatic systems can be initiated by both motivations and external information. If the intensity of motivations is high, external information may be disregarded. If intensity is low, the source of information often determines which automatic system is activated. For example, the same request for help by a sibling and by a friend will be processed within the kin-investment and reciprocation systems, respectively; in turn, different interpretations and responses will follow. In extreme situations, such as being suddenly confronted by a predator, one's eyes focus; nonrelevant noises and thoughts are disregarded; and one's body quickly prepares to flee or confront the danger. These responses point to the priority given to information processing exerted by the survival automatic system.

Within limits, automatic system priorities are learned and refined, and relative to motivational systems, a greater degree of cross-person variation is observed. Much of the learning and refinement occurs in association with emotions. A child may be unafraid of fire. He may play with it, burn himself, and experience pain. Subsequently, the importance he attaches to fire will be different.

Suboptimality and dysfunctionality of automatic systems have a major place in explanations of conditions. Suboptimality is implicated if automatic systems consistently function less efficiently and precisely than is typical for a person's age and sex, or if the systems are unresponsive to learning and refinement. Dysfunctionality is implicated if compromises are time-limited (state conditions). Clinical signs of compromised systems include inattention to goal-relevant information (e.g., failing to process biologically important signals), misprioritization of information (e.g., consistently viewing neutral events as dangerous), and insufficient information filtering (e.g., excessive intrusions of emotions or thoughts into consciousness). Within-family similarities in information processing point to genetic influences on automatic system function, and among families with a high prevalence of disorders, such as residual schizophrenia or schizoid personality disorder, these influences may be significant (Lenzenweger and Loranger, 1989).

Because automatic systems are responsible for the initial filtering, organization, and prioritization of information, they influence the functionality of the two infrastructure systems yet to be discussed (algorithms and functional capacities). And because automatic systems have physiological, psychological, and anatomic components, changes in one part of a system (e.g., psychological) may influence other parts. (This point explains in part the observation that the same condition can be associated with very different assessments [e.g., psychological, physiological] when features of a condition are viewed through the lenses of two or more models.) It follows that alterations in automatic systems may be achieved in a variety of ways; that is, psychological and pharmacological interventions may result in similar therapeutic outcomes (Baxter et al., 1992).

Mr. B

Mr. B was a 24-year-old divorced male who was referred for psychiatric evaluation by the court because he was physically abusive to women.

His history revealed a decade-long series of impulsive and violent acts; repeated conflicts with his parents and the law; fights with peers; periods of unexplained anxiety; and a six-year history of possessiveness and suspiciousness toward females. His education record was dismal, despite his performing well above average on IQ tests. He had no close male friends, and he lacked empathy.

Mr. B married at age 22. In the month following his marriage, he became preoccupied with the belief that his wife was sexually involved with other men. Wife beatings, which were preceded by periods of intense anxiety and anger, began soon thereafter. Four months into the marriage, his wife filed a civil complaint for spousal abuse and began annulment proceedings. Mr. B showed no remorse for his behavior and was convicted of spousal abuse and sent to jail. Three months after his release from jail, he was arrested for beating the woman he was dating. The arrest led to a court-ordered psychiatric evaluation.

Psychological testing and clinical interviews revealed that Mr. B viewed women as self-serving and dishonest; that he was highly opinionated and lacked insight; and that he misinterpreted neutral events (e.g., his wife's briefly talking to the next-door neighbor in the front yard). Physiological measures were within normal limits. Treatment consisted of anti-anxiety drugs, counseling, and group meetings with males with similar histories. Although medications resulted in a moderate reduction in his anxiety, treatment did not alter his belief that females were untrustworthy and likely to cheat sexually. After a year, treatment was considered unsuccessful and discontinued.

Proximate explanations of Mr. B's behavior include the misinterpretation of reproductive-related information because of automatic system suboptimality; frequent unpleasant emotions (anxiety and anger) associated with misinterpreted information; and a behavioral response (physical abuse) directed toward the perceived source of the unpleasurable emotions. The failure of the therapy was not unexpected. Conditions in which suboptimal automatic systems have a prominent role have proved to be difficult to treat, and in the models developed here, suboptimal automatic systems are a major contributing factor to chronic conditions (e.g., severe forms of paranoid, schizoid, borderline, and narcissistic personality disorders; sexual perversions; schizophrenia; and mental retardation).

The case of Mr. B is not without evolutionary precedent. In evolutionary context, males are predisposed to possess females for a number of reasons, including competitive advantage, predictable intimacy, the bearing of offspring, and reduced paternity uncertainty (Daly, et al., 1982). When possession fails, males often respond with anger. A limited capacity to empathize, which means a reduced capacity to think and feel as others, appears to be a key factor when anger turns to physical abuse and one is unable to modulate behavior that is predisposed (Miller and Eisenberg, 1988). Although such behavior is to be deplored, and although the legal and social consequences of physical abuse may reduce the frequency of such behavior, predisposed behavior is not always easily controlled, particularly in individuals whose automatic systems are suboptimal.

Mr. X

Mr. X was a 19-year-old, unmarried college sophomore at the time he was referred to a university psychiatric hospital for evaluation. Referral occurred two weeks following what his friends described as the onset of "strange behavior," "social withdrawal," "hallucinations," "intense fear and anxiety," and an "inability to make decisions."

Mr. X was the fourth of six children. As a child, he had been shy and often anxious. He had had few childhood friends. At age 8, he had been referred for a psychiatric evaluation because of persistent bed-wetting, marginal school performance, periods of compulsive behavior, and an inability to experience pleasure. Anti-anxiety medications were prescribed. They were moderately effective in reducing his anxiety and compulsive behavior, and there was a notable improvement in his school performance. He had continued taking medication until five weeks prior to his evaluation. There was no history of substance abuse. Two relatives, one each on the maternal and paternal sides of his family, had similar histories. Both had been diagnosed as "schizophrenic," and both were receiving antipsychotic medication.

Mr. X was admitted to a psychiatric inpatient facility and treated with antipsychotic medications. Initially, there was a reduction in his anxiety and hallucinations. However, other signs and symptoms persisted. Over the ensuing weeks, his mental condition became worse, and different combinations of medications failed to stop his deteriorating clinical course. Eventually, he was transferred to a long-term-care facility.

Most likely, Mr. X's diagnosis was schizophrenia. Schizophrenia is one of several conditions in which at least one automatic system, and probably more, is postulated to be severely suboptimal because of system deficits that result in incomplete and distorted processing and use of information. Mr. X's case also illustrates an important clinical point: Automatic system functionality seldom changes dramatically. Detailed histories of persons who seemingly develop severe disorders without a prior history of social difficulties often reveal signs of compromised function (Fish, 1987; Walker and Lewine, 1990; Fish et al., 1992). Shyness, bed-wetting, marginal school performance, excessive anxiety, compulsive behavior, and lack of friends are possible examples in the case of Mr. X.

As noted, automatic systems are the postulated basis of emotions. Descriptively, emotions are somatic-psychological states that last only a few seconds. Moods are somatic-psychological states that endure. Emotions and moods may be initiated by either internal or external information, including motivations, and through their automatic system effects, they modify the function of other infrastructural systems; for example, anxiety and depression lead to the reappraisal of information. In evolutionary context, emotions are evolved responses that have been selected because they assisted in solving adaptive problems (Plutchik, 1980, 1984a; Nesse, 1990a). Because they provide information and contribute to behavioral change, they can be considered forms of intelligence (Plotkin, 1994).

Chapter 5 offers a detailed discussion of emotions. Only two points will be mentioned here:

1. Emotions provide information more rapidly and influence behavior more forcibly than, say, objective reasoning. A strong emotion not only reduces the probability that other emotions will be experienced simultaneously but also increases one's focus on the perceived cause(s) of the emotion; for example, if another person is viewed as the cause of one's physical or emotional pain (e.g., shame), anger combined with retaliatory or escape strategies are likely.

2. Emotions provide information about the costs and benefits of past, present, and future behavior. Unpleasant emotions associated with a particular behavior are experienced as costly and fitness-reducing, and they mold subsequent behavior by reducing the likelihood that one will behave in the same way again (Nesse, 1990a). Conversely, pleasant emotions are experienced as beneficial and fitness-enhancing and increase the probability that one will engage in similar behavior in the future.

In summary, automatic systems are ultimately caused systems that have psychological, physiological, and anatomic components. Their primary functions are to filter, organize, and prioritize biologically important information and to mediate motivations. Different types and sources of information trigger different automatic systems, which are the basis of emotions. Suboptimal or dysfunctional systems are major contributors to conditions, as well as to many features of conditions. And the degree of automatic system functionality correlates inversely with condition probability and severity.

Algorithms

As noted in the preceding chapter, *algorithms* are ultimately caused, anatomical-physiological-psychological systems. They utilize information filtered by automatic systems (Cosmides and Tooby, 1987, 1989; Cosmides, 1989), and they are responsible for the mediation of behavior. Reasonable assumptions are that algorithms evolved somewhat later than automatic systems; that among humans they are elaborations of information-processing systems observed among nonhuman primates; and that they were selected because of their adaptive value. Like automatic systems, algorithms can be viewed as special forms of intelligence. Clinical experience suggests that algorithms are subject to greater degrees of variation than either motivations-goals or automatic systems; that they are subject to greater degrees of refinement; and that refinement occurs across the entire life span, although less so with advancing age.

Both special-purpose and general algorithms are thought to exist. Examples of special-purpose algorithms include assessing or estimating the costs and benefits of social exchanges (cost-benefit assessments), interpreting events and constructing causal models, developing novel behavior strategies, monitoring the effectiveness of one's behavior (self-monitoring), and empathy. (Special-purpose algorithms are discussed in detail in chapter 6.) Examples of more general-purpose algorithms include social learning and abstract concept formation. The total number of algorithms is unknown. The view that there are both special-purpose and more general algorithms conflicts with theories that assume that information is processed by a single all-purpose system. An all-purpose system seems highly unlikely, however. Indeed, it is no more probable that a single information-processing system could solve all the information-related adaptive problems an organism faces than it is that a single general-purpose organ could perform all the physiological functions required for life.

Algorithm functionality can be both facilitated or constrained by automatic system function. Facilitation occurs when automatic systems contribute to the accurate interpretation, unambiguous organization, and clear prioritization of information in ways that are orthogonal with motivations-goals and environmental contingencies. Constraint occurs when information is misinterpreted, poorly organized, or misprioritized. *Unlike automatic systems, individual algorithms appear to function in association with more than one automatic system; for example, the algorithms responsible for cost-benefit assessments and causal modeling function in association with each of the four automatic systems and only the information they manipulate differs.* (If this were not the case, efforts to identify cognitive rules might well be doomed.) Suboptimal and dysfunctional algorithms have their own clinical signatures and functional consequences. For example, a limited capacity to assess costs and benefits in dyadic relationships will lead to a high percentage of socially inappropriate behavior, and a limited capacity to self-monitor accurately will lead to repetitions of the same behavior despite negative outcomes.

Mr. Z

Mr. Z was a 27-year-old unmarried male who sought psychiatric treatment because of difficulties at work. On his initial interview, he was anxious and preoccupied.

Mr. Z was a foreman in a tool-and-dye factory, a job that he had held with several different companies. Despite his superior technical skills, his work history was one of average job performance and occasional costly mistakes. While at work, he suffered from anxiety and anger. These emotions were largely tied to interactions with his superiors whose behavior disappointed him and that he often criticized. Away from work, he appeared to function normally and was seldom symptomatic.

Mr. Z was the third son of a working-class family. His mother had died when he was 2 years old, and he had been raised by his father and two older brothers, who were strict disciplinarians and who punished him frequently for failing to complete chores, for pranks, and for marginal school performance. Following his graduation from high school, he had left home to seek employment in another town. He maintained minimal contact with his father and brothers and considered himself a loner.

Therapy consisted of medications to relieve his anxiety and psychotherapy to assist him to understand how his behavior contributed to his work-related difficulties. The medications were only partially effective. Psychotherapy revealed a host of findings characteristic of persons with similar histories, for example, his belief that he was misunderstood and unappreciated, his reliance on others' views of his job performance for his self-esteem, and his limited capacities to develop novel strategies for dealing with frustrating work situations. Despite these features, Mr. Z did not seriously misinterpret or misprioritize social information outside work, a point that suggests his condition was not due primarily to automatic system suboptimality, although intermittent dysfunctional periods were implicated because of work-related anxiety and anger. His awareness of the negative effects of his behavior on his supervisors suggested that his self-monitoring algorithm(s) was functional.

Psychotherapy continued twice a week for 16 months and focused on developing more realistic job expectations, alternative ways of interpreting the actions of his supervisors, and avoiding anxiety-provoking interactions. Moderate behavioral change followed. More dramatic changes occurred when Mr. Z "realized" that he had been "unknowingly" bringing past conflicts with his father and older brothers into the workplace. By the time therapy ended at 22 months, Mr. Z had largely resolved his conflicts with his supervisors. He was less frustrated and angry at work; he was experiencing anxiety-free periods without medications; his ability to take advice from his superiors had increased; and he was making fewer decision mistakes.

The case of Mr. Z illustrates two important points: (1) Misperceptions of one's social environment and socially inappropriate responses often accompany algorithm dysfunctionality (dysfunctional causal modeling and the development of novel behavior strategies in the case of Mr. Z), and (2), algorithm dysfunctionality can be treated by assisting individuals to develop novel models applicable to themselves and their social environment.

In summary, algorithms are evolved physiological, psychological, anatomic systems that manipulate information and put it to use. Individual algorithms function in association with more than one automatic system, and algorithm suboptimality or dysfunctionality can occur separately from, or in conjunction with, compromised automatic system function.

Functional Capacities

Functional capacities are evolved capacities to execute behavior. Their evolutionary origins lie in the adaptive problems associated with moving about and exploiting scattered energy sources (survival), locating mates, reproducing and protecting kin, and interacting with nonkin (Plotkin, 1994). Functional capacities are influenced by genetic information, and they are subject to greater degrees of variation than any of the three previously discussed infrastructures. They are mediated primarily by algorithms and are thus facilitated or constrained by algorithm functionality. They may also be constrained on their own: One can develop a behavior strategy yet lack the capacity to execute it, such as singing on key. Efficient function is contingent on capacities to enact functional capacities.

A list of functional capacities is presented in Table 4.2, and further analysis can identify the behaviors most often associated with each capacity. Examples are found in Table 4.4 and the remaining chapters.

It is unlikely that any of the capacities listed in Table 4.2 are unfamiliar. Literally all have been observed among humans across a variety of cultures and contexts (Eibl-Eibesfeldt, 1989; Ekman and Friesen, 1969; Ekman, 1971, 1993), and analogous behaviors are seen among nonhuman primates (e.g., Hinde, 1983). The initial expression of functional capacities, such as recognizing caretakers, avoiding strangers, and accurately communicating one's emotions, occurs early in life (Blurton Jones, 1972) and at a similar age among persons living in different social and physical environments (Freedman, 1974; Eibl-Eibesfeldt, 1983, 1989). Ethological studies of any one of the behaviors reveals tendencies toward context-specific uses (e.g., McGuire and Lorch, 1968; Hinde, 1974; Kendon, Harris, and Key, 1975: Coulter and Morrow, 1978). *The absence of a functional capacity is rare.*

The functional capacities listed in Table 4.2 are often apparent. Infants cry when they are hungry, in pain, or lonely, and specific behaviors (e.g., turning away) are associated with attempts by infants to avoid strange and fearful situations. As children

Table 4.2 Functional capacities.

Information processing: The reception and manipulation of information

Behaviors include:
- Memory
- Thinking
- Sense knowledge (the ability to use the basic senses, including sight, hearing, touch, pain, and taste, and to alter selectively the amount and kind of sensory information to which one attends)
- Observational learning (the ability to acquire new behavior patterns following the demonstration of these patterns by others)
- Active learning

Social understanding behavior: An awareness of the norms of interactions among the members of a group

Behaviors include:
- Understanding group spatial, postural, and interactional norms specific to the individual's age, sex, environment, and context
- Understanding the implicit and explicit group goals and goal-related behaviors
- Understanding others' motives, behaviors, and feelings (This behavior is limited to situations where one is in the presence of another. Critical information for this kind of understanding includes perceiving the actual of implied content of statements, behavioral intensity and frequency, the consistency between behavior and statements, and behavior in relationship to context)
- Anticipating the emotional, physical, and cognitive needs of others (This behavior differs from the behavior above in that it refers to inferences made from secondary sources or prior associations while the person in question is absent)
- Understanding the effects of one's behaviors on others (This behavior also includes understanding the effects of inaction when behavior is expected)
- Understanding the available social support systems
- Understanding one's available social options
- Monitoring others' behaviors (the ability to gain information by watching others without interacting with the person being monitored, as in a parent watching what a child is doing, a teacher determining if a student is studying, or a foreman determining if a laborer is working. One often attempts to disguise the fact that one is monitoring)

Social maintenance behavior: The preservation and continuation of behaviors that are useful in social interactions

Behaviors include:
- Verbal behavior (the ability to comprehend and to speak one's native language. Spoken communicative and interpretive competence, not reading or writing ability, are the behaviors indicated here. People often use bad grammar, have poor elocution, or utter incomplete sentences, yet what they say may be perfectly intelligible. Likewise, one may not understand the meaning of all one hears yet easily understands what is being said)
- Nonverbal behavior (the ability to comprehend and to communicate nonverbally through gestures, postures, facial expression, and movements associated with verbal communication. As in verbal behavior, style differences exist in this category, but they are not relevant unless they alter communication efficiency or precision)
- Overt behavior according to group spatial, postural, and interactional norms established through familiarity
- Overt behavior in accordance with the implicit or explicit goals of a group
- Communication of one's motives, thoughts, and feelings (Various means of information transmittal can be used, including verbal and nonverbal behavior, refusal to act according to others' expectations, and the use of special behaviors, the precise meaning of which is known only to friends)
- Toleration of motivational and emotional conflicts with others

Table 4.2 (continued)

Social manipulation behavior: Influencing the outcome of interactions with others to one's own advantage

Behaviors include:
- Utilizing one's available social options
- Using others' affective systems to alter their behavior (the ability to elicit feelings of guilt, sympathy, compassion, fear, anger, affection, joy, or cognitive changes that result in behavioral changes in others)
- Promising to others future emotional, cognitive, and material resources (the ability to communicate to others that they will be provided with emotional, cognitive, and/or material resources at some future time)
- Extracting from others future promises of emotional, cognitive, and material resources
- Display, territorial signaling, and/or territorial defense behavior (the ability to engage in postural, verbal, threat, and/or attack behaviors that convey information to others about one's attitudes, feelings, and/or potential behaviors as they relate to a given area, an idea, or a relationship with another person to which one believes one has special interactional, spatial, or reciprocity rights)
- Disguise of cheating (the ability to convey to others that one is providing them with emotional, cognitive, or material resources that are not being provided)

Social exchange behavior: Giving or receiving any commodity (e.g., material things, information, emotional support)

Behaviors include:
- Meeting others' emotional, physical, and cognitive needs
- Utilizing available social support systems
- Receiving altruistic behavior (the ability to accept beneficial assistance from others)
- Engaging in altruistic behavior (the ability to engage in behaviors that involve costs. In extreme cases, one may temporarily lower the probability of one's own survival to confer benefits on others. Helping people out of dangerous situations, sharing resources, and doing certain kinds of favors all belong to this category)

Self-understanding behavior: An awareness of one's own being and needs

Behaviors include:
- Feeling (the ability to experience and to alter different emotions)
- Anticipating one's own emotional, physical, and cognitive needs
- Identifing essential material resources (e.g., money, food)
- Understanding the operative characteristics of the material environment
- Anticipating one's material resource requirements (This category includes anticipating the requirements of those individuals for whom one assumes responsibility, e.g., children and elders)
- Monitoring one's own physical and psychological health (Correlational knowledge is sufficient in this category; e.g., a pain in the throat signifies throat infection)
- Understanding one's own motives and their relationship to one's thoughts, feelings, and behavior

Self-maintenance behavior: Providing for the preservation and continuation of one's own well-being

Behaviors include:
- Meeting one's basic physical, emotional, and cognitive needs
- Acquiring essential material resources
- Altering priorities
- Tolerating physical environment conflicts
- Enjoying certain emotions and behaviors
- Tolerating reciprocity imbalance
- Altering the material environment
- Physical adroitness
- Maintaining physical attractiveness

Note. The table is adapted from the McGuire and Essock-Vitale (1981), where the methods used in the selecting items in the table are discussed.

grow, parents try to shape and refine their capacities (e.g., teach social skills), but such efforts are only partially successful. Some capacities, such as effectively communicating one's feelings, are more important for achieving goals than others, although context is always an influencing variable. For example, in a safe environment, a limited capacity to accurately communicate one's emotions is likely to have greater consequences for social navigation than compromised physical adroitness, while the opposite point may apply in a physically dangerous environment. What often distinguishes persons with conditions from those without is a higher percentage of compromised capacities (see Figure 4.4) (Grant, 1968, 1969; Argyle, 1972a, 1972b). Said differently, the greater the number of compromised capacities the less one's social attractiveness (Gilbert and Allen, in press).

In summary, functional capacities and their associated behaviors are the means by which persons achieve biological goals, such as investing in kin, acquiring mates, and eliciting others' cooperation. Capacities may be suboptimal in their own right or dysfunctional because of automatic system or algorithm suboptimality or dysfunctionality.

Two points bring this section to a close. First, thus far, infrastructures have been discussed as if they operated linearly; for example, automatic system function influences algorithm function, which influences functional capacities, which influence function. As noted earlier, linear relationships are the exception in biological systems, in that system interactions are modulated by both internal and external information. For example, emotions provide information about the state of one's strategies, they alter others' behaviors, and they frequently lead to strategy changes. Second, the suboptimality or dysfunctionality of any of the four infrastructural systems reduces behavior plasticity, compromised automatic system functionality generally having the most significant effects, and compromised functional capacities having the least.

The Social Environment

Literally all species of diurnal primates spend the majority of their lives in social groups. *Homo sapiens* is no exception. The social environment includes those persons with whom and through whom one attempts to achieve biological goals, as well as the persons whom one attempts to avoid because of possible distasteful or unbeneficial interactions. Among psychiatry's prevailing models, only the sociocultural model accords the degree of importance to the social environment found in evolutionary biology. In evolutionary thinking, relationships between a person and the environment are bidirectional, and the environment *cannot* be excluded from explanations of behavior, normal or otherwise. Biological investments (costs) consist primarily of the expenditure of time, energy, and resources for the benefit of oneself, one's kin, and nonkin. If searching for nutrients is excluded, most of the biological benefits (e.g., repaid favors, increased probability of survival, offspring production) come from interacting with others. Environmental options and one's capacity to navigate the social environment correlate with the costs of behavior and the benefits that accrue. As a rule, the more benefits exceed costs, the less likely an individual is to develop a condition.

Discussions dealing with how the social environment is best characterized have a long and fascinating history (e.g., G. W. Brown, Bhrolcháin, and Harris, 1975; Krebs and Davies, 1978, 1987; Barton et al., 1992). This history will not be pursued here, except to note the following points: Persons differ in both their interpretations of and their responses to environmental events (A. B. Clark, 1987); adverse environments are associated with a high prevalence of somatic symptoms (I. Grant et al., 1981); culturally supported social systems, such as marriage, can be highly stressful during certain periods of ecological change (Low, 1990a, 1990b); and social behavior rules undergo change (e.g., infanticide was legally practiced in Europe until several centuries ago; Oliveric, 1994). Characteristics of the physical environment, such as light, nutrients, and poisons and their possible condition-triggering effects, are discussed in detail in psychiatric textbooks and will not be reviewed here. Instead, this section focuses initially on two features of the social environment that research and theory suggest interact with infrastructural functionality and condition probability: social structure and the rate at which new individuals enter one's social environment. We will then turn to a discussion of social support networks and social status systems.

Social Structure and Rate of Change among Members of the Social Environment

Social Structure

The structure of the social environment is a product of interactions between numerous factors, including the number of one's kin, same-sex preferences, and cognitive constraints. Findings from a recently reported study provide an unusually clear illustration of these interactions:

> Data on the number of adults that an individual contacts at least once a month in a set of British populations yield estimates of network sizes that correspond closely to those of the typical "sympathy group" size in humans. Men and women do not differ in their total network size, but women have more females and more kin in their networks than men do. Kin account for a significantly higher proportion of network members than would be expected by chance. The number of kin in the network increases in proportion to the size of the family; as a result, people from large families have proportionately fewer non-kin in their networks, suggesting that there is either a time constraint or a cognitive constraint on network size. A small inner clique of the network functions as a support group from whom an individual is particularly likely to seek advice or assistance in time of need. Kin do not account for a significantly higher proportion of the support clique than they do for the wider network of regular social contacts for either men or women, but each sex exhibits a strong preference for members of their own sex. (Dunbar and Spoors, 1995, p. 273)

The structure of the social environment, which may be defined as *the number, type, and attributes of persons with whom and through whom biological goals are achieved*, correlates with the probability of having a mental condition. Family composition provides an example. Children who grow up in households that include their two natural parents and one grandparent (usually a maternal grandmother) have an unusually low probability of developing a condition, in part, no doubt, because grandparents are less conflicted than parents with competing investment priorities. In contrast, children who grow up in families with only one biological parent (usually the mother), and often the mother's male companion, have an unusually high probability of developing conditions. For the single-parent–male-companion households, the referral rate of children for psychiatric care is seven times greater than it is for the two-biological-parent-one-grandparent household (Essock-Vitale and Fairbanks, 1979). Differences in referral rates are of course influenced by factors other than family structure, such as the capacity of household members to invest in one another. Nevertheless, the potential condition-contributing effects of structural features cannot be easily discounted, as persistent strong associations between these features and condition probabilities attest.

Structure and behavioral probabilities also interact. An obvious example is how most individuals behave in dyads compared to group situations. Persons with conditions are not an exception, as is illustrated in Figure 4.2.

Figure 4.2 shows that the percentage of socially atypical behaviors (e.g., peculiar postures, talking to oneself, repetitive behaviors) among persons hospitalized in a psychiatric inpatient facility inversely correlated with the number of people in functionally defined social areas (e.g., dining room, recreation room). When subjects in this study were alone in any of the social areas, the frequency of atypical behaviors reached its highest frequency. When the number of people in a social area increased, the frequency of atypical behaviors declined. The figure has two important implications: Despite their conditions, patients were responsive to the structural features of their environment (the number of individuals in this example), and an increase in social information (more people in a functional area) tended to normalize patient behavior.

Rate of Change among Members of One's Social Environment

The rate at which new members enter and leave one's social environment is associated with another set of findings and predictions. Figure 4.3 addresses these points by

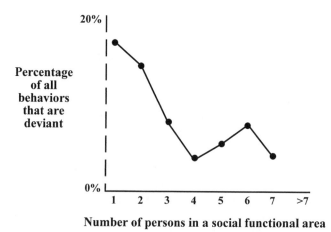

Figure 4.2 Percentage of atypical behaviors as a function of the number of persons in a social area. The figure is adapted from Fairbanks et al. (1977) and McGuire et al. (1977), where both the methods used in the study and the findings are discussed.

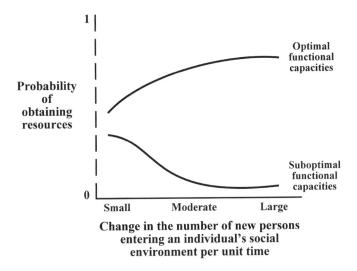

Figure 4.3 Probability of obtaining resources versus the number of new persons entering an individual's social environment per unit time.

plotting relationships between the number of new individuals entering (and leaving) an individual's social environment per unit time and the degree of resource acquisition among persons with different functional capacities.

Figure 4.3 makes the following assumptions: (1) Acquiring resources is contingent on interacting with others; (2) the degree of competition for resources is constant across the three conditions in the figure (small, moderate, and large changes in the number of new members); (3) persons seek out social environments in which they believe benefits will exceed costs; and (4) the cost of developing a new social relationship initially exceeds the benefits that accrue from such a relationship. The figure thus simplifies the reality of social environments where equilibriums differ because of the personalities and motivations of the participants and the local costs of achieving goals (Hirshleifer, 1978).

In Figure 4.3, the suboptimal-functional-capacities curve should be read as follows: When the number of new members per unit time in the social environment is small, the degree of resource acquisition is likely to be minimally influenced. The introduction of a few new members only slightly disrupts existing social relationships. However, as the number of new members increases, the probability of interacting with known others and of acquiring resources declines (Warburton, 1990). This decline largely reflects the costs associated with developing new social relationships, in particular the effort required to develop relationships to the point where they facilitate resource acquisition. The costs of developing new social relationships are assumed to be greater for individuals with suboptimal capacities than for those with more optimal capacities. Thus, for individuals with suboptimal capacities, there are advantages in staying at home, in supporting the *status quo*, and in trying to locate oneself in social environments into which few new members enter. This is the least costly or lowest risk strategy for such individuals. This type of behavior is characteristic of persons who are unusually dependent, as well as of those with chronic disorders, such as schizophrenia, a subset of whom experience significant difficulties in adjusting to changing social conditions (Johannsen, 1961) and who engage in restrictive territorial behavior during periods of social change (M. M. Singh, Kay, and Pitman, 1981). Put another way, rapid change in the members in one's social environment reduces one's social options, and (on average) for persons with suboptimal infrastructures, the fewer the social options, the greater the risk of both reduced goal achievement and the onset of a condition. There are some important exceptions, however. Persons with antisocial personality disorder are known to seek out environments where they are the new members because new membership reduces the chances that their deceptive strategies will be detected (McGuire et al., 1994).

A different set of relationships is predicted for persons with optimal social skills. Moderate rate changes among the members of the social environment will be associated with increased resource acquisition because membership change increases the chance of interacting with persons who can optimally facilitate resource acquisition. Nevertheless, the potential benefits of social change are limited even for persons with optimal capacities: When the number of new members per unit time is large, the probability of resource acquisition declines because the cumulative costs of developing new relationships offset potential resource gains. Possible somatic consequences are also relevant here. Findings from nonhuman primate studies demonstrate that frequent changes in the social composition of groups result in physiological alterations that positively correlate with both biochemical indices of stress and anatomical indices of disease (Kaplan et al., 1982, 1983; McGuire, Brammer, and Raleigh, 1986; Dillon et al., 1992).

Ms. Q

Ms. Q was a 27-year-old unmarried college graduate, who sought psychiatric consultation because of feelings of low self-esteem, fatigue, and intermittent periods of depression. Her signs and symptoms had been present for a year.

Ms. Q had a normal developmental history until age 25. Two years prior to seeking psychiatric help, she had taken a "high-pressure job" with a company that was underfinanced and poorly managed, and whose managers were insensitive to their employees. Although she worked an average of 65 hours per week and had been given increasing responsibilities, the high quality of her work had not been acknowledged, and neither her status in the company nor her salary had improved. Despite her numerous requests, the company managers had refused to meet with her and address her suggestions for improving working conditions and work productivity. After approximately a year on the job, she had begun to experience feelings of low self-esteem. Chronic frustration and depression followed, and the quality and efficiency of her work declined.

Ms. Q had a supportive family. Her parents had repeatedly advised her to change jobs and had offered financial support for a between-job transition period. Despite job offers from other companies, she had not changed her job.

Her psychiatric evaluation revealed a moderately depressed adult female who was fatigued, disillusioned, angry about her working conditions, and moderately withdrawn. Psychotherapy was recommended. Therapy focused on events contributing to her feelings of low self-esteem. As she became aware that her work-related expectations were reasonable and that the behavior of her employers was unlikely to change, she began considering alternative jobs. She sent her résumé to other companies, and within six weeks, she had obtained a new job. The new job had similar responsibilities but differed from her former job in that the management was both sensitive and well organized. Three months into her new job, she described herself as "almost back to normal." By four months, her signs and symptoms had disappeared. An interview with family members confirmed that there had been dramatic changes in her mood, her interest in her work, and her social life. Within a year, she received a major promotion at work.

The case of Ms. Q is a straightforward example of the effects of an adverse social environment on automatic systems and algorithm functionality and, in turn, functional capacities. While the details of how environmental information influences functionality are not discussed until later chapters, two points are worth emphasizing: (1) Optimal physiological and psychological states are contingent in part upon receiving specific types of social information, and (2) the absence of such information can contribute to dysfunctional states.

Social Support and Social Status Networks

There is far more to social support networks than having a few good friends. Social status is also involved, and status interacts with the type and the degree of support others will provide. An evaluation of how persons with and without conditions develop and maintain social support networks that are nested within social status systems illuminates some of the functional consequences of mental conditions.

Helping others and being helped by others are essential behaviors for developing and maintaining social support networks, and the functional capacities listed in Table 4.2 in the categories of social understanding, social maintenance, and social exchange are critical in determining the outcome of such efforts. The ontogeny of these behaviors can be traced to the early years of life. Children who are preferred by peers smile at, touch, and stay in close proximity to those who prefer them. Among friends, there is more frequent eye contact and special types of body orientation. Children with similar personalities are more likely to be friends and to share interests, although not necessarily attitudes (MacDonald, 1988a, 1988b). Similar points hold for body movements, as well as for conformity to rules applicable to coalition formation, resource allocation, and decision making, all of which are observed among preschool children (Charlesworth and LaFreniere, 1983; Hold-Cavell and Borsutzky, 1986). The refinement of these capacities continues through adolescence into adulthood, when, studies suggest, reciprocal interactions are tempered by cultural rules and context cues (Staub, 1974), and attention to rules and cues interacts with friendship and relationship quality (W. M. Brown and Palameta, 1995).

Social support networks are also social status networks (Argyle et al., 1970; G. W. Brown et al., 1975; Ellyson and Dovidio, 1985; Asher, 1990). Social status references one's abilities to acquire and control resources (e.g., property, others' time), abilities that are recognized by other members of one's social group (Stone, 1990). Status relationships are observed among the vast majority of primate and many nonprimate species (Wrangham, 1987; Ginsburg, 1991). Their nearly ubiquitous presence across the animal kingdom suggests that selection has favored the development of capacities for group living, in which members structure their relationships so as to facilitate cooperation and interaction predictability, foster moderate competition, and reduce certain types of conflict. Successful status-related behavior necessitates that one constrain extreme forms of competition, aggression, and deception; that one accept the behavioral constraints and social options associated with different status levels; and that one be sensitive to status-relevant information and expectations. Managing and putting such information to use can be a far from an easy task.

Like friendship, status relationships appear early in life. High social status at all ages of childhood is related to helpfulness, rule conformity, friendliness, and prosocial interactions (Coie, 1990; Coie et al., 1990). Popular children become leaders and set norms for the group. As children grow older, popular children tend to be above average in their academic and athletic achievements, and they are viewed as more attractive by others (Coie, Dodge, and Kupersmidt, 1982). Those who recognize status differences and follow status rules become members of social hierarchies. Those who fail to do so often do not, and they are likely to have smaller social support networks and to be viewed as less attractive socially. The behavior and attributes of high-status children can be viewed as a set of social assets for the child who possesses them, as well as for other children who value them, while the behavior and attributes of less-well-liked, low-status, and rejected children can be viewed as a set of social liabilities which are unattractive to others (MacDonald, 1988a).

Advantages and disadvantages are associated with status. For example, among high-status individuals, there are fewer social consequences for behaving atypically and perhaps also reduced expectations among lower status individuals of reciprocity for helping. Among lower status individuals, especially boys, there is an increased frequency of aggression, hyperactivity, marginal school performance, and social disruptiveness, as well as a smaller group of friends. Clinical experience is consistent with the view that socially deviant behavior declines with improved social status. Further, compared to persons with conditions, persons without conditions are more adept at developing and maintaining relationships, in which it is essential to act in accordance with status-related rules (Essock-Vitale and McGuire, 1990). Thus, the finding that high social status inversely correlates with the frequency of environmentally induced stress conditions is not unexpected (Gift et al., 1988; Murphy et al., 1991), nor is the finding that many individuals withdraw from social groups in which they have low status (Gilbert and Allen, submitted).

In summary, features of the social environment are essential elements in the evolutionary analysis and explanation of conditions. The importance of the social environment is not limited to the effects of adverse social conditions, such as poverty. Successful social navigation, which includes developing, participating in, and maintaining social support networks, and understanding and behaving in expected ways within status systems are equally critical.

Functional Analysis of Behavior

It is striking that, in an age of computerized axial tomographic scans, positron-emission tomography, and other technological advances in psychological measurement, the assessment of basic phenomenological hallmarks in psychiatry has remained elusive. (Kay, Wolkenfeld, and Murrill, 1988, p. 545)

Behavioral assessments [are] still the Achilles heel of biological psychiatry. (van Praag, Kahn, Asnis, Lemus, and Brown, 1987, p. 6)

We have tried the "specific disease, specific biology" approach for 40 years without much success. Let us consider a change. It's time to say "The emperor has no clothes." (Maas and Katz, 1992, p. 758)

In Chapter 2, we mentioned a number of the reasons for psychiatry's reluctance to engage in the detailed direct observation and analysis of behavior. This reluctance boils down to a preference for thinking of *DSM*-type disorder categories as valid and then to move to the study of either the causes of the putative disorder or the efficacy of interventions. We also mentioned that this reluctance has contributed to psychiatry's failure to seriously address the possibility that specific signs and symptoms have specific functions. That is, psychiatry avoids coming to terms with the fact that signs and symptoms are usually more amenable to scientific investigation than disorders (Costello, 1992; Slavney and Rich, 1980).

Three premises underlie the importance we place on the detailed assessment of behavior and function: (1) The study of conditions will be unnecessarily hampered unless functional questions are asked and answered; (2) the analysis of behavior and function offers psychiatry specific targets on which to focus; and (3) the analysis of behavior and function is an essential step in the development of evolutionary hypotheses dealing with the causes of conditions. The evolutionary approach to explaining conditions commences with an analysis of a patient's behavior and function. From this analysis, it progressively moves to the identification of possible contributing factors (McGuire, 1978; Dixon et al., 1989; Schelde, 1994). This approach differs from one of the stated aims of *DSM* classification, namely, decoupling taxonomic efforts from causal hypotheses. *Functional analysis yokes observation to condition classification and causal hypotheses, and it proceeds from the assumptions that conditions will be more meaningfully classified and their contributing factors more rapidly identified by such coupling* (e.g., Curio, 1994). Figures 4.4, 4.5, and 4.6 and Tables 4.3 and 4.4 illustrate features of this approach.

The findings in Figure 4.4 are for disorders as they were classified in *DSM-III* (APA, 1980). While some of the names and descriptions of these disorders have changed in *DSM-IV* (APA, 1994), these changes are not critical to the points being made here: If similar research was repeated by the use of *DSM-IV* categories, similar findings would be forthcoming.

In Figure 4.4, each of the disorders is plotted in terms of the categories used in Table 4.2; a score of 5.0 is the highest possible score. The mean scores for each category of control subjects are shown by a dot surrounded by a circle. Scores for persons with disorders are shown by bars, the top of each bar representing the mean functional capacity scores for each of the seven functional categories during remission. The bottom of each bar references the mean capacity scores during exacerbation. For all of the functional capacity categories, differences between remission and exacerbation scores are statistically significant (McGuire and Essock-Vitale, 1982); the findings in the figure are consistent with outcomes from numerous studies that demonstrate disturbances in social behavior among persons with mental conditions (e.g., Argyle, 1972a, 1972b).

Several points are suggested by Figure 4.4. First, and most obviously, persons with disorders had compromised functional capacities, and the type and degree of compromise differed across disorders. Although there is nothing new about these findings, it is worth emphasizing that functional capacities are not totally absent, even in

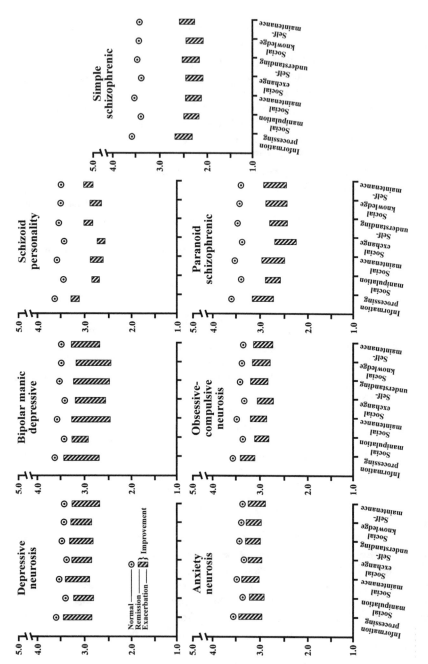

Figure 4.4 Disorders and functional capacities. The figure is adapted from McGuire and Essock-Vitale (1982), where the methods used in the study and the findings are discussed.

the most debilitating disorders. The finding that capacities are not totally absent is the primary reason behind the earlier mentioned decision to discuss adaptations on a positive (0–1) scale rather than to use an adaptive-maladaptive characterization. This finding is also the basis for our view that most conditions are best described as clusters of suboptimal and dysfunctional traits. Second, remission scores can be viewed as measures of suboptimal capacities, while exacerbation scores depict the summation of suboptimal capacities and dysfunctional states. Third, a noteworthy finding is that all of the scores for the control subjects were below 5.0. Within-trait variation among members of the control group is the most likely explanation.

Disorders are highly recognizable during exacerbation periods. They are also detectable during remission states. Figure 4.5 addresses this point.

Figure 4.5 can be understood in this way: During remission, 90% of persons with anxiety neurosis (AN) are unrecognized as persons with a disorder; for schizoid personality (SP), less than 10% are unrecognized; and so forth. To the degree that indices of disorders render individuals socially unattractive, persons with disorders should be socially disadvantaged (e.g., should have fewer social options). Studies are consistent with this prediction (e.g., Essock-Vitale and McGuire, 1990).

In a more detailed analysis, features that are recognizable during periods of remission should differ across disorders. Findings from a study testing this prediction are shown in Table 4.3.

In Table 4.3, for persons with paranoid schizophrenia (in remission), compromised social exchange, social manipulation, and information-processing capacities are most apparent to others. For simple schizophrenia, compromised social maintenance, information-processing, and social understanding capacities are most apparent. And so forth. The feature shared by the most seriously debilitating disorders in the table (paranoid schizophrenia, simple schizophrenia, schizoid personality disorder, and obsessive-compulsive disorder) is compromised information processing, which includes the functional capacities of memory, thinking, sense knowledge, observational learning, and active learning (Table 4.2). The most seriously debilitating disorders also have the lowest average remission scores in Figure 4.4. Similar findings have been reported from comparisons of persons with positive and negative signs of schizophrenia (e.g., Bellack et al., 1990).

Figures 4.4 and 4.5 and Table 4.3 have familiar clinical and evolutionary counterparts. Humans spend considerable time assessing others' capacities, and as a result of their assessments, they make judgments about when and how to relate to and invest in others (the same point applies to nonhuman primates; e.g., Berkson, 1973). To the degree that individuals with compromised functional capacities are viewed as poor

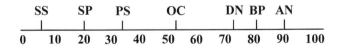

Figure 4.5 Probability of detection of persons with disorders in remission. The figure is adapted from McGuire and Essock-Vitale (1982), where the methodology of the study and details of the findings are discussed. SS = simple schizophrenia; SP = schizoid personality disorder; PS = paranoid schizophrenia; OC = obsessive-compulsive disorder; DN = depressive neurosis; BP = bipolar illness; AN = anxiety neurosis.

Table 4.3 Most apparent suboptimal capacities when disorders are in remission.

Paranoid schizophrenia	*Anxiety neurosis*
Social exchange	Self-maintenance
Social manipulation	Social exchanges
Information processing	
Simple schizophrenia	*Depressive neurosis*
Social maintenance	Self-maintenance
Information processing	Social exchanges
Social understanding	
Bipolar illness—manic type	*Obsessive-compulsive personality*
Self-maintenance	Information processing
Social maintenance	Social maintenance
Social manipulation	
Social exchanges	

Schizoid personality
Social maintenance
Information processing
Social understanding

Note. The figure is adapted from McGuire and Essock-Vitale (1982), where both the methods used in the study and the findings are discussed.

investment risks (McGuire et al., 1994), it follows that interactions with both kin and nonkin should occur less frequently than among persons without conditions. Literally every study that has asssessed this prediction has yielded findings consistent with this expectation (e.g., Henderson et al., 1978a, 1978b; Eisemann, 1984; Tantam, 1988; Essock-Vitale and McGuire, 1990).

Returning to the analysis of behavior in social settings, we find the diagnostic and potential explanatory implications of assessing functional capacities in social context illustrated by the findings in Figure 4.6.

Figure 4.6 shows the mean frequency per hour of approach-another-person and be-approached-by-another-person, two of the many behaviors that were recorded in a study of hospitalized patients with the *DSM-III* diagnosis of unipolar depression. Behavioral observations were made while subjects were in nonstructured social settings, for example, during recreational and free-time periods. Subjects whose functional capacities improved during the four-week hospitalization are identified by circles ($n = 9$). Those whose functional capacities did not improve are identified by triangles ($n = 16$).

Starting on the third hospital day, all subjects in this study began receiving antidepressant medications. Symptoms of depression declined among all members of the improved group, and among all but three members of the unimproved group. Symptom reduction did not statistically distinguish the two groups, a finding that other investigators have noted for persons with depression who receive medications but have significantly different clinical outcomes (e.g., Rosen et al., 1981a, 1981b). Changes in functional capacities did distinguish the two groups, however. For the two behaviors plotted in the figure, the improved and unimproved groups differed significantly ($p < .05$) at Week 3 of hospitalization, although differences were detectable within a few days (at times, a few hours) following hospital admission. Despite

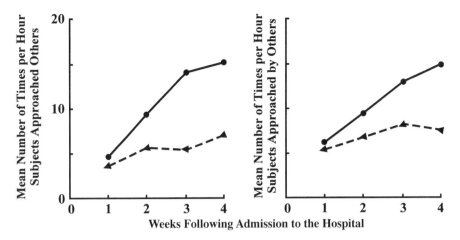

Figure 4.6 Social interaction behavior of persons who recover clinically compared to that of persons who do not recover. The figure is adapted from Polsky and McGuire (1981), where both the methods used in the study and the findings are discussed.

their symptoms, motivations and capacities to initiate social interactions, and to signal to others that they desired such interactions, were present among members of the improved group. Further, members in the improved group increased the frequency of their social behaviors over the course of hospitalization. Again, other investigators have reported similar findings: Persons with unipolar depression who improve clinically exhibit greater frequencies of talking, social participation, nodding, smiling, helping others, laughing, and gesticulating, while essentially the opposite set of findings applies to patients who do not improve (Schelde, 1994).

The findings in Figure 4.6 invite two types of interpretations. One is to view them from the perspective of infrastructural functionality. Members of the improved group had dysfunctional automatic systems but only moderately compromised algorithms, while members of the unimproved group had both suboptimal and dysfunctional automatic systems and algorithms. The failure of the unimproved group to increase the frequency of social interactions or to engage effectively in social behaviors even though their symptoms had declined (an automatic system change) may be understood in this way. (The clinical importance of this type of analysis is underscored by findings that show that persons with the diagnosis of unipolar depression who improve symptomatically yet fail to increase their socialization rate are at increased risk for suicide. [Schelde, submitted].) A second view is that members of the two groups suffered from different types of depression, distinguished in part by self-evaluation and their use of social strategies. In this view, the behavior of individuals who did not improve may have been due to strongly motivated escape and social avoidance strategies: Persons who suffer excessively from low self-evaluation (e.g., shame) often find that social interactions intensify their negative self-evaluations even though their symptoms have declined (Gilbert, 1992).

The behaviors in Figure 4.6 are explored in greater detail in Table 4.4, which examines clusters of behavior in time and space.

Table 4.4 Clusters of behavior among patients and control subjects.

Group A	Group B	Group C
Distance between subject and nearest neighbor .75 m or <		
Look-at-person		**Look-at-person**
Diagonal-in		**Diagonal-in**
Send-verbal	**Send-verbal**	**Send-verbal**
Receive-verbal	**Receive-verbal**	**Receive-verbal**
Task-nil	Task-nil	Task-nil
Head-even	Head-even	Head-even
Sit-support		Sit-support
	Slouch	
	Nonspecific-gaze	
Distance between subject and nearest neighbor 1.5–3.0 m		
Look-at-person		**Look-at-person**
Send-verbal		**Send-verbal**
Receive-verbal	**Receive-verbal**	**Receive-verbal**
		Head-on
Task-nil	Task-nil	Task-nil
Head-even	Head-even	Head-even
		Sit-support
Sideways	Sideways	
Diagonal-out	Stand	
Slouch		

Note. The figure is adapted from Polsky and McGuire (1980), where both the methods used in the study and the findings are discussed. Bold type = core package of behaviors characteristic of social interactions involving verbal exchanges among members of the control group. Underlined type = behaviors sometimes associated with core behaviors. Plain type = behaviors infrequently associated with social interactions among controls.

Table 4.4 shows social behavior clusters for three subject groups at two different interpersonal distances. For this study, clusters were selected for analysis if they contained either the behaviors of send-verbal or receive-verbal. All subjects discussed in Figure 4.6 (*n* = 25) were used in this study. Control subjects (*n* = 19) were added for comparison. Group A comprised patients whose social functioning improved during hospitalization. Group B comprised patients whose social functioning did not improve during hospitalization. Group C comprised staff members and guests of patients (the control group). The behavior categories in the table include task-nil (not involved in a task), diagonal-in (subject oriented in the direction of his or her nearest neighbor), diagonal-out (subject oriented away from his or her nearest neighbor), and nonspecific-gaze (eyes not focused on a person or an object. Other behaviors in the table are self-explanatory.

The clusters for the control group (Group C) serve as the basis for comparison with Groups A and B at the two distances shown in the table. The comparison illustrates how discrete behaviors associated with socialization are integrated in time and space and how far different clusters represent integrated behavior packages (molar behaviors), as, for example, when a person simultaneously speaks, gestures, and maintains eye contact while interacting . For both the near and far interpersonal distances, behaviors in bold type identify the core package of behaviors that were characteristic of social interactions involving verbal exchanges among members of the control group. Underlined behaviors are behaviors sometimes associated with core behaviors. Behaviors that are neither in bold type nor underlined were infrequently associated with social interactions among the control population.

At the low interaction distance (0.0 to 0.75 meters), the behavior cluster for Group A (improved group) is the same as the cluster for Group C (control subjects), while the cluster for Group B (unimproved group) differs from those for both Group C and Group A. For Group B, several of the behaviors observed in the Group C cluster are absent, and behaviors not seen in the Group C cluster (slouch and nonspecific-gaze) are present. Differences in behavioral integration are more evident at the greater distance (1.5 to 3.0 meters), Group A showing somewhat less integration than Group C, and Group B showing significantly less integration than the other two groups.

What kinds of conclusions do the preceding findings suggest? There are several: (1) Poorly integrated behavior, especially when one is motivated to interact, is a consequence of compromised functional capacities; (2) when lack of integration is persistent across social contexts, infrastructural suboptimality is implicated, and (3) as noted, an alternative yet less likely interpretation of the findings in Table 4.4 is that individuals are engaged in a social withdrawal strategy. What argues against this interpretation is that such individuals generally do not improve once they leave clinical care.

If the social withdrawal explanation is excluded, the findings in Tables 4.3 and 4.4 and Figures 4.2 to 4.6 illustrate a breakdown in social behavior patterning, and they raise both taxonomic and interpretive questions. For example, can psychiatric conditions be precisely characterized without detailed functional assessments? Our answer is no. Findings of compromised functional capacities frequently serve as the identifying features of conditions (McGuire and Fairbanks, 1977; McGuire and Essock-Vitale, 1982; Ploog, 1992). Further, the fact that patients in both the improved and the unimproved groups in Figure 4.6 had the same *DSM-III* diagnosis suggests that the failure to include functional assessments in diagnostic protocols invites diagnostic imprecision (Troisi et al., 1990, 1991; Bouhuys and Van den Hoofdakker, 1993). Do functional assessments have predictive value? The answer is yes. In those studies that have compared methods of predicting clinical outcome, detailed assessments of functional capacities predict the clinical course of disorders with greater accuracy than assessments based on either diagnostic categories or clinical impressions (Polsky and Mc-Guire, 1980; J. Pedersen et al., 1988; Schelde, 1994). Further, studies show that functional profiles predict the clinical effectiveness of drug treatment more accurately than *DSM-III-R* (APA, 1987) diagnostic categories (Troisi et al., 1989), and behavioral assessments independently validate characterizations of some disorders but not others; for example, the frequency of eye contact with interviewers is significantly higher among those patients who, by *DSM-III-R* criteria, belong to the good-prognosis subgroup of persons with schizophrenia than among the poor-prognosis subgroup (Troisi et al., 1991).

Are there constraints on the use of the functional approach? The answer is yes. As noted, time is required to collect behavioral data and to document the details of how persons navigate the social environment. Moreover, not all the behaviors observed in persons with conditions are relevant. These behaviors need to be identified, as do behaviors that reflect attempts to adapt when functional capacities are compromised. Because not all behaviors are tied to genes, some investigators have questioned the value of this approach (e.g., Jamieson, 1986). Both evidence and clinical experience refute this concern.

Concluding Comments

In this chapter, we have outlined an evolutionary-based theory of behavior in which behavior systems and their biological motivations-goals, automatic systems, algorithms, and functional capacities, and the social environment have critical and often equivalent roles. Overt behavior is the outcome of interactions within and between infrastructural systems and the environment. Suboptimal and dysfunctional automatic systems, algorithms, and functional capacities are associated with identifiable behavior profiles. The approach we have taken differs from the prevailing models, which focus primarily on disorder constructs rather than function. The need for coupling theory and behavior rather than decoupling them was stressed, as was the need for detailed functional analyses. What we have not done is to demonstrate that functional analyses can lead to novel causal hypotheses. This topic will be addressed in the ensuing chapters.

5

Mechanisms, Symptoms, and Affects

No topics in psychiatry or its related disciplines elicit more debate than those dealing with mechanisms and emotions. Thousands of publications now address these subjects and thousands more will follow—the debates are far from over. When an evolutionary perspective is introduced, more debates are likely, yet greater clarity may be the reward. In the first pages of this chapter, we spend some time looking at these debates in order to explore how an evolutionary analysis can change their format and content. We then turn to a discussion of physiological, behavioral, and psychological mechanisms or, as we refer to them, *states, traits,* and *events.* These terms, rather than *mechanism,* better describe psychiatry's current state of knowledge. In the last part of the chapter, we present evolutionary interpretations of emotions, moods, and affects. To remain consistent with evolutionary terminology, affects are defined as signaled emotions or moods.

Psychiatry's interest in condition-causing mechanisms can be traced back to the ancient Greeks and Hippocrates' (470–360? B.C.) humoral theory of mental illness (Zilboorg, 1941). More recently, it dates to at least 1868, when Erik Von Hartmann published *Philosophy of the Unconscious,* a book that anticipated many of the insights of psychoanalysis. Interest has remained intense ever since. Currently, advocates of psychiatry's prevailing models can be identified by the mechanisms they suspect are responsible for conditions and those that they attempt to influence through interventions.

Most psychiatrists subscribe to the view that conditions are a consequence of atypical or dysfunctional mechanisms, and literally all believe that an improved knowledge of mechanisms and their actions will lead to a better understanding of conditions and more effective interventions. Thus, the importance of mechanisms themselves is seldom debated. What clinicians and investigators argue about is how mechanisms are best defined and identified; which mechanisms explain which data; what type of mechanism-related research is important; and how interventions alter mechanism action.

A mechanism can be defined as the minimum number of elements required for a specific event. Elements can be molecular, physiological, psychological, behavioral, or environmental. Over the past three decades, possible condition-causing genetic and physiological mechanisms have commanded the greatest interest in psychiatry. For a few disorders, such as those due to a specific DNA configuration or a particular physiological state, studies show compelling associations between genetic profiles and physiological measures and phenotypes. In these instances, it is reasonable to assume that one or more condition-contributing mechanisms have been identified. This is the good news. The bad news is that a closer look reveals a host of problems associated with the definition and use of the term *mechanism*. Figure 5.1 serves to illustrate this point.

Figure 5.1 models image processing in the brain: (1) Light from an object enters the eye; (2) it is segregated at the retina; (3) segregated information is transmitted to different parts of the brain; (4) different parts of the brain organize and prioritize the information they receive; and (5) information from the different parts is integrated into images in yet other locations in the brain (after Zeki, 1992).

Can the definition of the term *mechanism* (the minimum number of elements required for a specific event) be applied to Figure 5.1? On initial reading, the answer appears to be yes. Events in the figure are identified, and each event is associated with a distinct function. A closer look raises questions, however. Each event in the figure is contingent on a host of other events, such as changes in ion channels and neurotransmitter interactions, not all of which are addressed. Defining the level of

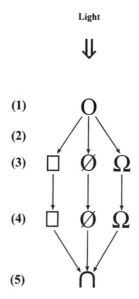

Figure 5.1   A model of image processing. Adapted from Zeki (1992).

analysis (e.g., molecular, physiological) at which explanations apply clarifies which elements should be included in explanations. But from another perspective, identifying the level of analysis simply acknowledges that the scope of an explanation will be limited.

To take the preceding view is neither to argue against mechanism-oriented research nor to oppose the introduction of findings from such research into clinical practice. Certainly, these activities should be carried out. Moreover, most investigators and clinicians are aware that the mechanisms they discuss are only a small part of a complex, multi-element puzzle. Limiting explanations to a subset of relevant variables is an expedient. Still, the fact that conditions reflect the action of many mechanisms often gets lost in the rush to identify the cause of a specific condition or to treat conditions by using drugs designed to alter specific mechanisms (e.g., a serotonin reuptake inhibitor).

One can arrive at essentially the same conclusion by considering the less than straightforward course of evolution. Evolution weaves back and forth, at times reversing itself, and more often than not, its products represent trade-offs between the competing internal systems and external factors that influence selection. Evolution is in part the story of new systems building on old, often either leaving rudiments of past adaptations and their associated systems in place or giving them new functions (MacLean, 1985, 1990; Mayr, 1988; Kavanau, 1990). It is this backward-forward-sideward process that is the likely basis for the fact that numerous parts of the brain are involved in vision (Figure 5.1), declarative memory (e.g., Zola-Morgan and Squire, 1993), language (e.g., Damasio and Damasio, 1992), and most other CNS functions. Many of the products of this process turn out to be well suited to specific purposes (e.g., acquisition of language, bonding to caretakers, rapid withdrawal from painful stimuli) but not for others, such as recognizing subtle cheaters, quickly discontinuing an intense emotion once a frightening stimulus is no longer present, or predicting one's feelings several months in advance.

When new systems build on old, a number of things happen. Physiological and anatomic systems turn out to be connected sometimes directly, sometimes indirectly (e.g., via intermediate messengers such as hormones). Systems compete with one another (e.g., systems responsible for fatigue, which compete with systems responsible for attention). The same molecule (e.g., serotonin) can serve different functions in the brain (neurotransmitter) and in the peripheral blood system (blood-clotting constituent). And hormones, such as estrogen, have different somatic and psychological effects in males and females (Nyborg, 1994). It is these many possibilities that in part underlie the often significant differences in genetic, physiological, and behavioral measures that are observed among individuals with and without conditions (e.g., Berger et al., 1987; Joyce, Mulder, and Cloninger, 1994). In short, there is little in the current understanding of evolution to suggest that mechanisms have evolved either separately or in an orderly fashion, or that hypotheses that are limited to putative actions of a single or even a few mechanisms should dominate causal explanations.

The preceding discussion may seem unnecessarily skeptical. It may also be viewed as evidence-based and realistic (Yuwiler and Freedman, 1987; S. Rose, 1995; Yuwiler, 1995). Despite frequent calls for a biomedical imperative in psychiatry (Guze, 1989, 1992; Stampfer and German, 1988), not only are unambiguous mechanism-condition relationships few and far between, but how mechanisms are thought to explain conditions begs for reevaluation. Hence, the view that we have staked out: Observe behavior and its functions; introduce ultimate cause concepts; use comparative (cross-species) data as well as laboratory and clinical research findings to assess infrastructural functionality; infer the effects of genetic information, physiological changes, and environmental variables on infrastructural function and behavior; and develop and test hypotheses dealing with interactions between putative condition-contributing states, traits, or events and conditions. *In this view, events that are usually discussed in terms of separate genetic, physiological, and psychological mechanisms turn out to be subparts of infrastructures that contribute to specific functions.* When this approach is applied to Figure 5.1, considerable interpretive headway is made if each of the events in the figure is viewed as part of a functional system in which selection has favored capacities to develop visual images that facilitate such functions as moving efficiently, tracking predators, and locating nutrients. Given this view, subparts of the system—such as rapidly acting biochemical systems for processing visual information, the neuronal structure permitting three-dimensional images, and biochemical processes that minimize afterimages—take on new meaning. Research may require that individual states, traits, or events be studied separately or in conjunction with a few interacting variables. But making optimal sense of findings from such studies requires both a concept of function and an appreciation of functional units.

One of the outcomes of this approach is that it introduces alternative explanations of conditions. For example, consider the not infrequently encountered clinical situation in which a married mother of children is in her mid-30s, depressed, and no longer happy with her marriage yet feels that leaving the marriage will compromise her offspring's development. In addition to her psychological state, physiological correlates may be found. However, a major factor contributing to her condition may be her inability to act on one of two ultimately caused yet conflicting goals: to act in her own short-term self-interest by leaving the marriage or to remain in a situation that benefits her offspring.

Selected physiological, behavioral, and psychological findings are discussed in the following pages. For the most part, they are separated into familiar categories and are described as if events within each category occur independently. Although the point is not emphasized in this chapter, trait variation (e.g., different baseline hormone and neurotransmitter profiles) is applicable throughout the discussion. We stress that each example describes only small parts of highly complex, multidimensional systems.

Physiological States, Traits, and Events

Physiological states, traits, and events exert their effects somewhere between genes and overt behavior. Findings from studies of serotonin and social hierarchies among vervet monkeys (*Cercopithecus aethiops sabaeus*) illustrate this point and hint at some of the complexities involved in interpreting physiological findings even among species that are assumed to be less complex than *Homo sapiens*.

Hierarchical or status relationships are characteristic of most Old World primate species, and among nonhuman primates, including vervets, status relationships may remain stable for years (McGuire, Raleigh, and Johnson, 1983). In natural settings, vervets usually live in groups that include several adult males and females and their subadult male and female offspring. Males migrate to neighboring groups when they become young adults. Females remain in their natal groups throughout their lives. Among literally all groups, there is only one high-status or dominant male and one high-status or dominant female. We will focus primarily on males.

Dominant and subordinate male vervets behave in very different ways. Compared to subordinate males, dominant males are more attentive to events outside the group (e.g., watching for migrating males and predators), more tolerant of infant play, less often initiators of aggression, and more often initiators of affiliative behavior. Dominant males have priority access to desirable perches and food; copulate more frequently; and more frequently intrude in fights among adult females. Dominant and subordinate males also differ physiologically. For example, in stable groups dominant males have peripheral serotonin levels 1.5 to 2.0 times higher than those of subordinate males (Raleigh et al., 1984).

Dominance relationships can be manipulated by withdrawing a dominant male from his group and allowing the remaining subordinate males to compete for dominant status. When one of the subordinate males wins the competition—animals compete for high status, not low status—both his peripheral serotonin levels and his behavior change to values and profiles characteristic of dominant males. Moreover, if the removed dominant male is not returned to his group within two weeks, his peripheral serotonin levels will decline to levels characteristic of subordinate males. Peripheral serotonin levels also correlate with individual test performances; for example, dominant males solve maze-escape problems and learn novel tasks more rapidly than subordinate males (McGuire and Raleigh, unpublished data, 1989).

It is possible to induce changes in CNS serotonin activity by administering small doses of selected drugs (e.g., serotonin agonists, reuptake inhibitors) or nutrients (e.g., tryptophan, the amino acid precursor of serotonin). Increases in CNS serotonin activity correlate with increases in the frequency of affiliative behavior and with decreases in the frequency of aggressive behavior in both dominant and subordinate males (Raleigh et al., 1991). However, many behaviors, such as recognition of kin, group members, and one's social status are unaffected.

Dominant and subordinate males also differ in their responses to drugs and nutrients. Weight-adjusted doses of tryptophan or serotonin reuptake inhibitors result in significantly greater changes in the frequency of serotonin-influenced behaviors in dominant males than in subordinate males (Raleigh et al., 1985). Said differently, dominant males are more responsive to induced changes in CNS serotonin activity. Thus, in this species, peripheral measures of serotonin are predictive of CNS serotonin sensitivity.

Other studies show that an increase in CNS serotonin activity facilitates subordinate animals' achieving dominance status. These studies are performed by removing a dominant animal from his social group and treating a randomly selected subordinate male with a drug that increases his CNS serotonin activity. Treated animals become dominant literally 100% of the time, and in large part, they do so by increasing the frequency with which they engage in affiliative behavior with females, which join them in coalitions and thereby help them to establish their dominant status. Conversely, if a dominant male is removed from his group, and a randomly picked subordinate male is given a drug that reduces CNS serotonin activity, the treated animal becomes more aggressive and also reduces the frequency of his affiliative behavior toward females. Such animals do not become dominant (Raleigh et al., 1991). Finally, there are known limits to the degree to which alterations in CNS serotonin activity (e.g., triple drug doses) affect those behaviors that serotonin is thought to influence most. This finding is simply an indirect confirmation of the well-known fact that serotonin is only one of many neurochemical systems involved in behavior that can be influenced by pharmacological manipulations. For example, epinephrine-norepinephrine ratios differ among dominant and subordinate vervet males (Dillon et al., 1992), as do the CSF metabolites of dopamine and norepinephrine; under certain conditions, cortisol concentrations also differ (McGuire et al., 1986). Similar status-related findings have been observed in baboons, whose adrenocortical function, social rank, and personality are known to interact (Sapolsky, 1983, 1989, 1990a, 1990b). Opioid activity is another candidate, although studies designed to clarify possible links between social status and opioid function have yet to be reported.

The preceding findings could be taken as support for the view that physiological states and their influence on specific behaviors should be the primary focus of psychiatric investigations. As noted in Chapter 2, this approach has been advocated by a number of investigators (e.g., van Praag, 1989; Iny et al., 1994), primarily as an antidote to the view that *DSM*-type classification can obfuscate searches for the causes of conditions. Yet, at its heart, this recommendation is no more than a reaffirmation of the biomedical model partially liberated from the constraints of *DSM*-type classification. Studies of interactions between physiological variables and specific phenotypic features, rather than conditions, might lead to improvement in our understanding of behaviors mediated by specific neurochemical systems. But such a focus will fall far short of what is possible if physiological findings are integrated with other condition-interacting variables and the integrations are viewed from a functional perspective. For example, differences in both peripheral serotonin levels and aggressive behavior among dominant and subordinate males disappear several weeks after females are removed from a group (McGuire and Raleigh, unpublished data, 1989), and serotonin differences tend to disappear when animals are crowded.

The vervet-serotonin story does not end here. There is another important chapter dealing with the influence of female behavior on spontaneous male status changes. From time to time in stable groups, females will approach a subordinate male, and together, they will form a coalition. The coalition then challenges the reigning dominant male. Reigning dominant males turn out to be a poor match for such coalitions, and over a period of several weeks, they are displaced, and the female-selected subordinate male becomes dominant (Raleigh and McGuire, 1989). These types of status changes rarely involve physical contact. Rather, they are characterized by ritualized displays, threats, and submissive behavior. In displaced dominant males, peripheral and CNS measures of serotonin, as well as status-related behavior, change to values characteristic of subordinate males, while the opposite changes occur in subordinate males that become dominant. Why females act to influence the choice of dominant males is unknown, but a strong possibility is that they do so in order to alter the genetic makeup and immunological competence of their offspring.

The preceding discussion of serotonin in vervet monkeys can be summarized as follows. Available evidence suggests that serotonin plays an important role in the mediation of a variety of social behaviors. Equally important, both social structure (e.g., dominant-subordinate relationships in males, the presence of females in groups) and dynamic features of the social environment (e.g., occasional female coalition formation with a subordinate male) interact with adult male serotonin activity and function. It is not suggested that serotonin molecules change during these events; rather, the effects of serotonin molecules differ across social conditions because of events in the social environment, as well as their interaction with physiological systems other than the serotonin system.

There is yet another and instructive way to look at the preceding findings. Consider what might have happened had the initially observed broad range in peripheral serotonin measures been interpreted by means of a model that disregarded social behavior and function. The most likely outcome is that the range would have been interpreted as an indication of individual variation, perhaps due to differences in the genetic makeup of animals. Further, high and low measures would have been tied to specific behaviors (e.g., correlations between low serotonin measures and the frequency of aggressive behavior). Once a physiology-behavior relationship is identified, it is usually followed by more focused within-animal studies to identify the detailed features of states, traits, or events that might explain the influence of serotonin on specific behaviors. While such studies are important, they usually occur without an appreciation of the many nonphysiological factors that may be instrumental in altering physiological states, for example, social status change. It was the use of evolutionary models that took this research down a different path.

Among humans, differences in peripheral serotonin levels have been observed, and in most studies, high-status males have higher levels of peripheral serotonin (Madsen and McGuire, 1984; Madsen, 1985). However, some findings have been reported among individuals with different personality types that suggest a different direction (Madsen, 1986). Moreover, personality types are thought to be associated with different ratios of the neurotransmitters serotonin, dopamine, and norepinephrine (Cloninger, 1986). There are also a multitude of findings in which conditions or condition features correlate with serotonin measures. To cite only a few examples from an extensive literature that builds primarily on CSF 5-HIAA findings (low CSF 5-HIAA implies low CNS serotonin activity) and CNS challenge studies: CNS serotonin activity is low in some but not all persons who engage in specific types of impulsive behavior, such as arson (Linnoila et al., 1983; Coccaro, 1989; Kruesi et al., 1990); in some but not all persons who commit suicide (e.g., Mann, McBride, Brown, et al., 1992; Stein and Stanley, 1994); in a subset of clinically depressed persons (Mann, McBride, Anderson, and Mieczkowski, 1992); and in persons with a variety of other disorders (Zohar et al., 1987; Coccaro et al., 1989; Meltzer, 1989; D. L. Murphy et al., 1989). However, low CNS serotonin activity is also observed among a subset of persons without conditions. All the above may be said without the consideration of a host of other potentially relevant points; for example, there are currently thought to be approximately two dozen different subclasses of serotonin receptors, each of which may be associated with different functions or behaviors, as well as differentially influenced by the actions of nonserotonin biochemical systems.

Given the preceding points, it is not surprising that it is yet to be established that serotonin is the primary causal factor in any of the conditions for which it has been so implicated. In addition, the findings on the relationship between serotonin levels and states in vervet monkeys and humans and the findings of the relationship between human serotonin and personality do not easily fit with the view that serotonin can be discussed as a separate physiological system or mechanism responsible for specific behaviors or conditions. Further, the fact that animals that change status do not alter their interpretation of many kinds of information (i.e., kin recognition), yet do alter their behavior strategies and behavior, suggests that automatic system or algorithm function changes accompany alterations in CNS serotonin (and related neurochemical) activity, and that these changes have adaptive consequences. *Selection may have favored the capacity to change how automatic systems and how algorithms process information in response to changes in social information. That is, infrastructural changes are in part contingent on alterations in CNS serotonin activity, and CNS serotonin activity is in part contingent on others' social signals.* It follows that our understanding of serotonin changes and their consequences may improve significantly if serotonin changes are viewed in terms of their influence on and interactions within infrastructures, and vice versa.

In summary, the key points in this section are these. First, the function of physiological systems, such as the one(s) that controls serotonin, will be incompletely understood unless both psychological variables (e.g., social information), function, and interacting physiological systems (e.g., norepinephrine, Sulser, 1987; oxytocin, Kalin and Shelton, 1989) are taken into account. And second, viewing neurochemical systems as elements of infrastructures rather than as independent mechanisms results in alternative ways of thinking about physiological systems and, in turn, explaining conditions.

Behavioral States, Traits, and Events

For the majority of behaviors, the underlying operations of infrastructures are transparent. This is the case when a clinician asks a question of a patient and the patient provides an appropriate answer. Such exchanges occur in familiar sequences punctuated by both verbal and nonverbal signals. Participants are aware of the products of the sequence (e.g., questions and answers in social context), but not of the underlying events contributing to the interactions. These must be inferred. What often goes unnoticed during interactions is that the spoken word is only part of the interaction story: Nonverbal signals both modify the meaning of what is said and initiate responses in others.

Table 5.1 lists familiar examples of nonverbal initiation-response relationships.

The nonverbal signals listed in the left-hand column of Table 5.1 nearly always elicit specific types of responses in others (e.g., Ekman, 1993). Because the signals and responses in the table are observed so often, reasonable assumptions are that predispositions for both signals and responses have been favored by selection; that selection has occurred because the behaviors and responses facilitate the solving of adaptive problems; and that signals and responses reflect the operation of infrastructural systems.

The findings in Table 5.1 inform our understanding of conditions in a number of ways. When infrastructural systems are compromised, normal responses to signals are often absent or atypical. A response may be delayed, exaggerated, or accompanied by unexpected postures or gestures (recall the cluster data in Table 4.4). As a rule, the more typical and less obvious a response sequence, the better the clinical prognosis and the greater the likelihood that infrastructures are minimally compromised. Con-

Table 5.1 Signaling and others' responses.

Identifier's signal	Recognizer's response
Adult smiles at infant	Infant orients toward adult
Offspring looks unhappy	Parent orients toward offspring
Sibling acts deceptively	Other sibling initiates deception-detection strategies
Attractive female appears	Male's attention focuses on female
Attractive male appears	Female's attention focuses on male
Identifier behaves atypically in a nonthreatening way	Observer interprets the behavior
Identifier threatens	Observer flees or fights
Identifier doesn't act submissively when expected to	Observer is irritated, perhaps angry

Note. Adapted from McGuire (1988).

versely, the more atypical and obvious a response sequence, the worse the clinical prognosis and the greater the likelihood of severely compromised infrastructures. The latter point was illustrated in Figure 4.6, which shows that a group of patients with the diagnosis of unipolar depression did not increase the frequency of either initiated or received social signals over the four-week course of hospitalization: Their capacities to initiate and respond to social signals remained largely unaltered, even though their symptoms declined significantly in response to antidepressant medication.

While several reasons for associating atypical behavioral sequences with a guarded clinical prognosis have been mentioned, one is worth reemphasizing here: Interpersonal interactions tend to normalize behavior. That is, social interactions influence behavior in the direction of predictable, socially typical interaction norms, which, for example, specify the range limits of behaviors such as closeness, assertiveness, submissiveness, and social comparison (discussing others' traits; Gilbert and Allan, 1994). Socially typical responses by others decrease the psychological and physiological effects of uncertaintythat, to some degree, always accompany social interactions, and others respond positively to socially typical and context-expected behavior, just as they respond hesitantly or negatively to behavior that varies from age, sex, and context norms. These effects are not limited to ongoing social interactions and often extend to possible future encounters; for example, there is enhanced memory for the faces of cheaters compared to noncheaters (Mealey et al., 1996). A clear implication of this line of reasoning is that the importance of social norms and socially appropriate behavior should not be underestimated (Wenegrat et al., 1996a, 1996b). Socially atypical or negative behavior by others has the opposite effect, and as will be discussed in chapter 7, such behavior can trigger both infrastructural changes and the onset or worsening of a condition.

The preceding examples bring us back to the subject of function. Thus far, function has been discussed primarily as if it applies only to individuals; for example, one is efficient or inefficient in certain functions. What Table 5.1 suggests is that the concept of function can be extended to social units: A does X, B does Y, and these interactions serve specific functions, such as increasing the responsiveness of parental care for distressed infants, thereby reducing infant distress.

If we view Table 5.1 another way, behaviors and responses that repeatedly differ from those shown in the table permit inferences about which infrastructures may be compromised provided behaviors and responses do not reflect adaptive strategies. For example, if environmental information is misperceived or atypically organized and prioritized, compromised automatic systems are implicated. If environmental information is not misinterpreted, but causal modeling or behavior strategy development is atypical, compromised algorithms are implicated. If neither of the above occurs, yet behavior is strategically ineffective, compromised functional capacities are implicated.

In summary, the key points to be taken from the discussion of behavioral states, traits, and events are that (1) others' behavior initiates infrastructural activities that have physiological, psychological, and behavioral consequences; (2) interaction sequences can be viewed as functional units; (3) atypical interaction sequences are often consequences of compromised infrastructures; and (4) inferences about infrastructure functionality can be made through the observation of behavior and its functions. Most clinicians would agree with Point 3. However, Points 1, 2, and 4 are only sporadically a part of clinical thinking and, even then, far from center stage where we would place them.

Psychological States, Traits, and Events

Understanding the content of what others say, drawing inferences from their behavior, anticipating how they will behave under different conditions, planning and organizing one's activities, constructing mental maps of the environment, and memory are examples of psychological events that have been and are studied by a variety of disciplines. Psychologists concern themselves with the rules of cognition. Biologists investigate how physiological systems interact with cognition and emotion (e.g., Näätänen, 1990; Ungerleider, 1995). Linguists focus on deep structures and the organization and uses of language (Chomsky, 1957; Pinker, 1994). Psychoanalysts study the effects of intrapsychic distortions and ego functions. And evolution-oriented psychiatrists have offered novel interpretations of many of the psychological phenomena that have interested other disciplines (e.g., Nesse and Lloyd, 1992) and have also identified parallels between psychoanalytic and evolutionary ideas (e.g., Slavin and Kriegman, 1992; Badcock, 1994). It would be prohibitive to review this voluminous literature. Thus, we will focus on only two examples: the possible adaptive functions of psychic defenses and of circumscribed delusions. (Chapter 6 offers a detailed discussion of the psychological features of key algorithms.)

The terms typically used in psychoanalytic discussions of psychic defenses include *denial, isolation, regression, rationalization, intellectualization, reaction formation, projection, reversal, splitting, repression,* and *sublimation.* Defenses are thought to be products of the psychic processes responsible for information manipulation and distortion and to have their own clinical signatures. Their excessive use is observed frequently in persons with conditions. However, they are sometimes also observable in persons without conditions, particularly during periods of stress, fear, disappointment, or fatigue. Defenses may be traits or states. When they predate the onset of conditions and extend beyond condition remission, they are most likely traits. When they are confined to the period of a condition, they are most likely states.

While psychoanalysts have focused their inquiries largely on the negative consequences of the exaggerated or prolonged use of defenses, a few investigators have inquired into whether defenses represent evolved strategies that can have adaptive functions (Slavin, 1985; Nesse, 1990b; Gilbert, 1992, 1993; Nesse and Lloyd, 1992; Schore, 1994). Denial, rationalization, and repression will serve as examples. Denial may be understood as a form of information filtering that reduces the amount of encoded external information. Its potential adaptive value is to help the individual to set priorities and develop strategies during periods in which external information is either ambiguous or negative. For example, disregarding others' potential anger if one wins a competitive encounter, as in acquiring a highly desirable resource or mate, may be adaptive if sufficient benefits are to be gained by winning. Rationalization combines selective information filtering with excessive (but inaccurate) causal modeling of one's behavior, in effect, the development of a plausible but invalid explanation of why one is behaving in a particular way. While a failure to recognize one's mistakes may follow, rationalizations may also increase one's capacity to deceive others, which, from an evolutionary perspective, may be adaptive. And repression, which is the selective filtering of internal information, limits the amount of potentially conflictual information available to awareness and may facilitate carrying out self-interested behavior strategies without a recognition that they conflict with social rules (cf. Vaillant, 1971).

Gilbert (1993) elaborated on the function of defenses by emphasizing interactions between defense, safety, and subordination. In his view, the defense system "evolved to allow a signal-detection system and provide a menu of response options to potential threats, to instigate self-protective behaviour and enact strategies. But the defence system did not evolve to 'cause psychopathology'" (p. 149). Because defense systems are easily activated and are thought to have evolved on a better-safe-than-sorry principle, they are often associated with selective information filtering, information manipulation, stereotyped responses, and negative affects (Gilbert, 1993). This concept can be extended to submissive and subordinate behavior as forms of social defense (Gilbert and Allan, submitted), to involuntary subordination or dependence as a factor in depressive vulnerability (Gilbert, Allan, and Trent, 1995), and to self-esteem as an interpersonal monitor of others reactions; for example, low self-esteem alerts individuals to the possibility of social exclusion (Leary et al., 1995).

Compartmentalized delusions provide the second example. The common ingredient of this type of delusion is a false belief that is maintained in the face of strong evidence or of conventional thinking that suggests alternative interpretations. Clinically, compartmentalized delusions may manifest themselves as atypical thoughts about the self, others, inanimate objects, or ideas. They may be persecutory, somatic, systematized, or bizarre (Manschreck, 1989, 1992). These delusions seldom extend to all forms of cognition or emotion. That is, some types of information processing are relatively normal.

A number of explanations have been offered for delusions: The steps involved in computing events are in error; delusions represent extreme forms of social attribution (e.g., assigning the causes of one's behavior to events in the environment); delusions are a consequence of attempts to deal with highly conflicted intrapsychic ideas or emotions (Maher, 1992); and delusions reflect dysfunctional physiological systems.

The preceding hypotheses may explain features of delusions. However, a more compelling explanation emerges from an evolutionary analysis. Looked at in an evolutionary context, compartmentalized delusions can be understood as extreme forms of information encapsulation (Gilbert, 1989), the primary function of which may be to protect the functionality of other infrastructures (a function analogous to isolating an infection). One may be aware of encapsulations yet still be unable to alter them, as is the case with visual illusions of which one is aware: One's logic will not alter one's perception (Gilbert, 1989). That there may be physiological correlates of delusions is not in question here. Physiological correlates exist for all states. A clinical vignette illustrates this interpretation.

Mr. C

Mr. C was a professor who completed writing a book approximately every two years. He attended psychotherapy sessions once every two weeks. In the months prior to completing a book, he isolated himself, worked long hours, became increasingly anxious, and developed a delusion about the local police, who he believed were conspiring to interfere with his writing. As the delusion developed, he engaged in what he referred to as "countertactics," which amounted to leaving his home at unscheduled times, taking alternative routes to the market and to his office, not answering his phone, and carrying his manuscript with him at all times. He was aware that he was suffering from a delusion, but this recognition did not relieve his fears. Moreover, the delusion did not interfere with his writing in any identifiable way. He met his publishers' deadlines, and his books were well reviewed. Following the completion of each book, the delusion slowly disappeared over a period of three to four months.

From the perspective of functional units, psychic defenses have several critical features in common: (1) They minimize the effects of certain types of information; (2) they preserve functional systems and facilitate strategy execution in the presence of informationthat, if recognized, might compromise strategy enactment; and (3) they increase the possibility of social participation and goal achievement. Thus, psychic defenses can be viewed as internal functions for controlling and limiting the potentially negative or destabilizing effects of information. Compartmentalized delusions may be viewed in a similar way: They represent a "last-ditch" effort to maintain infrastructural and functional integrity when one or more infrastructures is compromised.

In summary, different psychological states, traits, and events have clinical signatures; we can make inferences about infrastructural functions and their possible contributions to conditions by observing these states and traits; and both psychic defenses and compartmentalized delusions, which are usually thought to be indices of conditions, may have adaptive value.

Emotions, Moods, and Affects

Following the publication of Darwin's *Expression of Emotions in Man and Animals* (1872), nearly a century passed before psychiatrists began to think seriously about emotions, moods, and affects in the framework of evolutionary theory. Along the way, psychologists, behaviorists, philosophers, and novelists developed their own views. Despite the divergence of approaches, there were and are many points of agreement. For example, most clinicians and investigators believe that emotions are more revealing than words and that the intensity of unpleasant emotions and affects (signaled emotions) provides a rough approximation of condition severity. There are also points of disagreement. These have often centered on such issues as emotion-cognition interactions and the contribution of physiological system(s) to emotions. The positive side of these disputes is that they have fostered research that has expanded our knowledge of neural pathways and cerebral nuclei, neurotransmitter and hormone functions, and the physiological and psychological consequences of experience. The negative side is that new research findings have accumulated so rapidly that they have seldom been fully tested or critically evaluated. At this time, it is probably fair to say that most clinicians believe emotions are evolved traits. Yet, with the exceptions discussed below, psychiatry has seldom turned to evolutionary theory to interpret emotions or affects.

As was noted in chapter 4, emotions are somatic states that last a few seconds. Moods are sustained emotions. We will use the terms interchangeably. Affects are signaled emotions. Within limits, affects are effective means of altering the behavior of others (see Table 5.1).

In evolutionary context, emotions and affects are evolved traits, not things primarily learned or taught, and they provide information about behavior and events. In the first moments of postuterine life, emotions largely reflect internal states. Infants fuss and cry when they are hungry or when they desire to be held. With development, emotions remain responsive to internal states and become increasingly sensitive to environmental information. Some events elicit positive emotions, such as joy when a caretaker appears or pride when one is praised. Other events elicit unpleasant emotions that become apparent at different points during development: Disgust and distress are present at birth, sadness and fear appear during the last half of the first year, and guilt and shame appear during the second year (MacDonald, 1988b).

Throughout the average day, one normally experiences a variety of emotions. If they are not overly intense or prolonged, and if they do not require major adjustments in living, they are usually taken as normal. Under certain circumstances, normal emotions are intense or prolonged. Anxiety in response to a frightening experience, such as an earthquake, and grief in response to the loss of an important other are familiar examples. As a rule, strong emotions shape behavior more than rational thoughts; for example, they "energize" or "deenergize" beliefs and behavior and increase the probability of approach, avoidance, or a specific action. They also have different long-term consequences. Strong emotions of anger or love are not easily ignored nor easily forgotten, while cognitive assessments of these states are often fleeting. When emotions are understood as adaptive traits, what often appears to be an excessive emotional response makes evolutionary sense: The costs of an excessive response (e.g., intense fear of a predator) are minimal compared to the potential consequences of an inadequate response (Nesse, personal communication, 1995).

After Freud, Jung, and Meyer, A. J. Lewis (1936) and Price (1967) were among the first psychiatrists to discuss emotions in an evolutionary context. However, psychiatry had to wait until 1980 and the work of Plutchik for its first comprehensive evolutionary interpretation of emotions. In Plutchik's models, emotions are physical-psychological states that integrate and organize internal and external information and reflect attempts by the organism to achieve control over survival-related events (Plutchik, 1980, 1984a, 1984b, 1991). One need not always be aware of emotions, which in Plutchik's view are more primary than feelings. Yet, one is aware of feelings, and feelings presuppose emotions. Capacities for emotions and feelings are ultimately caused, although at any moment in time, their presence and their intensity can be explained by proximate events. Different emotions are associated with eight species-characteristic motivational systems that overlap with the four behavior systems and their motivations-goals, as discussed in chapter 4: incorporation, rejection, destruction, protection, reproduction, reintegration, orientation, and exploration (Plutchik, 1980). Combinations of emotions occur more frequently than single emotions, in part because several motivations may be associated with behavior, and in part because environmental contingencies are seldom unitary.

Following Plutchik, Nesse (1987a, 1990a, 1991a, 1991b) is the psychiatrist who has most often analyzed emotions in evolutionary context. His focus has been on the functional features of emotions, which he characterizes as specialized states that can increase individual fitness in at least two ways: by highlighting information from both the internal and the external environments, and by altering the probability of behaviors that have potential evolutionary payoffs. Emotions inform an organism if its investment strategies are working. Unpleasant emotions lead to attempts to alter or eliminate the source of the emotions; for example, they initiate strategy changes. Pleasant emotions lead to attempts to prolong the behavior that contributes to the emotion (see also Westen, 1995).

The works of Plutchik and Nesse are the source of four basic evolutionary axioms: (1) Emotions evolved for the quick recognition of biologically important events and contingencies, (2) emotions reflect the preferential selection, organization, and prioritization of information; (3) the information provided by such systems is easy to interpret; and (4) emotions influence both ongoing and future behavior so as to render it more adaptive. Normal emotions qualify on each of these points. Debilitating moods often do not.

Several recent explanations of emotions overlap with the models discussed here. Affect theory deals with "innate affects" or, in the terminology we are using, innate emotions. Innate emotions are viewed as primary physiological systems that at birth are minimally responsive to external events. Most innate emotions are thought to retain their internal roots, yet to become increasingly responsive to external events (Nathanson, 1993). There are exceptions, however: Experiences also tie memories to emotions, and emotion programs "take over" remote structures that have evolved for other purposes, the result being unexpected states in which emotions and behavior are combined (e.g., fear associated with piloerection). Thwarted-action-state-signaling theory has its origins in the ethological analysis of emotions and the attempt to provide a "realistic identification, classification, and explanation of the phenomena of emotion" so as to facilitate and recognize associated brain systems and processes (Salzen, 1991, p. 48). Still to be determined is the number of emotions. Plutchik suggests eight, as do others (e.g., Salzen, 1995), but estimates range from six upward (Plotkin, 1994; see also Buck, 1988; Schore, 1994).

In chapter 4, we noted that automatic systems are responsible for emotions. The point deserves elaboration. Automatic systems are closely integrated anatomic-physio-logical-psychological systems, and it is this integration—namely, the influence of each part on the other parts—that accounts for the multiple effects of automatic systems: One's somatic state, what one thinks, and how one behaves never occur completely independently. Similar points apply to symptoms, which, along with emotions, appear to have at least three sources:

1. One source reflects an extreme of normal automatic system function, as in instances where the effects of an unexpected threat linger: A threat initiates automatic system activity; an unpleasant (yet informative) emotion is experienced; but the emotion does not decline in intensity in the expected time period.
2. A second source is automatic system dysfunctionality. This is the likely basis for intermittent moods that may or may not be tied to specific events. The fact that condition-related emotions have features similar to normal emotions suggests that many of the same infrastructural processes mediate both types of states. Further, the fact that pharmacological agents often reduce the unpleasant effects of undesirable moods is consistent with the biomedical view that dysfunctional physiological systems are an integral part of moods.
3. A third source is automatic system suboptimality, of which two types can be inferred from clinical data. One type is an inability to process information in ways that lead to the discontinuation of existing moods. Features of chronic mood conditions and unremitting cases of psychogenic pain are examples. The other type is an inability to selectively filter information that initiates emotions. Excessive emotional sensitivity to others' signals is an example.

There are also important relationships between unpleasant emotions and the development of novel behavior strategies (an algorithm function). Among persons without conditions, unpleasant emotions usually initiate strategies to alter the perceived source of the emotions and to reduce their undesirable effects. Examples are seeking out a friend when one is bored or mildly depressed, exercising when one is anxious, and changing tasks when one is unable to concentrate. Such behavior is expected from an evolutionary perspective: Changes in behavioral strategies should occur in response to information that one's behavior is minimally adaptive. Persons with conditions often attempt to employ similar strategies although they frequently fail. Figure 5.2 explores these points further.

Figure 5.2 is an inverted U curve depicting relationships between the intensity of unpleasant emotions and the likelihood of developing novel behavior strategies. Up to Point X, unpleasant emotions increase the probability that persons will try to develop novel strategies, for example, to discard unfaithful friends, change jobs, or take a holiday. However, if emotions become overly intense, the likelihood of novel strategy development declines. This occurs to the right of Point X: The accumulated cost deficits associated with unpleasant emotions serve to inhibit novel strategy development because additional strategy-related costs will be incurred before benefits can be achieved. In effect, prolonged symptomatology and its associated costs curtail algorithm functionality. At Point Z, emotion decoupling (discussed below) can occur.

There are some exceptions to Figure 5.2, but these occur primarily among persons without conditions. For example, in certain instances, unpleasant emotions may not

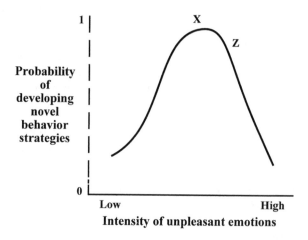

Figure 5.2 Emotions, novel behavior strategies, and decoupling.

lead to novel strategy development in the area to the right of Point X. This may happen when the costs of altering a strategy or adopting a new but questionable strategy are expected to exceed the costs of continuing with the original strategy. The frustrations often associated with dealing with kin with severe personality disorders is an example.

A related point concerns the inability to discontinue or turn off emotions. Evolutionary reasoning suggests that systems should have evolved to limit the duration of moods, particularly when mood-initiating contingencies are no longer present. The frequently observed close association between conditions and undesirable moods, which is seen, for example, in mood disorders and posttraumatic stress disorders, can be explained in part as a consequence of an inability to discontinue adaptive warning systems that focus attention on some anticipated or actual malfunction of the organism or on environmental contingencies that threaten reproductive or survival potential (Nesse and Williams, 1994). Automatic system suboptimality or dysfunctionality is implied when warning systems fail.

Mood-behavior decoupling is also observed. When the costs associated with moods become excessive (Point Z, Figure 5.2), decoupling may be the least costly way of reducing debilitating effects, and in ways analogous to compartmentalized delusions, decoupling can be viewed as a last-ditch strategy to reduce the effects of undesirable emotions. What is involved in decoupling is modeled by nonhuman primates when infants are separated from their mothers. Their initial response is protest, which is followed by a period of depression-like behavior. Eventually, the infants become immobile. Such behavior is consistent with the view that the cost of initiating a novel strategy (e.g., search for mother) will soon become greater than an animal can tolerate without increasing the intensity of a painful mood.

Among persons without conditions, decoupling is observed in extremely stressful or threatening situations, such as torture, loss of a loved one, the destruction of personal possessions (e.g., in a fire or an earthquake), or frightening information (e.g., a diagnosis of cancer). Among persons with conditions, decoupling is seen in *la belle indifférence,* where a highly atypical somatic state is not accompanied by the expected emotional response; in dissociative states such as amnesia; in alexithymia when there is decoupling between the appraisal and the expression of moods (Sifneos, Apfel-Savitz, and Frankel, 1977); and in incongruent mood (socially inappropriate mood) in schizophrenia.

The remainder of this chapter deals with features of the emotions and affects of anxiety, depression, frustration-anger, pleasure-satisfaction-joy, pain, and power-control-elation. Much of what is said builds on the ideas of Price (e.g., Price et al., 1994), Gardner (e.g., 1982), Sloman (e.g., Sloman and Price, 1987), Gilbert (e.g., 1989, 1992), Ploog (1989b), Nesse (e.g., 1987a, 1990b, 1991a), Marks (1987), Frank (1988), Marks and Nesse (1994), Plutchik (1980, 1984a, 1984b, 1991), and Salzen (1995). Each emotion or affect may be associated with each of the four automatic systems discussed in chapter 4. Trait variation, which is characteristic of mood states, is not emphasized. For each emotion, a feeling state description is given, followed by a sampling of clinical and research findings, and then by an evolutionary interpretation. The discussion is organized in textbook form so that it may be quickly accessed.

Anxiety

The Feeling State

Anxiety is an unpleasant emotional state associated with the sense that undesired or dangerous events will occur. Dyspnea, palpitations, chest pain, dizziness, tension, apprehension, visual difficulties, increased vigilance, and/or a sense of uncertainty about familiar settings and persons may be experienced. In mild forms, the emotion is irritating, but usually manageable. In more intense forms, it becomes increasingly central to one's experience and behavior, and cognitive functions (e.g., concentration) are often compromised. Extreme and debilitating forms of anxiety are seen in panic, of which fear may be a major component.

Clinical and Research Findings

There is an extensive literature dealing with putative causes of anxiety, much of which is the subject of debate (e.g., Klein and Klein, 1989; Kagan and Schulkin, 1995). In the prevailing models, the postulated causes include a sense of impending physical harm; the possibility that high-priority goals may not be achieved (e.g., performing well on school examinations, successfully going through a surgical operation); intrapsychic conflicts; a sense that one will lose control; and dysfunctionality of one or more neurochemical systems. Findings from a number of studies point to predisposition influences (e.g., Torgersen, 1985). Other studies suggest that predispositions manifest in different ways at different points during the life cycle. For example, chronic, generalized anxiety and adult agoraphobia are reported to occur more frequently among persons with a history of separation anxiety during childhood (Bowlby, 1973; Breier, Charney, and Heninger, 1986; Zitrin and Ross, 1988).

Neuropsychological studies implicate the septohippocampal system and its contributions to behavioral inhibition as possible factors (Gray, 1982). PET measurements implicate basal ganglia involvement (Wu et al., 1991). Physiological responses, including increased systolic blood pressure, shorter cardiac interbeat intervals, and elevated norepinephrine and cortisol activity, have been reported (Nesse et al., 1985; Igbal, Bajwa, and Asnis, 1989; Charney et al., 1990; J. F. Thayer et al., 1996), as have reduced levels of thyroid stimulating hormone (Corrigan et al., 1992), altered serotonin activity (Kahn et al., 1988), possibly altered dopamine activity (Pitchot et al., 1992), and opioid system dysfunction (Kalin and Shelton, 1989). When anxiety blends into panic, both physiological and psychophysiological measures are atypical, and multiple studies implicate endocrine changes (Nesse et al., 1985). Cerebral blood flow is decreased during periods of intense anxiety, although apparently not during periods of mild anticipatory anxiety (Zohar et al., 1989). Electroencephalogram (EEG) and computerized-tomography (CT) findings have been reported as normal across a range of anxiety conditions (Lepola et al., 1990). Reports also suggest that anxiety has specific autonomic system correlates (Thayer et al., 1996), and that it is a frequent component of many conditions, including conditions in which depression is the most prominent symptom (Breier et al., 1984; Stein et al., 1990; Lydiard, 1995).

Studies of nonhuman primates suggest that specific cues elicit distinct types of anxiety-fear and defensive behavior (Kalin, 1993). Once elicited, fear and defensive behavior may not abate. Findings from experimental studies from a variety of disciplines are compatible with the formulation that

> human and nonhuman primates have what could be called "evolutionary memories" that play a role in determining which objects and situations are most likely to become the objects of fears and phobias. It seems likely that these evolutionary memories underlie selective associations in fear conditions, which in turn mediate the nonrandom distribution of fears and phobias seen clinically. . . . Memory processes affecting the acquisition and retention of fears serve to promote maintenance and overgeneralization of fear with the passage of time . . . That is, through natural selection fear may have come to be associated with conservative cognitive biases which, under ordinary circumstances, are more likely to promote the reinforcement, enhancement, or overgeneralization of fear rather than the forgetting of fear. (Mineka, 1992, p. 162)

Evolutionary Views

Anxiety is thought to have evolved as a system warning that high-priority biological goals may be jeopardized (Nesse, 1990a; Marks and Nesse, 1994; Stevens and Price, 1996). It is a future-oriented emotion. This interpretation is consonant with the view that selection has favored systems that rapidly provide behavior-influencing information dealing with impending dangers (Nesse, 1991a). In situations of perceived external danger and panic, both automatic systems and algorithms contribute to the flight-or fight-features of anxiety, including physiological responses, the amplification of external vigilance, and the overriding of other potentially competing goal priorities (Cameron and Nesse, 1988; Friedman et al., 1993).

There is also the issue of internal models that can alter emotional states; for example, humans can become sexually or aggressively aroused or calmed by fantasies or daydreams in the absence of external signals. Imagining a moment of bonding may be temporarily helpful to a person who is lonely, just as imagining a failure at bonding may be temporarily unhelpful. Persons also ruminate and generate fantasies and images that maintain a mood. Feedback is relevant as well: Anxiety may increase threatening thoughts, which may then increase anxiety (Gilbert, personal communication, 1996). Thus, some internal management of these systems is essential.

In mild forms, anxiety often correlates with the development of alternative behavior strategies. If these strategies are effective, a potential negative cost-benefit outcome may be offset, and the mood may dissipate. If not, the mood is likely to remain, and possibly to increase in intensity. Males and females are thought to differ with respect to the types of situations that are most frequently associated with anxiety; for example, stronger correlations between anxiety and anticipated affiliative and reproductive failures are expected in females, while in males, stronger positive correlations are expected between anxiety, resource acquisition, status, and reputation-related failures.

A coevolution of both the affect of anxiety and its recognition is probable. Under most circumstances, feelings of anxiety are accurately signaled and accurately recognized by others. Signaling may warn others of potential external dangers. Anxiety also initiates assistance by others. Others may be empatheic, particularly when somatic signs, such as sweating, hyperventilation, and speech alterations are obvious. Viewed this way, the affect features of anxiety can be understood as an adaptive strategy within the context of often debilitating emotions. In repeated social encounters, appearances of anxiety may reduce one's competitive advantage. This point may explain in part why chronically anxious individuals frequently attempt to disguise their feelings in social encounters.

Time-limited, external-stimulus-initiated anxiety may be viewed as a normal response. Conversely, extended anxiety without a stimulus can be viewed as minimally adaptive (Marks and Nesse, 1994). Both types of responses are automatic-system-mediated, and compromised automatic system functionality is associated with a reduction in the frequency with which novel anxiety-reducing strategies are developed. The often rapid anxiety-ameliorating effects of drugs point to physiological dysregulation as a key factor in dysfunctionality. Other types of conditions, particularly the absence of anxiety in situations in which it is expected, require other explanations, such as infrastructural deficits (suboptimality). Deficits may explain the responses of certain criminals and persons with antisocial personality disorder who have below-average arousal in response to anxiety-provoking stimuli (Eysenck, 1964; Raine, Venables, and Williams, 1990). The possibility that the incapacity to develop anxiety may have adaptive features is considered in chapter 8.

Looking to the future, we anticipate that different types of anxiety will be identified, each relating to a different type of adaptive problem. For example, separation anxiety in children should correlate with their fear that an important person may not be available in the future (survival anxiety); social anxiety should correlate with one's status (reciprocal anxiety); and potential loss of a kin should correlate with the fear that one will be restricted in kin investment options (kin selection anxiety). Nonhuman primate data are consistent with the preceding points. For example, in macaques, a mother's emotional concern over a potential threat to her own well-being (survival anxiety) is largely independent of her concern over a potential threat to her infant (kin investment anxiety; Troisi et al., 1988).

Depression

The Feeling State

Mild to intermediate depression is an unpleasant emotional state associated with dull pain, usually in the upper body; a sense of heaviness; a loss of interest in external events; a reduced sense of motivation; pessimism; self-depreciating thoughts; an inability to experience pleasure; a sense that routine activities require extra effort; and a sense of impending doom (Gilbert, 1984; APA, 1994). Severe forms of depression include the signs and symptoms noted above, as well as reduced appetite, insomnia, psychomotor retardation, agitation, loss of energy, and condition-characteristic postures and affects.

Klein and his colleagues (1980) provided a particularly insightful picture of the emotion:

> The various pleasurable life activities can be considered either consummatory or appetitive in nature. Consummatory pleasures are those directly related to satisfaction of biological drives, such as occurs with eating, drinking, and sexual intercourse and, perhaps, sleeping . . . In addition to consummatory activity . . . there is the pervasive appetitive activity; the sort of activity that gets the animal into position to enjoy consummatory activity . . . This activity may directly precede consummatory pleasure, e.g., sexual foreplay; may be further removed, as is the case with hunting or sports; or may be still further removed, as with intellectual activity or seductive behavior. Our point is that such activities are primarily rewarding, not simply secondarily rewarding because of their contiguous relationship to consummatory activities . . . From this standpoint the behavior of the person with endogenous depression can be understood as attributable to severe inhibition of both consummatory and appetitive pleasures . . . The situation of the demoralized person, however, is more easily understood as an inhibition of only the appetitive pleasure response, the consummatory pleasure responses remaining intact. (pp. 230–231)

In this formulation, both forms of inhibition would be due to the consequences of compromised automatic systems.

Clinical and Research Literature

The clinical research literature provides support for each of the prevailing model hypotheses. There is epidemiological evidence pointing to the importance of predispositions and gene-environment interactions (vulnerability) (e.g., Whybrow, Akiskal, and McKinney, 1984; Akiskal, 1985; Kendler, Heath, et al., 1993; Kendler, Pedersen, et al., 1993). Depression is reported to be more frequent among offspring whose parents are depressed (Weissman et al., 1984; Rice et al., 1984) and among women than among men (Weissman and Klerman, 1977; Weissman et al., 1984). Physiological explanations have focused on the dysregulation of the norepinephrine (Schildkraut, 1965; Igbal et al., 1989) and serotonin systems (Siever and Davis, 1985; vanPraag, Kahn, Asnis, Lemus, and Brown, 1987; van Praag, Kahn, Asnis, Wetzler, and Brown, 1987; Mann, McBride, Anderson, and Mieczkowski, 1992; Mann, McBride, Brown, et al., 1992), and some investigators have noted that the clinical and norepinephrine features of atypical depression differ from more standard forms of depression (Asnis, McGinn, and Sanderson, 1995). Platelet studies are reported to distinguish controls from persons who are depressed (Ellis and Salmond, 1994). Slow encoding of stimuli and prolonged processing of stimulus-response compatibility are reported for retarded as well as anxious-agitated and impulsive depressed patients (Pierson et al., 1996). Other investigators have focused on possible neurochemical links between depression, anxiety, and stress, and still others have suggested that specific types of neurochemical dysfunction may be associated with specific symptoms irrespective of diagnosis (Iny et al., 1994); for example, temperament type correlates with the hypercortisolemia observed in some depressed patients (Joyce et al., 1994). Recently, dopamine system dysregulation has been implicated as a possible contributing factor in depression (Kapur and Mann, 1992).

Among psychoanalysts, depression is usually viewed as distinct from grief, in that grief, more than depression, is associated with personal loss and vivid memories of what has been lost. Some authors have suggested that complicated grief- and bereavement-related depressions are distinct conditions (Prigerson et al., 1995), while others fail to make such distinctions. These views are consistent with findings showing that dimensionality ratings of depression are more consistent with a multidimensional than with single-dimensional structures (Suzuki et al., 1995).

Other findings focus primarily on social and psychological factors. The frequency of depression positively correlates with stressful and depriving environments and marital dissolution (Stack, 1980; Cochrane and Stopes-Roe, 1981; Sturt, Kumakura, and Der, 1984), as well as reduced life control (Gilbert, 1992). Depression has been associated with parental representations: Perceptions of shaming by parents and being a nonfavored child can contribute to shame and depression vulnerability (Gilbert, Allan, Ball, and Bradshaw, 1996). Characteristic personality features (Cofer and Wittenborn, 1980; Hirschfeld et al., 1983), attitudes toward others (Howes and Hokanson, 1979; Hokanson et al., 1980), and thought patterns (Donnelly et al., 1980; Silberman et al., 1983; Silberman, Weingartner, and Post, 1983; Powell and Hemsley, 1984), as well as low self-esteem, are often observed, although exceptions have been reported (Silverman, Silverman, and Eardley, 1984). Seasonal factors are also implicated (N. E. Rosenthal et al., 1984; J. Pedersen et al., 1988). Ethological studies demonstrate that certain types of nonverbal behavior correlate with depression (Ellgring, 1989; Bouhuys et al., 1987; Schelde, 1994), and that different behavior profiles predict clinical outcome more accurately than clinical assessments based on diagnostic categories (Schelde et al., 1988; Schelde, 1992; Sachdev and Aniss, 1994; chapter 4). A number of ethological studies have failed to identify the subtypes of depression found in *DSM-III-R* (Troisi et al., 1990) and suggest that many features of depression have a closer fit with a symptom-continuum model (i.e., mild to severe depression).

Evolutionary Views

Depression is thought to have evolved as a somatic indicator that biological goals have not been or are not being achieved, that is, that one's fitness has been or is being compromised (Nesse, 1991b). Time-limited depression due to loss of another, or to repeated external setbacks, can be explained as a consequence of normal automatic system function (Gilbert, Allan, and Goss, 1996b). Clinical findings are compatible with this explanation: The close correlation between depressed feelings and the sense that life has not turned out as desired (because of, e.g., the loss of a valued other or resources or a decline in social status) is too striking to ignore. Further, in instances in which depression appears to be a response to a failed attempt to achieve a goal, it often disappears after the goal is either achieved or given up. Severe and chronic forms of depression imply automatic system disruption. As predicted, grief intensity ratings were found to be higher among monozygotic than among dizygotic twins when one twin dies (Segal and Blozis, 1995).

In mild forms, depression often correlates with the development of alternative behavior strategies. If the strategies are effective, a negative cost-benefit ratio may be offset, and the mood may dissipate. If not, the mood tends to remain, and even to increase, until alternative high-priority biological goals influence behavior. However, severe depression is seldom significantly modified by novel strategies or changes in goal priorities. The different situations that are likely to increase the probability of depression in males and females are noted in chapter 7.

Possible adaptive functions of depression have also been mentioned and include the increased probability that others will provide help (Lewis, 1936); that help need not be reciprocated (Sloman and Price, 1987); that one can remain a member of one's social group while temporarily avoiding group-related participation, including competitive encounters (Price, 1967); and that one can alter others' behavior in ways that are beneficial (Nesse, 1990a, 1991a, 1991b). The physiological and psychological slowdown associated with depression may conserve energy and reduce both current and possible future costs (Engle, 1980), and ruminations may reduce the probability of taking on new and possibly costly projects (Nesse, personal communication, 1995). Recently, it has been suggested that some of the biochemical abnormalities associated with depression represent adaptive responses to features of the condition (Post and Weiss, 1992). Like anxious persons, persons who are depressed often attempt to hide their state from potential competitors, yet because of nonverbal features, the emotion remains difficult to conceal. Attempts at concealment may be due in part to shame (Gilbert, personal communication, 1996). It is worth emphasizing that persons select different environments in which to become depressed. Some do so in the presence of known others, while others do so in private. Very different implications follow: Depression in the presence of known others implies that persons hope that others will provide assistance; depression while alone implies the view that known others are hindrances to clinical improvement.

Finally, different types of depression are likely to be associated with failures in solving different types of adaptive problems. For example, depression associated with the loss of a significant other may have a stronger affective component than depression associated with failure to acquire resources.

Frustration-Anger

The Feeling State

Frustration and anger are associated with feelings of physical tightness, tenseness, discomfort, diffuse pain, and a heightened focus on the person or situation that is perceived to be responsible for the emotion (which may include oneself). Autonomic nervous system changes (e.g., increased heart rate and blood pressure) are often present, and aggression toward others or objects may occur.

Clinical and Research Literature

The postulated causes of frustration and anger range from actual events in the social and physical environment, to misperceptions of such events, to predispositions to act impulsively, to neurochemical dysregulation. Nearly all instances of what might be called normal frustration or anger are associated with identifiable external or internal stimuli, such as being cheated or realizing that one has acted stupidly. Extended states of frustration sometimes accompany depression. Pathological anger may or may not be associated with an actual failure to achieve goals (McGuire and Troisi, 1989a, 1989b). Related conditions include pathological jealousy (Mullen, 1991) and clinical conditions in which anger becomes both chronic and an all-consuming mood and behavior-influencing trait.

Neuroanatomical investigations point to the importance of the limbic system (MacLean, 1990), particularly the amygdala (Kling, 1986; Kling and Brothers, 1992). Neurochemical theories have usually emphasized the involvement of the norepinephrine or serotonin systems, and CNS serotonin activity is known to correlate inversely with aggression toward others and the self (reviewed in Masters and McGuire, 1994). Recently, gamma-aminobutyric acid (GABA) and dopamine have been implicated as possible contributing neurochemical systems. In comparisons with persons without conditions, a higher frequency of both anger and aggression have been observed in persons with paranoid schizophrenia, antisocial personality condition, and some organic brain syndromes (P. McC. Miller et al., 1985; McGuire and Troisi, 1989a; Tardiff, 1992). However, not all disorders have a higher than average incidence of aggression (Troisi and Marchetti, 1994).

Nonhuman primate studies are compatible with the view that animals experience feelings analogous to frustration and anger prior to aggression; that aggression is goal-directed; and that its primary function is to alter the behavior of others to the advantage of the aggressor. Anger frequently occurs in response to threats by other animals or following resource-acquisition disputes. Aggression also varies with certain biological and social states. For example, it is more frequently observed among lactating females (Troisi et al., 1988), as well as among low-status compared to high-status males (McGuire et al., 1983). Negative correlations between aggression, CSF monoamine metabolite levels, and measures of adrenal activity have been reported. Pharmacological manipulations of the serotonin, norepinephrine, and dopamine systems sometimes, but not uniformly, alter the frequency of aggression. These effects differ across social settings (Olivier, Mos, and Brain, 1987; McGuire and Raleigh, unpublished data, 1997).

Evolutionary Views

Frustration and anger are automatic system-mediated somatic states that develop in response to perceived high-cost situations. Anger is a likely emotion when resource access and one's control over resources are threatened, when expected reproductive options are constrained, when one is the victim of deception, or when one has actively pursued a high-cost strategy without beneficial results. Mild anger may improve reasoning (Ketelaar and Clore, 1995), but the opposite outcome usually applies to extreme anger. The capacity for anger appears to have coevolved with the capacity to accurately calculate the short-term consequences of actions directed at others, as well as with the capacity to constrain such actions; we direct aggression toward others less often than we might like (McGuire and Troisi, 1989a). However, the capacity to accurately estimate the long-term effects of anger, particularly possible negative social consequences, appears to be less well developed.

In mild forms, frustration correlates with the development of alternative behavior strategies. If the strategies are effective, a negative cost-benefit balance may be offset, and the emotion may dissipate. If not, the emotion tends to remain, and even to increase in intensity, and extended periods of anger and resentment may follow. Chronic anger, even when there is a known precipitant, implicates automatic system suboptimality.

Others may provide assistance when one is frustrated or mildly angry, particularly if the anger is not directed toward the potential helpers. Mild frustration can be hidden from others. Hiding intense anger is another matter. *Most important, the influence of anger on the behavior of others cannot be too strongly underscored: Whatever else anger may be, it is one of the most effective ways of rapidly altering others' behavior. Indeed, with the possible exception of intense pain, no other affect has such a predictable impact.* Again, different types of anger may be associated with different types of behavior. For example, future studies may distinguish between anger designed to intimidate others and displaced anger.

Pleasure-Satisfaction-Joy

The Feeling State

Pleasure, satisfaction, and joy, which are difficult to separate, are warm and desirable emotions, coupled with a sense of well-being, safety, security, optimism, and interpersonal closeness. These feelings may occur when one is alone (e.g., as a result of the pleasure of reading a humorous story), during bonding, or in groups (e.g., as a result of membership participation). Normally, pleasurable emotions are closely tied to achieving biological goals, but there are exceptions, such as the pleasure of relaxed contentment when one feels that one no longer has to achieve. Unlike unpleasant feelings, which may become chronic, pleasure, satisfaction, and joy are usually short-lived. Feelings associated with hypomania are not thought to be pleasurable in the sense discussed here, but to be more akin to feelings of power-control-elation, which are discussed below.

Clinical and Research Literature

On average, persons with mental conditions experience less pleasure than persons not suffering from conditions. Either compromised automatic systems or an inability to achieve biological goals is implicated. Psychological theories of pleasure and happiness applicable to persons without conditions are reviewed elsewhere (Argyle, 1987; Tiger 1992). Compared to the psychiatric literature discussing anxiety, depression, and anger, that discussing the possible causes of pleasure-satisfaction-joy is limited largely to the kinds of findings discussed below for drugs. These findings suggest that the CNS has several pleasure centers.

In reviewing the clinical and research literature dealing with drug-induced states, it is important to distinguish between the states of pleasure that result from actions associated with achieving goals and those that result from self-induced chemical alterations, such as heroine use. While the pleasure experienced with successful bonding and the use of certain drugs may be similar, it is unlikely that the excessive self-induction of pleasure by using drugs (induction to the point where reality is distorted) is an evolved trait.

Two distinct neurochemical systems are thought to mediate pleasure and reward: the opioid and the dopamine systems (Kosten and Kosten, 1991). The opioid system will be discussed first.

Among normal volunteers, high doses of naloxone, a drug that blocks opioid receptor sites, results in increases in tension, anxiety, irritability, and depression (M. R. Cohen et al., 1983). These feelings are analogous to those that are associated with social loss. For persons who lack social skills, and who are addicted to drugs influencing the opioid system, such as heroin, the self-induction of pleasure by means of drugs may be a way of attenuating persistent feelings of personal isolation (Panksepp, Siviy, and Normansell, 1985). Clinically, the opposite point also holds: Persons who are addicted to pleasure-inducing drugs are less inclined to engage in bonding behavior than persons not using these drugs. For example, new mothers who are addicted to opiates bond less intensely with their offspring. These findings permit the following interpretations: Endogenous opioids are part of the neurochemical reward system for social attachment (Schino and Troisi, 1992); the pleasurable quality of optimal social relationships is associated with endogenous opioid release; and the subjective feelings of distress associated with minimally adaptive or undesired social events (e.g., social ostracism) are associated with CNS opioid dysregulation. The street drug "ecstasy" might also be mentioned here. Users report that it increases a sense of general social affiliation as contrasted to dyadic affiliation, and this feeling of "group-belonging" is one of the principal attractions of the drug.

While bonding may be viewed as the ideal social model for achieving pleasurable states, it is not without its costs and uncertainties. Not only does successful bonding usually require considerable time and effort, but it also necessitates that partners be motivated and receptive to achieving similar ends. Thus, costs may be considerable in advance of benefits, and benefits may not be immediately forthcoming. In this context, the attractiveness of drugs such as heroin is understandable because they result in predictable feelings of pleasure with less interpersonal cost than may be associated with bonding. In effect, the desire for the pleasure associated with successful bonding appears to remain intact among persons who use pleasure-inducing drugs, while capacities to achieve successful bonding may be compromised.

Findings from experimental studies of nonhuman species are consistent with the preceding interpretations. Opiate withdrawal with the use of receptor-blocking techniques (e.g., naloxone) increases affiliative behaviors (Meller, Keverne, and Herbert, 1980; Fabre-Nys, Meller, and Keverne, 1982), and the acute administration of nonsedating doses of morphine decreases the degree and frequency of affiliative behavior (Miczek et al., 1981). Separation from mothers with the concurrent administration of naloxone to infants increases the distress vocalizations emitted by the infants (Kalin and Shelton, 1989). And the frequency and intensity of isolation calls when adult animals are separated from peer groups also increase with naloxone administration (J. C. Harris and Newman, 1988).

When we turn to the dopamine system, it appears that it influences different behavioral dimensions and responds to different features of fitness than opioid-influencing drugs. Cocaine is a dopamine-influencing drug, as is amphetamine. Still, unlike opiates, which are seldom associated with hostile or violent behavior during intoxication, cocaine and amphetamine are often associated with competitive and agonistic social interactions. Persons who use cocaine usually continue to engage in socially competitive behavior. These findings permit further speculation. Compared to the effects of heroin, the stimulation of the dopamine system appears to be more related to feelings of power-control-elation than to those of pleasure-satisfaction-joy. Further, the distinction noted above, in the discussion of depression, dealing with appetitive and consummatory responses may correspond to distinctions between the dopaminergic reward system (appetitive) and the opioid reward systems (consummatory). The dopaminergic system and its association with increased competitive activity may account for the inhibition of social bonding associated with cocaine use. A variety of syndromes, including psychotic depression, Parkinson's disease, frontal lobe syndrome, Type II schizophrenia, pseudohyperparathyroidism, neuroleptic-induced akinesia, and amphetamine or cocaine withdrawal, are thought to result from the hypofunction of dopaminergic systems. Clinically, each of these syndromes reveals signs of diminished motivation, as well as lack of emotion, interest, and concern for others (Marin, 1997).

Evolutionary Views

Pleasure, joy, and satisfaction are automatic-system-mediated emotions that are thought to have evolved not only as indicators that biological goals are being or will be achieved (i.e., that benefits are exceeding or will exceed costs), but also as a way of increasing the likelihood of similar emotion-producing behaviors in the future (Nesse, 1990a). The fact that there is an abiding cultural preoccupation with achieving states of pleasure is not surprising (Tiger, 1992), nor is the fact that persons attempt to induce pleasure in a vast number of ways, through sex, jokes, empathy, social gatherings, and holidays. The usually short-lived nature of pleasurable feelings may protect persons against a long-term reduction in vigilance, which often accompanies pleasure. The feeling state also appears to have coevolved with the capacity to accurately signal one's feeling state to others. Within limits, signaling pleasure may make one more attractive (e.g., a "winner") and thereby decrease competitive encounters. Although role and individual differences apply, as a rule the persistent identification of pleasurable feelings has the opposite effect, in that others may take offense if one remains happy for too long. That different types of pleasure are associated with different types of competitive victories is a strong possibility.

Pain

The Feeling State

Physical pain is associated with sharp, hurtful and sometimes dull, unpleasant somatic sensations. Externally induced pain results in rapid withdrawal from the source. Internally induced pain (e.g., a broken leg, a gall bladder attack) leads to behavior to minimize its effects (e.g., a reduction in activity). Psychological pain is experienced as physical pain, as in phantom limb syndrome, or as a diffuse but hurtful sensation, as in psychogenic pain syndrome (Troisi and McGuire, 1991).

Clinical and Research Literature

Although pain is ubiquitous in the human condition, it continues to defy precise definition. Few investigators would argue that it is not part of the body's defense system, and the most frequent defensive action is to interrupt the activity or events responsible for the pain. Physical pain caused by others often leads to retaliation and aggression. Psychogenic pain associated with events such as being raped or severely cheated can have similar effects. If one is helpless to act so as to alleviate pain or is unable to retaliate, depression often follows. Personality traits color the behavior of individuals in pain (Bond, 1978), and shame has features of pain, hence its inclusion here: Shame and fear interact and have the common feature of fear of negative evaluation (Gilbert, Pehl, and Allan, 1994). And when shame is intense, it is a powerful behavior-altering experience (Gilbert, in press).

A number of physiological and psychological systems have been implicated in explanations of externally induced pain. Gate-control theory postulates that the transmission of nerve impulses beginning in the periphery result in CNS-initiated alterations in arousal and autonomic modulation (Blackwell, 1989). Multiple pain representations in the CNS have been reported (Talbot et al., 1991), and representations appear to be specific to different types of pain, including auditory, visual, and somatic. Selected medical syndromes associated with the reduced response to pain are due to the absence of pain fibers, a condition that is modeled by automatic system deficits. Several neurotransmitter systems are thought to modulate pain, the serotonin system being the most frequently mentioned. Clinically, increasing CNS serotonin activity reduces the perception of pain and, perhaps, the effect of pain on infrastructural functionality.

Reports suggest that pain thresholds change during development (Izard, Hembree, and Huebner, 1987), and that they are increased in persons with anorexia nervosa, bulimia (Lautenbacher et al., 1991), anxiety, and unipolar depression (Adler and Gattaz, 1993). It is not as yet clear if pain thresholds are reduced in persons with psychogenic pain disorder. That specific emotional states can temporarily override pain seems clear from reports of people who are injured and frightened: Often, the pain from the injury is not perceived until the frightening experience is over.

Evolutionary Views

For mental pain, pain-mediating systems are likely to have evolved as indicators of minimally adaptive anatomic and physiological states, as well as of states in which fitness is being compromised (N. W. Thornhill and Thornhill, 1989). If there is a perceived external source of pain, it is the focus of attention, and anger directed at the source is usually an accompanying feature (N. W. Thornhill and Thornhill, 1990a, 1990b). Thus, the experience of pain may quickly shift to aggression. Capacities to generate behavior strategies for rapidly exiting from pain due to an external physical source are well developed. Capacities to behave in ways that reduce mental pain seem less well honed. In situations in which mental pain is extended, it compromises both algorithm function and functional capacities. Thus, chronic pain is unlikely to be a selected trait, even though it may signal others to offer help. When the source of pain is psychological, evolutionary interpretations may provide critical insights, as in instances of rape. Rape is a fitness-reducing event for women, and evolutionary reasoning predicts that the psychological pain associated with rape will be greatest among reproductive-age women. Findings support this prediction (N. W. Thornhill and Thornhill, 1990a, 1990b, 1990c). Evolutionary reasoning also predicts that events or situations associated with pain will be avoided in the future. Findings also support this prediction for a variety of experiences, ranging from painful social relationships to unpleasant working conditions.

The affect of physical pain is accurately signaled and recognized by others, and recognition generally results in rapid and helpful responses. The psychological pain associated with experiencing unexpected loss or being cheated generates similar responses. However, the chronic expression of psychological pain generally leads to the opposite outcome, particularly when there is no identifiable source: Others' sympathy and willingness to help diminish over time, in part because the source of the pain is difficult to identify, and in part because the recognizers are helpless to alter the pain.

Power-Control-Elation

The Feeling State

Power-control-elation is associated with a sense of mastery over one's physical or social environments, a sense that desired objectives are being achieved, and a sense that one has attained a competitive advantage. Ambiguity and uncertainty are reduced, and a sense of pride may be present. As noted, this feeling state is sometimes mistakenly associated with pleasure or positive emotions (e.g., those resulting from a massage or a nice meal with a friend). Power-control-elation is generally a far more energizing emotion.

Clinical and Research Literature

Clinically, the emotion is associated with winning a competitive encounter, perpetrating a successful deception, or experiencing an upward shift in the conditions of reward (Strongman, Wookey, and Remington, 1971). Under normal conditions, feelings of power-control-elation are short-lived, perhaps with good reason: In the real world one's competitive advantage is often fleeting and only a prelude to renewed competition with others. However, these points do not deter persons from attempting to extend the duration of these feelings. Persons with psychiatric conditions, except hypomania, do not appear to suffer from an excessive sense of power-control-elation, although certain personality conditions (e.g., narcissistic personality disorder) have been interpreted otherwise.

In the psychiatric literature, there are few references to this emotion. Hypomania is one exception, and the prevailing view is that this feeling state is a consequence of major physiological and psychological dysregulation and that environmental information is relatively unimportant. Possible involvement of the dopamine reward system was mentioned earlier. A recent report suggests that CNS serotonin is *not* altered in manic states (Yatham, 1996). This finding raises questions about interpreting mania as a status-related behavior (cf. Gardner, 1982). Whether hypomania is an extreme state of a normal emotion or a separate emotion remains an unanswered question, although the former seems more probable.

Evolutionary Views

Evolutionary explanations of power-control-elation have focused on interactions between achieving a competitive advantage and rising in one's social hierarchy (e.g., Price, 1967; Gardner, 1982). Findings from studies of vervet monkeys discussed earlier serve as a model of how social events interact with physiological states. Desired emotional states may be instrumental in influencing animals to continually seek to improve their status. It is also likely that capacities to control the expression of this emotion have evolved. While the signaling and recognition of this emotion are generally accurate, there are often good reasons to disguise it; otherwise, others' competitiveness may increase.

Concluding Comments

The six emotions can be integrated with the concept of functional systems in the following ways. Emotions provide information that functional systems are in a cost-deficit state (depression, frustration-anger, pain), will be in a cost-deficit state (anxiety), or are in a benefit-excess state (pleasure, power-elation-control). For the actual or anticipated cost-deficit states, strategies are usually developed to alter the deficit. If the deficits are too great, decoupling, which amounts to a strategy to reduce further cost deficits, may take place. Conversely, for the benefit-excess states, efforts to repeat or perpetuate the emotions are common. Viewed in this manner, emotions turn out to be critical information sources that contribute to behavior modulation and strategy modification, primarily in the service of achieving goals. The fact that a large percentage of the population without conditions frequently experiences varying degrees of the six emotions described here suggests that both modulation and strategy modification are unending events.

We understand moods and affects more clearly when we view them as adaptive functions. Automatic system dysfunctionality can contribute to emotions, which, in turn, become moods through the failure to accurately process emotion-terminating feedback information. In conditions in which a mood(s) is a prominent feature and there are no obvious associated strategic advantages, the mood is likely to be a consequence of automatic system suboptimality or dysfunctionality, for which there are a variety of sources: deficits, state changes, and so on. But the preceding analysis has only scratched the surface of emotions; for example, in addition to the points already discussed, the emotions of shame and pride may be strongly influenced by reactions to the opinions of others (Fessler, 1995) and may thus be motivating factors in cooperation.

6

Information Recognition
and Signaling

To discuss mental conditions is also to discuss information recognition and information signaling. Over half of the mental status examination performed by psychiatrists is devoted to evaluating these capacities, and the outcome influences diagnosis, explanation, and intervention. *Tangential thinking, poor recall, delusions,* and *flat affect* are terms familiar to clinicians of every theoretical persuasion. Nevertheless, explanations of recognition and signaling remain far from satisfactory. This is so for good reasons. Information is complex. The ways in which it is processed and transmitted are equally complex, and the details of these complexities are not easily studied (Hinde, 1985). Therefore, it is reasonable to ask if our understanding of both normal and condition-related recognition and signaling can be improved by an analysis of these behaviors in an evolutionary context. Our answer is *"Yes, provided function is included in the analysis."* This is the approach we have taken. The chapter begins with a discussion of general features of information processing. It then turns to an analysis of recognition and self-deception. Signaling, deception, and recognition of deceivers are the subjects of the final section.

General Features of Information Processing

Information filtering-organizing-prioritizing, causal thinking, scenario development, behavioral strategies, and *self-monitoring* are five clinically identifiable information-processing functions. Numerous other processing functions could be discussed but in our view, these five are essential in the assessment of every patient. Information filtering consists of filtering, organizing, and prioritizing biologically important information. Causal thinking is modeling and attributing causes or explanations to behavior and events. Scenario development is the imaginary rehearsal of behavior and events. Behavioral strategies are scenarios turned into action. And self-monitoring is the process of auditing the cost-benefit effectiveness of one's behavioral strategies. Most likely, each of these capacities is a product of selection. Information filtering was first discussed in chapter 4. It is an automatic system function. Algorithms are primarily responsible for the remaining four functions. One may be only partly aware of these functions. Genetic information and trait variation influence each of the functions, and within limits, each can be refined.

Evolutionary discussions of information processing cover a broad range of topics. Details can be found in studies dealing with the evolution of the brain (MacLean, 1990); consciousness (Crook, 1980); intelligence (Stenhouse, 1973; Byrne, 1994); interactions between social structure and brain function (Chance, 1988); modular concepts of brain function (Gazzaniga, 1989, 1992); algorithms (Cosmides and Tooby, 1987, 1992; Symons, 1990; Tooby and Cosmides, 1990a; Barkow et al., 1992; Wang, 1997); emotion-cognition interactions (Plutchik, 1980; Tooby and Cosmides, 1992); and memory.

Factors that are likely to have influenced the selection of sophisticated and flexible information-processing capacities have been noted by Plotkin (1994), who points out that in species that have extended periods of development (e.g., humans), there is a time lag between receiving genetic material and putting the products of such material to use free of the guidance and protection of elders. In humans, this is the period between conception and adolescence. During this period, the environment may change. If information-processing systems are inflexible, a person may be ill adapted to deal with the changed environment. Capacities to utilize others' causal models, to learn from observing, to integrate different experiences in the service of developing novel behavior strategies, to self-monitor, and to alter one's behavior on the basis of both monitored information and information provided by others are consistent with this view.

Within the ranks of investigators who look at the human brain from an evolutionary perspective, there is general agreement on a number of points. The brain is a highly complex system that not only processes internal and external information but is also responsible for physiological and psychological homeostasis and the mediation of behavior. It has numerous self-correcting (feedback) systems, and it is capable of manipulating information in a nearly endless variety of ways (Rimé et al., 1985; Marks, 1987; Gilbert, 1993). Findings from studies of closely related species support this view. Nonhuman primates of all types share many of their information-processing capacities and functions with humans, although there are clear species differences. For example, nonhuman primates are capable of developing efficient food-searching strategies, distinguishing between friend and foe, anticipating the behavior of conspecifics and predators, and understanding the behavior rules associated with hierarchical behavior (e.g., Cheney and Seyfarth, 1990). When the net is given a broader cast, it is clear that each species responds preferentially and differentially to particular types of information. Both similarities and differences provide insights into past adaptive problems and their solutions, as well as hint at adaptive potentials in current environments.

When discussions turn to questions about which infrastructure might be responsible for which functions, there are fewer points of agreement. Debates over algorithm function, which were first discussed in chapter 3, illustrate this point. Some investigators favor the view that learning is highly flexible (general-purpose algorithms), while others take the opposite view (special-purpose algorithms). A clear implication of these distinctions is that special-purpose systems are pre-prepared to selectively process certain types of information, while general-purpose systems are far less constrained (and thus more difficult to study). As noted in chapter 5, our view is that both general- and special-purpose systems are integral to the models developed throughout the book, although in this chapter, the discussion focuses on systems that carry out specific functions.

The special-purpose concept is consonant with many findings from neurophysiology which demonstrate that specific cells in the temporal lobe respond selectively to faces (Desimone, 1991); that facial information is interpreted modularly (Harries and Perrett, 1991); that different parts of the face ("local elements") are preferentially processed in the brain's left hemisphere (Hillger and Koenig, 1991); and that specific parts of the brain selectively process social stimuli (Brothers, 1990a, 1990b; Brothers and Ring, 1992, 1993). Similar specificity is likely for the auditory, tactile, and olfactory systems (Ploog, 1989b, 1992), and perhaps also for social structure information, such as the detection of coalitions (Kurzban, Tooby, and Cosmides, 1995). In short, "The modern human is a bundle of special-purpose systems that allow us to communicate, evaluate facial expressions, make inferences, interpret feelings, moods, behaviors, and all the rest" (Gazzaniga, 1992, p. 203). At least, this is true of normal individuals, who seemingly process information effortlessly. Not only do individuals with mental conditions experience various degrees of difficulty, but compromised information processing is probably the most frequent observation in clinical psychiatry (e.g., McGuire and Lorch, 1968; McGuire, 1979b; Mithen, 1995).

Because their operations are largely transparent, it is easy to overlook both the complexity and the utility of special-purpose systems. However, to do so is to invite misunderstanding of conditions and their consequences. For example, consider a situation in which two persons exchange favors involving different "currencies." Person A provides several hours of physical assistance to Person B (e.g., moving Person B's furniture). Person B repays the favor with an informed discussion about the upcoming possibilities of the stock market. And both A and B feel that the helping debt has been repaid. How is this possible when the currency of helping and repayment differ? In our view, the most parsimonious answer is that algorithms have evolved to make currency translations (e.g., Cosmides, 1989; Fiddick, Cosmides, and Tooby, 1995). The alternative interpretation—namely, that currency translations are learned—is not only unwieldy, but also hard to envision, if only because no two reciprocal exchanges are ever quite the same.

Information Filtering

The initial filtering of information is an automatic system function involving the identification, selection, and prioritization of biologically important information. (Figure 6.1, in a later section, provides a visual representation of this process.) As noted, one usually filters information without being aware of the filtering process, although one is aware of some of the products of filtering, as when one unexpectedly meets and recognizes a friend or focuses attention on a rapidly approaching car and quickly moves across the street. Depending on the circumstances (e.g., novel versus nonnovel stimuli), there are differences in the amount of attention paid to information as well as the intensity and rapidity with which it is processed. Nonnovel social stimuli generally command less attention than novel stimuli.

While most adults selectively filter information on the basis of motivational priorities, filtering is tempered by experience. The young child who sticks himself with a pin has to learn not to do so again. Emotions have a central role in this educational process. Built-in biases or information-processing constraints are also likely. For example, the brain appears to favor a better-safe-than-sorry bias. An animal that is in a field and hears an unfamiliar noise usually finds it safer to run away than to investigate, although there might be no real danger; that is, the brain may be designed to minimize the cost of making potentially dangerous mistakes and not always to collect all potentially relevant information (P. Gilbert, personal communication, 1996). Over the course of development, experience gradually builds a "library" of responses to and priorities for specific stimuli, and within limits, experience can override or replace biases (e.g., learning to like snakes or to dive off high cliffs).

Mental conditions frequently have their own characteristic filtering signatures, and as a first clinical approximation, the more obvious the signature, the more guarded the clinical prognosis (e.g., Fairbanks, McGuire, and Harris, 1982). Amnesia is associated with excessive filtering. Residual schizophrenia is associated with the selective filtering and the selective enhancement of certain types of information (Gaebel and Wölwer, 1992)—every clinician is familiar with situations in which ostensibly neutral words (e.g., *store, tree, car*) trigger strong emotional reactions in persons with this disorder (Minami, Tsuru, and Okita, 1992). Paranoid schizophrenia is associated with information misperceptions and distortions: Information that normally receives minimal attention, such as two friends talking in a hallway next to their offices, may be selectively processed, assigned high priority, and given a low-probability interpretation. Alexithymia is associated with altered capacities to process one's own emotions and to interpret others' affects (Troisi et al., submitted, 1997). Dyslexia and hypomania are yet other examples. And a number of rarely encountered conditions, such as Ganser's syndrome and Cotard's syndrome, appear to be consequences of suboptimal information filtering. That condition-characteristic information processing is in part the result of dysregulated physiological states is an obvious inference from experimental findings designed to identify interactions between neurochemical states and information processing (e.g., C. R. Clark, Geffen, and Geffen, 1987a, 1987b). Still, not all conditions have identifiable information-processing signatures. Mood conditions and selected personality conditions often fail to reveal consistent patterns; for example, mild states of depression and anxiety can be associated with near-normal filtering of much internal and external information, and persons with antisocial personality disorder who are socially successful often filter information accurately.

Four special-purpose information-processing algorithms are discussed below. For clarity, the discussion proceeds as if each of the algorithms functions separately and its effects are linear. Exactly the opposite is the case, however. Algorithms may function and interact concurrently, and feedback influences are ever-present.

Causal Modeling

The term *causal modeling* refers to capacities to model one's own behavior, others' behavior, and events, and to attribute causes (e.g., Bickhard, 1980; J. R. Anderson, 1991). A critical function of this type of modeling is to reduce uncertainty and ambiguity by giving meaning to events.

Causal modeling capacities are not easily divorced from language (a dictionary of memorized symbols and a set of generative and symbol-processing rules; Pinker, 1994). Compared to persons with below-average language capacities, persons with above-average capacities have a greater facility for interpreting and manipulating information. Atypical language patterns are often present in persons with disorders (e.g., residual schizophrenia), where poverty of speech, tangentiality, speech derailment, and illogicality are observed (Andreasen, 1979a, 1979b). It is unlikely that these patterns have no effect on thinking. However, in other disorders, such as mild forms of hypomania, anxiety, somatoform disorders, and narcissistic personality disorder, language capacities may be indistinguishable from those typical of persons without disorders.

Modeling capacities are refined during infancy, a period in which most children devote considerable time not only to figuring out why their parents, siblings, and others behave as they do, but also to testing and assimilating their models. Culture concurrently contributes to model refinement: All cultures have models dealing with behaviors that are believed to have fitness-reducing outcomes, just as all cultures value and disseminate information about fitness-enhancing behavior. But assimilation can cut two ways, as is sometimes the case for children who grow up in environments in which significant others are seldom present and, when present, are rejecting. These experiences can become the basis of a child's models of others' behavior. When such models are carried into adulthood, they often lead to socially unattractive behavior designed to disconfirm expected interaction outcomes, such as making excessive demands on others to test the degree to which they care.

Causal modeling is influenced by a variety of predisposed biases (Forsyth, 1980; Forsyth, Berger, and Mitchell, 1981; Nesse and Lloyd, 1992). For example, persons tend to overvalue their altruistic acts and to favorably interpret their own intentions (Essock-Vitale and McGuire, 1985b), and they usually view close kin, especially offspring, in the best possible light. Risk-sensitive models (the assignment of risk to specific conditions) vary as a function of biologically important variables (e.g., degree of relatedness) and age (Wang, 1997). Such biases are likely to serve specific functions; for example, the overvaluation of one's altruistic acts may reduce the chances of helping potential nonreciprocators by making one more attentive to others' strategies. Other types of biases are also apparent: All-or-none thinking, overgeneralization, and excessive personalization are examples, and like atypical language, they contribute to suboptimal modeling (Beck, Freeman, and Associates, 1990). There is also a bias—indeed, an apparently strong bias—favoring the use of already existing models in preference to developing new models, especially when persons are confronted with information that is novel and difficult to interpret. The studies described here, which were conducted over three decades ago, provide an amusing illustration of this point as it applies to persons without conditions.

The studies in question involved undergraduate students at premier U.S. universities. The students were asked to engage in a conversation through a teletypewriter that was connected to a remotely located computer. While these studies took place during the early days of computer development (the late 1960s), it was public knowledge in both universities and the national media that interactive computers had been developed and that they could carry on conversations. When subjects began the study, they were told only that they "would be communicating through a Teletype." During the experiment, subjects typed statements into the Teletype to which the computer generated replies. The computer software program was designed to simulate human typing, and it had approximately 100 software rules that served as the basis of computer-generated responses. For example, if a subject typed a sentence in which the word *if* appeared, such as, "I plan to go to the beach tomorrow *if* it doesn't rain," the computer rule was "Disregard all of the words in the sentence prior to the word *if*, repeat the word *if* and all that follows, and add the phrase *tell me more*." In this example, the computer would have replied, "If it doesn't rain, tell me more." Subjects were paid to participate in the study, and they were told in advance that their performance would not affect their pay.

Described out of context, both the rule above and its computer-generated reply seem embarrassingly simple. Nevertheless, subjects experienced the computer replies as if they were communicating with another person. After an hour-long test period and approximately 60 subject-computer exchanges, the subjects were interviewed. A key interview question was "Were you communicating with a person or a computer?" Ninety percent of the subjects answered, "A person." When subsequently asked, "Is it possible that you could have been communicating with a computer?" over 80% answered, "No."

The focus of research then changed to identifying the minimal number of computer rules that would result in at least 50% of the subjects' responding that they were communicating with a computer. A new group of subjects was used for this study. Over the course of the study the software rules were progressively degraded. The result was computer-generated replies that omitted key words (usually verbs), made obvious grammatical errors, and were sometimes nonsensical. Through each degradation stage, the majority of subjects continued to believe that they were communicating with another person, not a computer. Eventually, the software program consisted of only one software rule: "No matter what is typed by a subject, reply with 15 A's (*AAAAAAAAAAAAAAA*)." After an hour of communicating with this program, 70% of the subjects still believed they were conversing with a person. When asked if they could be communicating with a computer, over half of the subjects still said, "No," and would then volunteer explanations such as, "Someone was trying to convince me that he is a computer."

A third study was then conducted with another group of subjects. The computer program was explained and demonstrated. The subjects were then asked to develop programming strategies that would lead persons at a remote Teletype to conclude that they were communicating with a computer. The subjects made a variety of suggestions, of which the most frequent were the same rules the investigators had used: omitting key words, generating nonsensical statements, and so forth. In effect, the subjects repeated the same "mistakes" as the investigators by imagining that certain types of computer-generated responses would lead to the development of new causal models, that is, that one's interaction partner was a computer, not a person.

There are many ways of explaining these findings, and many are discussed elsewhere (McGuire, Lorch, and Quarton, 1967; Quarton, McGuire, and Lorch, 1967). The subjects entered a novel experimental environment. In response to their typed input, a machine produced replies similar to what might be expected if another person were typing at a remote location. Thus, in retrospect, it is not surprising that the subjects initially thought they were communicating with another person. It is also not surprising that the subjects in the third study recommended the same software degradation strategies that the investigators had used. There were a limited number of options. Further, the subjects were not selected (in the evolutionary sense) to communicate with computers, and for several reasons, it is this point that makes the findings all the more interesting:

1. The subjects' participation in the study had the same financial benefit irrespective of their performance; that is, all subjects were paid the same fee. Thus, the subjects were free to experiment, and many did so by typing in tongue twisters or nonsensical statements.

2. It was conceivable that another person was typing replies, and there was no foolproof strategy the subjects could adopt to disconfirm this possibility. Disconfirmation was further complicated by the fact that nonverbal elements were absent. In short, no matter what the subjects typed, they remained uncertain if they were interacting with a person or a computer, and they explained the information they had by accommodating the model they had brought to the experimental situation to the information available to them.

The preceding points can be instructively applied to modeling others' behavior rules, that is, to understanding others well enough to predict how they will behave in novel situations (MacKay, 1972; Brothers and Ring, 1992; Leslie, 1995). Such modeling often goes under the rubric *theory of mind*. With experience, capacities to model others' rules are refined, yet modeling is never so accurate that it is errorless. Others sometimes withhold information, deceive, or unexpectedly shift behavioral strategies without indicating that they are doing so. How one obtains information about others is also important. Consider a situation in which one is attempting to determine if another person will repay a favor. At least three kinds of information-gathering strategies can be identified. In the probe, or trial-and-error strategy, one does a small favor for another and awaits a response. If the response is not forthcoming, a likely conclusion is that the person benefited is not an altruist. This method has the advantage of minimizing the costs of determining if another person is an altruist, but it has the disadvantage of not accurately predicting whether bigger favors will be repaid. In the experience strategy, one assesses others over time and by multiple information routes, some direct (e.g., interacting, observing) and some indirect (e.g., third-party information). In terms of time and effort, this is the most costly method, but it is also the method most likely to result in accurate assessments of others' behavior rules (McGuire et al., 1994). In the third, or symbol-language, strategy, others communicate information about their behavior rules (e.g., motives, values) or information from which their rules may be inferred (e.g., how one responded to a particular situation). The cost of listening to others is relatively small, but the chances of being deceived, or mistakenly equating what others say with how they will act, are large.

Introspection and empathy bear on the preceding points. Introspection amounts to using the products of one's information-processing systems as a source of information for modeling one's intentions, biases, and priorities. There are good theoretical (Trivers, 1985) and empirical reasons (Nisbett and Wilson, 1977; Gazzaniga, 1989) for doubting the accuracy of introspections; the workings of most infrastructural systems are not available to awareness despite often heroic efforts to make them so. Nevertheless, it is reasonable to argue that a capacity that can so influence how we think, feel, and act is unlikely to have appeared by chance (e.g., Hermans, 1996). Most persons introspect; most give some credence to their introspections; and introspections often trigger strong emotions (e.g., shame). Thus, the possibility that the capacity has been selected deserves consideration.

It is our view that introspection evolved because it serves two functions. One is to gain *partial* access to one's own behavior rules—determining the importance of a goal, assessing one's motivations, and so forth. The other is to model others' behavior rules, a state psychoanalysts and others refer to as *empathy*. In psychoanalytic writings, empathy is described as the temporary suspension of one's emotions and thoughts while thinking and feeling as another. The possible neural bases for such behavior has been discussed elsewhere (Brothers, 1989, 1995). The fact that persons without conditions are likely to have similar motives and infrastructural systems permits the inference that causal models of one's own behavior rules provide reasonably accurate estimates of others' behavior rules. Viewed this way, sensing that one knows how others' feel and will behave (empathizing) is an attribution function of causal modeling (Forsyth, 1980), one in which introspections are used to model others. Because of greater genetic closeness, this postulate should apply more to kin than to nonkin, yet even among nonkin, genetic similarity may make it more cost-effective to initially favor the use of one's own models than to try to interpret others' behavior rules from scratch or from insufficient information.

Modeling the behavior rules of persons with conditions is another matter, in part because conditions and inaccurate modeling often go hand in hand (Nesse and Lloyd, 1992; Trower and Chadwick, 1995), and in part because one's introspections are often not applicable to persons with conditions. Model rigidity, excessive model revision, and motivation-induced invalid models explain some of the clinical findings. Model rigidity is observed among persons with paranoid character disorder, paranoid schizophrenia, social phobia, obsessive-compulsive disorder, and agoraphobia. For these individuals, chronic mood states appear to freeze (inhibit) the development of alternative models or to constrain the revision of existing models of others' minds even when disconfirming information is available. Excessive (frequent) model revision results in a different clinical picture, as is seen in persons with borderline personality disorder and in intensely anxious or confused persons, whose interpretations of others' rules or events may change rapidly. Motivation-induced invalid models are seen in what psychoanalysts refer to as *transference,* as well as in hypomania and erotomania, the latter being characterized by sexually skewed misinterpretations of others' motivations and behavior. A similar process is characteristic of fixed ideas and extended periods of anger when they are coupled with demeaning thoughts of others. A common thread in the preceding examples is that models are not viewed as working constructs that are subject to change in response to new information. Rather, they are taken as facts. Further, the persistent misreadings of events and others' behavior rules can have both physiological and psychological consequences. Studies showing that dysphoric ideas alter CNS glucose metabolism (Pardo et al., 1993) and blood flow (George et al., 1995) and that fantasies and meditation alter the activities of different physiological and hormonal systems (McGuire and Troisi, 1987b) point to this conclusion.

Scenario Development

Scenarios are imaginary rehearsals of one's own and others' behavior as well as events, such as imagining how one can engage in beneficial social interactions, avoid anxiety-provoking encounters, influence the behavior of others, improve one's social attractiveness, or reduce environmental uncertainty (Gilbert, submitted, 1997). *Role enactment* is the term sometimes used to characterize scenarios, and implicit in the scenario concept is some sense of self-other differentiation (Gilbert, 1995). Scenarios range from those that are straightforward to those that are deceptive, as when one develops elaborate plans to acquire resources at another's expense. Studies suggest that there are species-characteristic perceptual biases that lead to misassessments of risk (e.g., Lichtenstein et al., 1978; Johnson and Tversky, 1983). Somewhat surprisingly, these biases are not amplified in conditions where amplification might be expected, such as panic (Nesse and Klass, 1994).

Scenarios may depict events or experiences in the past, present, or future; and the amount of time persons spend developing and refining scenarios is often revealing; for example, excessive rehearsals of low-probability events are observed in many socially isolated individuals. Scenarios may appear to represent a special case of causal modeling, yet there are important differences: Scenarios are usually understood to be hypothetical, to be tentative, and to be subject to change if contingencies change; at least, this is true of most persons without conditions. In persons with conditions, tentativeness is often absent, and both the number and the type of scenarios are often constricted. Said another way, persons without conditions usually choose from several scenarios in their efforts to achieve goals, while persons with conditions (antisocial and borderline personality disorders possibly excepted) choose from a limited and often minimal array.

Differences in male and female scenarios are reported, and from an evolutionary perspective, such differences are expected. For example, males are more inclined to develop scenarios of competitive victories, while females are more inclined to develop scenarios of interpersonal closeness and possession (Ellis and Symons, 1991). Characteristic sex- and age-related scenarios are often absent among persons with conditions (Beahrs, 1990).

Cost and benefit estimates are integral to scenario development, and persons with and without conditions differ in how they estimate. Optimists lower expected costs and raise expected benefits. Pessimists do the opposite. Within limits, these biases are not so atypical that they are considered indices of conditions. Persons with conditions cover the entire range of possibilities. Hypomania is associated with the strong tendency to minimize cost and maximize benefit estimates; persons with social phobia usually do the opposite; and persons with somatoform conditions appear to make relatively accurate short-range cost-benefit assessments with respect to obtaining help and attention from others.

Behavioral Strategies

Behavioral strategies are scenarios turned into action. Behavior may include inaction if the expected costs exceed the expected benefits, or if the outcomes are uncertain. Developing strategies often involves memory, and specific algorithms may be tied to specific types of memory, for example, proprietary memories (Tooby and Cosmides, 1995). While simple strategies (e.g., carrying out familiar acts) are usually efficiently executed, this is less often the case when strategies are contingent on others' behavior. Contingent strategies usually require midcourse revisions. Thus, within limits, which strategy one adopts may be less important initially than is usually assumed, provided one is capable of midcourse strategy alterations. Moreover, more than one strategy can be effective in achieving a goal (Thorngate, 1980). It follows that paying close attention to how strategies unfold is likely to improve cost-benefit outcomes. This is the topic of self-monitoring discussed below. It is also in part the topic of emotions: Anger, upsetness, depression, anxiety, and jealousy inform one that a strategy is in trouble.

The psychiatric literature contains numerous references to the idea that individuals utilize "secondary-gain" strategies, as when a person acts as if he or she is ill in order to increase the possibility that others will provide care (Henderson, 1974). This idea is discussed in chapter 9, but one point should be noted here. While persons with conditions (e.g., hypochondriasis) often act and are often treated by others in ways that seem consistent with the secondary-gain hypothesis, an evolutionary analysis suggests that secondary gain is an unlikely explanation for such behavior. In an evolutionary context, conditions and attempts to act adaptively go hand in hand, and attempts to adapt are not primarily volitional. One does not easily settle for second best when one has a condition. Instead, one attempts to achieve goals within the constraints of one's condition. In short, the difficulty with the secondary-gain hypothesis is that there is little that is secondary about it if it is allowed that conditions like hypochondriasis and suboptimal infrastructures go hand in hand.

Self-Monitoring

Self-monitoring is an algorithm function that provides an answer to the question: Have the costs and benefits of my actions changed relative to my expectations? There is an extensive psychological research literature dealing with this function, which goes under such names as *self-focused attention, self-regulation,* and *role assessment* (e.g., Duval and Wicklund, 1972; Carver and Scheier, 1981; Pyszczynski and Greenberg, 1987). In the classical formulation of this process, focusing attention on the self initiates self-evaluations that are associated with self-standards. If one meets or exceeds one's standards, positive self-assessments follow. If one fails to exceed one's standards, negative self-assessments follow (Duval and Wicklund, 1972), and negative assessments serve a self-correcting (and thus a potentially adaptive) function. Debates continue over whether negative assessments are aversive, whether they motivate individuals to alter their behavior, and whether certain conditions are associated with self-focus avoidance (e.g., Carver and Scheier, 1981; see also Frijda, 1993). It is not critical to resolve these debates. What deserves attention is an important difference in the formulations above and our own: Our view of self-monitoring is that it is far more utilitarian than the theories mentioned here. In effect, just as one is *locked in* to the social environment to achieve critical goals, one is also *locked in* to self-monitoring in terms of cost-benefit assessments. What often differentiates persons with and without conditions is the degree to which negative self-assessments are avoided. Psychic defenses have been mentioned as one means of avoidance. Self-deception, discussed later in this chapter, is another.

At times, the "How am I doing?" question can be answered straightforwardly. If one sets out to repeat a familiar strategy, such as having lunch with a friend, and if one enjoys the lunch, it is clear that the strategy has worked. For more complex strategies, information filtering, causal modeling, and emotions are involved, as when one is monitoring another's response to acts of helping or coercion. If unfolding events compare favorably with strategy-related cost-benefit estimates—that is, if there is a close fit between what is expected and what is assessed—one usually views one's efforts as worthwhile, and the strategy continues to unfold (Nesse, 1990a). On the other hand, if the comparison is unfavorable, strategies may be altered or aborted. Emotions are relevant to this formulation in the following way: If the fit between expectation and outcome is not close, frustration, anxiety, or depression may indicate that strategies should be changed. As expected, suboptimal and dysfunctional self-monitoring are frequent trademarks of persons with conditions. In addition, a cardinal sign of many conditions is the failure to use self-monitored information in self-beneficial ways, as when persons disregard the effects of their behavior on others. Paranoid, borderline, antisocial, histrionic, and narcissistic personality disorders frequently qualify on this point, as do hypomania and obsessive-compulsive disorder.

To summarize some key points: (1) Each of the preceding information-processing functions is integral to efficient (cost-effective) behavior; (2) compromised information processing is a frequent but not a pathognomonic sign of conditions; and (3) compromises of one or more information-processing systems go hand in hand with cost-benefit deficits, reduced functional efficiency, and reduced goal achievement.

Information Recognition and Conditions

The topic of compromised information processing is not new to psychiatry. For at least a century, psychiatry's textbooks have devoted approximately half of their pages to the information-distorting features of conditions (e.g., poor reality testing, delusions). Clinical findings have been well documented elsewhere, and they will not be reviewed here. Rather, structural models designed to facilitate the interpretation of condition-related information distortions are introduced, along with the question: Are different kinds of information distortions best interpreted as examples of infrastructural suboptimality or dysfunctionality?

Structural Models of Recognition

Structural models serve as frameworks for assessing the contributions of traits and states to conditions. Persons with and without conditions often differ significantly in their capacities to process, recall, manipulate, and utilize information. Specifying whether capacities are suboptimal or dysfunctional is important in both diagnosis and intervention. For example, treating dyslexia (the absence of a structural connection; see below) with drugs makes little sense, while treating anxiety-related dysfunctional information processing with either an environmental alteration or a drug does make sense.

Figure 6.1 contains five structural models of information recognition. The models assume that there is a positive correlation between nondistorted recognition and the number of intact connections depicted in the models, and that there is a negative correlation between the number of absent or atypical connections and both the costs and the accuracy of recognition. The figure should be read as follows. Information from others and the physical environment enters at the top level and flows downward (1). Automatic systems function at the second level, where both internal information (e.g., motivations, memory, emotions) and external information are filtered and prioritized (2). Algorithms function at the third level (3). At the fourth level, one becomes partially aware of processing outcomes (4). The arrows at the bottom of each of the structures indicate that information is transmitted to other systems (e.g., memory). For

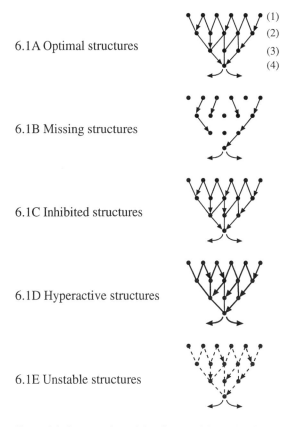

6.1A Optimal structures

(1)
(2)
(3)
(4)

6.1B Missing structures

6.1C Inhibited structures

6.1D Hyperactive structures

6.1E Unstable structures

Figure 6.1 Structural models of recognition. The figure should not be interpreted as a design of brain architecture or brain circuitry.

clarity, the models are depicted as if processing events occur in a linear fashion, but again, feedback systems are integral to elaborated forms of the models.

Optimal connections (Figure 6.1A). In this figure (continuous lines), all automatic system and algorithm connections are optimal; that is, a person has the capacity to receive and process external and internal information without constraints or misinterpretations brought about by system suboptimality or dysfunctionality. For multiple reasons (e.g., trait variation, prior experience, self-deception), this figure characterizes only a small percentage of individuals (see, for example, Figure 4.4, chapter 4).

Missing connections (Figure 6.1B). In this figure, automatic system and/or algorithm connections are missing (missing lines). This is a suboptimal or deficit condition. Deficits may be genetic, physiological, or psychological. Suboptimal conditions may be lifelong, or they may develop during one's life, as occurs in dementias due to neuronal death. Limited information is available for encoding, and the information that is available is subject to misinterpretation. Misperceptions and high-cost, low-benefit judgments and actions often follow. The figure models features of disorders such as Alzheimer's disease and other forms of dementia, mental retardation, and possibly also disorders in which misinterpretations of external and internal information are chronic (e.g., paranoid character disorder and dyslexia). Interventions are ineffective in restoring connections, but secondary consequences can often be treated (e.g., special reading techniques for dyslexia, drugs for anxiety reduction in dementia).

Inhibited connections (Figure 6.1C). Inhibited connections (solid lines) have information-processing consequences similar to those caused by missing connections, although generally, the effects are less severe. This is a dysfunctional condition that may sometimes affect anyone (e.g., during periods of unusual stress). Inhibited connections model features of atypical recognition observed in persons with, for example, temporary states of acute anxiety or dissociative disorders. These states are often amenable to treatment.

Hyperactive connections (Figure 6.1D). This configuration (downward-pointing large arrows) models the overvaluation and misinterpretation of information that is observed in persons with hypomania, borderline personality disorder, delusions, and hallucinations; in persons who experience periods of intense fear after the source of the fear has long disappeared; in persons whose responses to internal information are excessive, as is the case in those with hypochondriasis; and in persons who experience information overload and disorganization during acute phases of schizophrenia or following devastating losses. Hyperactive connections may be due to either suboptimal traits or dysfunctional states. If connections are dysfunctional, they are often amenable to treatment.

Unstable connections (Figure 6.1E). Unstable connections (downward-pointing arrows and interrupted lines) model fluctuating recognition capacities. This structure combines features of Figures 6.1C and 6.1D. Recognition among persons with schizoaffective disorder, borderline personality disorder, hypomania, and acute forms of schizophrenia are examples. Chronic unstable connections are due to suboptimal and intermittently dysfunctional infrastructures.

In different ways, missing, inhibited, hyperactive, and unstable connections compromise both automatic system and algorithm functionality. Missing and inhibited connections lead to minimally adaptive emotions and reduce the information available to causal modeling, while hyperactive and unstable connections either result in excessive amounts of information or lead to frequent model changes. Functional inefficiency is the unavoidable outcome.

Recognition Traits, States, and Conditions

Suboptimal or dysfunctional recognition is a feature of most conditions. Suboptimal traits are implicated when atypical recognition precedes condition onset and remains unchanged or worsens during periods of condition exacerbation. Dysfunctional states are implicated when recognition distortions are limited to periods of condition exacerbation. Trait and state distortions can be conveniently subdivided into three categories: recognition traits associated with conditions; recognition trait complexes classified as disorders; and recognition states associated with conditions.

Recognition Traits Associated with Conditions

This category includes information-encoding traits associated with a variety of conditions; yet, when considered alone, they are seldom sufficient to meet disorder diagnostic criteria. Examples are chronic pessimism and low-level fear, tangential thinking, and invariant interpretations of others' behavior (e.g., model rigidity). In this category, traits do not appear for the first time when persons are first given a diagnosis, only to disappear when disorders remit. Rather, traits are there all along and often become more obvious during exacerbation. In some instances, interventions may minimize their effects; for example, the anxiety component of chronic social phobia often responds to drugs. However, in such instances, traits do not disappear. Instead, they are present but chemically managed, as is suggested by their often reappearing if medication is discontinued.

Recognition Trait Complexes Classified as Disorders

Disorders such as schizophrenia, paranoia, hypochondriasis, agoraphobia, and obsessive-compulsive personality are in large part identified and classified on the basis of suboptimal or dysfunctional trait recognition.

Schizophrenia is characterized by both missing and hyperactive connections (Figures 6.1B and 6.1D). Paranoia is characterized by the persistent misinterpretation of others' motives and external events. Both missing and hyperactive connections (Figures 6.1B and 6.1D) model these misinterpretations. For persons with hypochondriasis, unrealistic fears or a belief that one is suffering from a disease without any apparent evidence is usually present. Some studies suggest that persons with this condition engage in what may be termed *somatosensory amplification,* a concept implicating hyperactive connections (Figure 6.1D) with an associated reduction in the ability to efficiently recognize and process internal and external information that could disconfirm the belief that one is ill (Barsky et al., 1988; Barsky and Wyshak, 1990). Chronic agoraphobia is characterized by a marked fear of being unable to escape from public places. Hyperactive connections (Figure 6.1D) are implicated. Obsessive-compulsive personality disorder is characterized by persistent ego-dystonic ideas and fears. Hyperactive and missing connections (Figures 6.1B and 6.1D) are implicated. A further possible entry in this list is alexithymia if replication studies confirm findings of a relative lack of imaginative activity (Vogt et al., 1977). As in the preceding category (recognition traits associated with disorders), traits may become more apparent when disorders are exacerbated.

Recognition States Associated with Conditions

A heightened sensitivity to external events or internal information is usually an indication that a disorder is present (Billings and Moos, 1984). Recognition changes that are confined to periods of exacerbation, such as intense anxiety, fear, or depression, are included in this category. Inhibited and hyperactive connections (Figures 6.1C and 6.1D) most frequently model state conditions, and both may be associated with striking recognition distortions (e.g., amnesia).

One feature of state-influenced recognition distortions deserves emphasis, namely, the influence of moods on historical information. This influence is perhaps most obvious during periods of depression when persons negatively interpret past events relative to their interpretations during nondepressed states. This observation is consistent with the view that automatic system dysfunctionality influences information interpretation. Still, it needs to be emphasized that event reinterpretation is not necessarily an index of a condition. Ongoing reassessments are predicted in evolutionary models. New information is expected to continually modify interpretations of prior experience (rewriting one's history) so as to render such experience more valuable in developing future strategies. It is the sudden change in valence of interpretations in association with moods that implicates dysfunctionality.

Self-Deception

In evolutionary context, self-deception is the failure to recognize and process selected information (Trivers, 1985; Alexander, 1979; Lockard, 1988; Lockard and Paulhus, 1988). It is a different concept from the psychoanalytic postulates of repression and denial, and its implications extend beyond those of psychic defenses.

In a discussion of information access constraints, important distinctions must be made between the types of information that are available to consciousness. First, there are limits to the amount of information available to consciousness; that is, one can think about only a limited number of things simultaneously. Second, studies repeatedly suggest that individuals do not have access to higher order cognitive processes (Nisbett and Wilson, 1977; Power and Brewin, 1991). These access constraints apply to persons with and without conditions, although probably differentially. Repression and denial imply something different, namely, that certain information and/or processes *could* be available to consciousness, but that psychic processes have somehow blocked their availability. A key feature of this formulation is that repression and denial are not uniformly tied to goal achievement or social interactions, although at times they may be; for example, one may repress or deny information about one's childhood that, if conscious, would *not* alter one's current choice of strategies. Self-deception has another explanation: It occurs specifically in relationship to goal-related activities and is limited largely to those motives that would conflict with carrying out a particular strategy.

In evolutionary thinking, self-deception has two main functions: to keep selected information out of awareness so that deceptive strategies can be carried out efficiently, and to facilitate both kin investment and nonkin altruism. In the first function, a person who is able to conceal his or her motives from himself or herself may have a better chance of concealing them from others (Trivers, 1985), and fewer signaling mistakes or nonverbal indicators of "deception leakage" are likely to occur (Ekman, 1971, 1988). Viewed this way, self-deception may be adaptive because it facilitates carrying out self-interested strategies that might otherwise be constrained or detected. The behavior of persons with chronic pain syndrome and hypochondriasis is consistent with this idea when the purpose of their deception is to obtain others' help and consideration. Similar interpretations apply to self-deception among persons with antisocial and histrionic personality disorders. Reproduction-related fears among persons with anorexia nervosa may also qualify. Normally, these fears are not available to consciousness. As to kin and nonkin altruism, self-deception may reduce the estimated costs of investing in kin and thereby increase the likelihood that such investment will occur; conversely, overestimating the cost of helping nonkin may result in a more discriminating approach to helping. Empirical data are consistent with both of these predictions (Essock-Vitale, McGuire, and Hooper, 1988).

Information Signaling and Conditions

Signaling is a term biologists use when discussing the transmission of information from one person or animal to another. *Communicating* is a more familiar term outside biology.

Evolutionary views on signaling and its functions are relatively straightforward. As with recognition, capacities have been shaped by selection, and signaling is a highly complex process. Accurate signaling (which may include deception) is a prerequisite for successful social navigation (Dawkins and Krebs, 1978; Guilford and Dawkins, 1991). And the refinement of signaling goes hand in hand with maturation and socialization. Everyday experience is consistent with these views. Persons discuss their motives, goals, and feelings, and their capacity to do so improves over time. They persuade and manipulate. They sometimes say too little, sometimes too much. And so forth. In addition to words, signaling has nonverbal features, some of which are understood universally, for example, hand gestures signifying *you, me, stop, yes, and good-bye*. Raising one's eyebrows signifies that one understands (Ekman and Friesen, 1969; Ekman, Sorenson, and Friesen, 1969; Grant, 1969; Grammer et al., 1988; Ekman, 1993) although in certain contexts, it signals fear or astonishment. Characteristic facial expressions are associated with states of happiness, anger, sadness, surprise, and fear, and although there is cross-person variation, the meanings of these expressions are widely understood across cultures (Ekman, 1971; Argyle, 1972a; Eibl-Eibesfeldt, 1989). Further, depending on context and motives, nonverbal signals often communicate more biologically important information than words. Similar points apply to the animal kingdom, where signaling may be visual, vocal, chemical, or contextual (Altmann, 1967; Smith, 1977; Beecher, 1982, 1989; Belcher et al., 1986; Hopkins and Savage-Rumbaugh, 1991).

Signaling capacities and rules are refined during development; for example, children gradually learn context-specific rules dealing with how long to talk and when to interrupt. Among adults, rules serve to guide and bracket conversations, and they influence what one signals as well as how signals are interpreted (McGuire and Lorch, 1968; Innes and Gilroy, 1980). Signaling and status also interact. Higher status individuals show an increased frequency of self-referencing behavior (Hold-Cavell and

Borsutzky, 1986), more complex uses of language, and greater degrees of self-deception (McGuire and Troisi, 1990); lower status individuals reference themselves less often and deceive others more often. Sex differences are also observed. Males and females posture differently, and their preferred dress, appearance, shape, and odor differ (Buss, 1989; Feierman and Feierman, 1992). Each of these signals carries its own message.

Structural Models of Signaling

The evolutionary approach to conditions inevitably leads to the study of features that make people susceptible to conditions and to their identification and characterization (Nesse and Williams, 1994). Like recognition models, structural models of signaling, which are presented in Figure 6.2, are useful because they provide a way of characterizing and thinking about condition-related signals. Five structural models are presented in the figure, the last four of which model conditions.

In Figure 6.2, the apex of each structure (1) represents a molar behavior (e.g., greeting or threatening another). Level (2) references individual behaviors (e.g., pos-

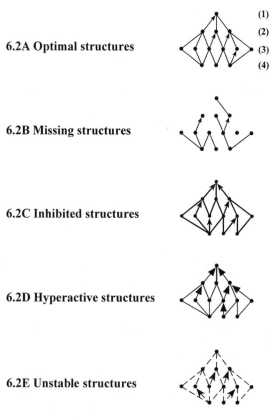

6.2A Optimal structures   (1) (2) (3) (4)

6.2B Missing structures

6.2C Inhibited structures

6.2D Hyperactive structures

6.2E Unstable structures

Figure 6.2 Structural models of signaling. The figure should not be interpreted as a design of brain architecture or brain circuitry.

tures, gestures, facial expressions, words) that together make up a molar behavior (recall the behavior clusters in Table 4.4). Algorithms are operative at Level 3 (3), and automatic systems at Level 4 (4). Assumptions applied to Figure 6.1 apply here as well. The greater the number of intact connections, the greater the possible number and diversity of signals; the greater the possibility of precise signaling; and the less the cost per signal. Atypical signaling can occur because of suboptimality or dysfunctionality at any of the levels in the figure.

> *Optimal connections* (Figure 6.2A). The figure depicts an optimal structure: the greatest number of intact connections; the greatest possible diversity of signals; and the least cost per signal. Continuous lines signify that connections are intact. Optimal connections are present when individuals communicate precisely, without constraint, and at minimal cost. The structure has clinical implications in that signaling capacities may be close to optimal among persons with disorders (e.g., antisocial personality disorder).

> *Missing connections* (Figure 6.2B). Missing connections are identified by missing lines relative to Figure 6.2A. The absence of connections leads to a reduction in both the options for and the precision of signaling. Missing connections are assumed to be irreversible and to represent suboptimal traits. They can be inferred in situations where expected behaviors do not occur or, if they do, where they are incompletely executed. Missing connections model degenerative disorders in which the capacities to execute signals have declined or disappeared. They may also model chronic features of disorders, such as impoverishment of speech and discontinuity of discussion, which are often observed in persons with chronic forms of schizophrenia (Frith, 1979). When connections are missing, signals increase in cost because of the need to use alternative but less efficient ways of communicating. Interventions result in minimal clinical improvement, although secondary salutary effects may be achieved.

> *Inhibited connections* (Figure 6.2C). Inhibited connections (thick lines) are transiently inoperative connections present from time to time among all persons, as when one's capacity to express oneself accurately is compromised during periods of anxiety. Inhibited connections are a consequence of dysfunctional systems, and when they are present, the clinical picture may be similar to that observed with missing connections. Both pharmacological and psychotherapeutic evidence is consistent with the view that inhibited connections are influenced by infrastructural function (e.g., infrastructural alterations may improve function). The figure models signaling associated with intermittent or state features of disorders: The costs of signaling increase, and the use of less relevant signals associated with intact structures may increase. The structure will shift toward Figure 6.2A when interventions are effective or when disorders remit spontaneously.

> *Hyperactive connections* (Figure 6.2D). Hyperactive connections (upward-pointing large arrows) are associated with short-duration, rapidly changing molar, or incomplete behaviors. Such behavior may represent a state or a trait. Because connections are preactivated, the short-term costs of signaling will be less than in Figure 6.2A Reduced short-term costs are the likely basis of high-frequency and often exaggerated behaviors observed in hypomania as well as states of hyperarousal, intense anxiety, fear, and overresponsiveness to minor yet familiar events. Reductions in short-term costs are offset by the costs associated with continual preactivation. As with Figure 6.2C, pharmacological and psychological evidence is consistent with the view that overactive connections have both physiological and psychological components.

> *Unstable connections* (Figure 6.2E). Structural instability (combined arrows and dashed lines) results in unpredictable signaling due to combinations of inhibited and overactive

connections. Figure 6.2E models features of disorders in which emotional states, thoughts, and overt behavior frequently change valence without an apparent external cause, as among persons with borderline and volatile or impulsive personalities, persons in psychotic states and delirium, and persons who are in transition states between disorder remission and exacerbation. Changing infrastructural dysfunctionality is implicated in each of these disorders. Chronically unstable connections implicate suboptimal infrastructures.

Factors other than those mentioned here influence structural functionality. Circadian rhythms affect alertness (Monk et al., 1989). The right and left sides of the brain process information differently (Gazzaniga, 1989), and persons differ in the degree of dominance associated with the two sides. Different parts of the brain are active in different emotional states (George et al., 1995). And male-female processing differences may exist (Buss, 1994). In addition, the fact that most persons not suffering from conditions exhibit transient forms of the kind of signaling modeled in Figures 6.2C to 6.2E suggests that the capacity for temporary structural change is a ubiquitous feature of the human brain. While behavior modeled by specific structures may serve as disorder signatures (e.g., hyperactive connections with hypomania, inhibited connections with depression), combinations of suboptimal or dysfunctional structures are more commonly observed clinically.

Whether individuals can voluntarily change the state of their structural connections without using drugs is still an unanswered question. Clinically, voluntary change appears to occur infrequently when the social environment and motivations are stable. However, some degree of self-determined mood and information modulation seems probable because emotions are tied to energy and motivation (Thayer, 1989). Moreover, environmental contingencies are known to alter brain physiology (see chapter 8), and persons do have the capacity to seek out social environments that are associated with observable structural shifts, as when they seek out a calming friend or engage in highly emotional group activities such as attending a rock concert.

Signaling Traits, States, and Conditions

Categories similar to those used in discussing recognition and conditions are applicable to signaling and conditions.

Signaling Traits Associated with Conditions

Signals are the information provided to others about one's motivations, traits, and states. The same signal may differ in its meaning depending on the context (e.g., large versus small groups, familiar versus unfamiliar individuals; Trivers, 1985), and nonhuman primates, as well as humans, are known to adjust their signals in response to features of the social environment (Elowson and Snowdon, 1994).

Certain signals lead to social acceptance or, if not, neutrality on the part of others (Gilbert and Allan, 1994). Other signals lead to negative responses or may imply that something is wrong (Ploog, 1992). Signaling traits characteristic of disorders fall into this category. Repetitive movements are observed in obsessive-compulsive disorder (Hollander et al., 1990); deficits in the spontaneous display of negative affect are observed in persons with alexithymia (McDonald and Prkachin, 1990); and combinations of meticulousness, excessive vigilance, pessimism, self-centeredness, shyness, and aggressiveness are characteristic of other conditions. Chronic emotional states also qualify. During condition exacerbation, traits may be more obvious than during periods of remission, and some traits, such as those first observed during the onset of chronic schizophrenia or dementia, remain chronic.

Traits in this category inform our understanding of both infrastructural function and functional strategies. Gaze aversion in infantile autism provides an example. At first glance, gaze aversion appears to be minimally adaptive. However, an analysis of this behavior suggests that it reduces eye-to-eye contact, which autistic children misperceive as a form of interpersonal aggression (Hutt and Ounsted, 1966; Richer and Coss, 1976; J. Pedersen, Livoir-Petersen, and Schelde, 1989). Survival system suboptimality is implicated in these findings. A related finding suggests that autistic children use object-centered strategies in preference to the person-centered strategies characteristic of children without autism (Philips et al., 1995). Interestingly, autistic children locate themselves closer than average to others, as well as engage in a high frequency of physical contact (J. Pedersen et al., 1989), two findings that suggest a desire for bonding. A variant of this interpretation—namely, that autism is associated with an absence or a diminished capacity to read others' behavior rules—has been suggested (Tooby and Cosmides, 1995).

Signaling Trait-Complexes Classified as Disorders

Signaling trait-complexes may be classified as disorders. Examples are somatization disorder, histrionic personality disorder, psychological pain disorder, hypochondriasis, Munchausen syndrome, and antisocial personality disorder. Likely candidates are general neurotic syndrome (Andrews, Stewart, Morris-Yates, et al., 1990), paraphilias (e.g., fetishisms), and psychosexual dysfunction disorders (e.g., inhibited sexual desire). Traits associated with hypochondriasis and Munchausen syndrome are listed in Table 6.1 and serve as examples of trait complexes.

For both of the disorders in Table 6.1, the trait complex is usually present long before the disorder is diagnosed clinically. Treatment for an actual or assumed medical illness usually leads to diagnosis. Milder forms of these disorders may go unnoticed or may occur intermittently; for example, a large percentage of the adult population may manifest transient forms of hypochondriasis during periods of prolonged stress (Barsky and Wyshak, 1990).

Trait complex disorders, such as somatization disorder, psychological pain disorder, histrionic personality disorder, hypochondriasis, and Munchausen syndrome, have overlapping features, including chronicity, general social disability (e.g., reduced capacities to utilize and benefit from normal social options, constricted social functioning), and diverse symptoms and complaints (Sigvardsson et al., 1986; Zoccolillo and Cloninger, 1986; Troisi and McGuire, 1991). Characteristic signaling profiles are associated with each. For example, women with histrionic personality disorders often initiate relationships with men by signaling that they are good prospects for intimate relationships. Persons with hypochondriasis, psychological pain, and somatization disorders signal that they are ill, and they engage family, friends, and institutional personnel in caretaking relationships, often to such a degree that they have been characterized as engaging in "morbid care-eliciting behavior" (Henderson, 1974). As noted

Table 6.1 Clinical features of hypochondriasis and Munchausen syndrome

	Primary trait condition	Consequences of secondary trait-state conditons
Trait-identification	Organic or hypochondriasis related disease	Excessive diagnostic procedures; missed organic diseases
Mechanism	Involuntary	Physiological changes often occur
Consequence	Manipulate others; receive attention and care	Social withdrawal by others; depression and chronic anger among those afflicted
Treatment strategy	Within limits, improve strategies for interpersonal relationships	Minimize medical interventions
Relationship to Figure 6.1	Primarily due to missing structural connections	Periods of inhibited connections
Munchausen Syndrome		
Trait identification	Organic or related disease	Inability to care for self
Mechanism	Voluntary induction	Physiological changes as a result of treatment
Consequence	Manipulate others; receive attention and care; excessive diagnostic procedures	Anger and rejection among those who recognize the condition
Treatment strategy	Allow persons to express dependent behavior	Minimize medical interventions
Relationship to Figure 6.1	Primarily due to missing structural connections	Periods of inhibited connections

earlier, our view is that such behavior reflects a primary strategy—the best they can do—among persons with constricted capacities, not behavior motivated by secondary gain. While it may be true that the behavior of persons with these disorders irritates others and renders such persons socially unattractive, it may also be true that relating as they do is their least cost-expensive strategy, as well as one that reaps benefits (others' attention and assistance) with enough frequency so that it is continued.

An unanswered question for hypochondriasis, psychological pain, and somatization disorders is whether persons with these disorders are overly responsive (e.g., hypersensitive to somatic information). Our view is that overresponsiveness is a secondary feature, and that it is frequently observed in individuals who are highly constricted in their capacities to achieve goals. Persistent failure in goal achievement correlates not only with symptom intensity and increased sensitivity to information, but also with attempts to gain others' assistance.

From almost any perspective, the benefits associated with the preceding list of disorders are likely to be less than the psychological, physiological, and social costs; for example, reduced social attractiveness and social options, limited resource access, possible social ostracism, and limited goal achievement. Because these costs are likely to have changed little over time, it is doubtful that these disorders have been selected. Further, the earlier analysis implicated missing structural connections and it is hard to imagine how missing connections could improve adaptive capacities. *In this view, an argument can be made that these disorders represent attempts to adapt by persons with suboptimal infrastructures. In effect, what is classified as a disorder may be understood as an attempt to minimize the competitive disadvantages that accrue from limited capacities to navigate socially.* Support for this interpretation is found in the excessive use of capacities that are minimally compromised, such as drawing attention to oneself, seductive behavior, and the overinterpretation and frequent use of symptoms to gain support and assistance from others. It follows that there is no reason to suspect dramatic improvement in response to interventions irrespective of their type. This is the usual clinical outcome.

States and State Signaling Associated with Conditions

This category deals with the presence of states and state signaling that is confined largely to periods when conditions are exacerbated. For example, the onset of depression is usually associated with the appearance of specific postures, facial expressions, and pessimistic utterances. These behaviors may be absent prior to condition exacerbation and may disappear following remission. Dysfunctionality at any one of the structural levels in Figure 6.2 may be implicated (Bouhuys and van den Hoofdakker, 1991, 1993; Bouhuys, Jansen, and van den Hoofdakker, 1991).

How do the preceding points explain compromised function? Suboptimal or dysfunctional recognition or signal systems compromise function in two critical, but different, ways. Recognition distortions mean that scenario and strategy development, as well as self-monitoring, are compromised. Signal distortions mean that behaviors are inefficient, as well as misinterpreted by others. These consequences are not easily avoided: To the degree that persons are *locked in* to participating with others to achieve goals, the probability of goal achievement declines, while that of conditions increases.

Deception

Deception enters the discussion at this point for the following reasons: It can be an adaptive trait, and it is a common feature of persons with conditions.

Deception is defined as a discrepancy between a perceived signal and the state of the displayer or events (Redican, 1982; Trivers 1985; Mitchell, 1986). In evolutionary models, it is viewed as a product of self-interested motivations, not accidents, and awareness of one's motivations is not implied. Thus, deception does not include conscious engagement in deceiving (e.g., white lies). In social interactions, the person who deceives and does so effectively may have an initial competitive advantage.

Evolutionary explanations dealing with why capacities for deception have evolved build on two ideas: Systems of animal communication are not necessarily systems for the dissemination of the truth, and members of all mammalian species try to alter the behavior of other animals (Trivers, 1985). In doing so, they may signal correct information, misinformation, or both (Hyman, 1989; McGuire and Troisi, 1990). An extensive literature addresses the conditions that are likely to have favored the selection of this behavior (Whiten and Byrne, 1988), its many uses (Redican, 1982; Whiten and Byrne, 1988), and its detection (Ekman, Friesen, and Scherer, 1976; Hocking and Leathers, 1980). Deception has morphological and behavioral components, and both are observed throughout the animal kingdom. Examples are nonpoisonous snakes with markings similar to those of poisonous snakes, sham rage and piloerection during aggressive displays among nonhuman primates, and alarm calls and displays by birds that divert the attention of predators from the location of nests and valued resources. Among humans, concealed ovulation may also be an example: Overt continual sexual receptivity in females without external signs of ovulation correlates with increased male parenting (Alexander and Noonan, 1979), an outcome that is likely to be advantageous in female reproductive strategies. As might be expected, male counterstrategies to possess and attract females (e.g., deception dealing with one's parenting intentions and one's capacities to acquire resources) appear to have evolved.

Among humans, postural, verbal, and social appearance deceptions are common. It is likely that all humans deceive and that many do so frequently. Selection should favor capacities to deceive if deception contributes to achieving biological goals at less than average cost. Data documenting the frequency of deception among humans, as well as its long-term consequences, are meager. Could it be otherwise? Status is also important; studies suggest that, compared to low- and high-status males, middle-status males are more likely to deceive females to gain sexual access (Ast and Gross, 1995). A reasonable expectation is that the frequency of deception will be relatively low in familiar settings, that is, settings in which the individuals, the interaction rules, and the behavioral expectations are known to all participants (Whiten and Byrne, 1988). Conversely, the frequency should increase where persons are less well known to each other. And of course, there are different degrees of deception, ranging from elaborate forms of fraud to subtle interpersonal manipulations (Kligman and Culver, 1992). Further, different forms of deception should be associated with different goals and their associated functional systems. The failure of many attempted deceptions suggests that not all persons who deceive are good deceivers (Ekman et al., 1976). Conversely, the success of many deceptions suggests that those who are deceived are imperfect detectors.

Even though very little is known about the details of deception, some general points should apply. Perhaps the most obvious is that deception can be a far from simple or cost-free undertaking (Kligman and Culver, 1992). For example, if one is deceiving another about the value of a resource, success is likely to hinge on a number of factors, such as misrepresentation of the value of a resource to the person being deceived, the costs of acquiring the resource through deceptive versus nondeceptive means, and one's capacity to develop and execute behavior strategies and to accurately estimate both the probability of detection and its possible consequences. Figures 6.3 to 6.6 elaborate on these points as they apply to persons with and without conditions. The assumptions applicable to each of the figures are that the cost of achieving a goal using nondeceptive behavior remains constant and that the variables discussed in the figure are not conscious.

If we look at the cost-benefit curve for deception (Figure 6.3), which plots the probability of deception as a function of a potential deceiver's cost-benefit estimates of deception, a revealing pattern emerges. The probability of deception increases as a function of decreasing benefits relative to increasing costs. This decrease begins to occur to the right of Point X. Situations in which Figure 6.3 apply include deceiving others about the state of one's intentions, such as promises of cooperation, or the state of one's resources so as to gain such benefits as valuable information. (Academic deception, stock market fraud, and the adventures of bigamists also qualify.) To the

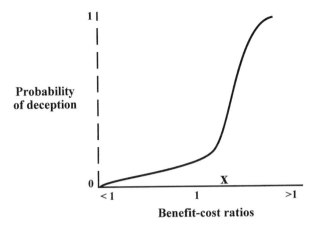

Figure 6.3 Probability of deception versus cost-benefit estimates. Adapted from McGuire and Troisi (1990).

degree that one misreads events or others' behavior rules, or that one miscalculates costs and benefits, deviations from the curve should occur.

As noted in chapter 3, the degree of genetic relatedness is a pivotal idea in evolutionary explanations of behavior (W. D. Hamilton, 1964), and the degree of relatedness should influence deception. Figure 6.4 explores this possibility.

Figure 6.4 depicts the probability of deception in relation to two coefficients of relatedness: $r = 1/2$ (by common descent, the deceived shares half of the deceiver's genes) and $r = 1/8$ (by common descent, the deceived shares one eighth of the deceiver's genes). The curve designated $r = 1/2$ (in which only the deceiver benefits) plots the likelihood of deception when the deceiver shares one half of his or her genes with

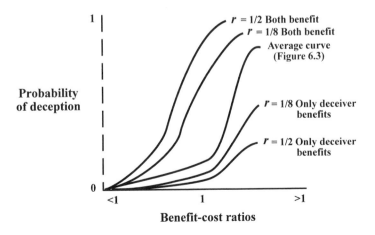

Figure 6.4 Probability of deception versus degree of relatedness and potential benefit. Adapted from McGuire and Troisi (1990).

the deceived and the deceived is unlikely to benefit from a deception. A successful deception, while potentially beneficial to the deceiver, may be costly to the deceived and may lead to a reduction in the deceiver's fitness. In this condition, the probability of deception is low. The curve designated $r = 1/2$ (in which both benefit) plots the likelihood of deception for an $r = 1/2$ relative who, along with the deceiver, is likely to benefit from the deception. This is a win-win situation. For example, a parent may bias information about an offspring's potential wealth so that she or he will be selected as a mate by a person who has above-average resources. The curves designated $r = 1/8$ (in which only the deceiver benefits) and $r = 1/8$ (in which both benefit) depict similar predictions for $r = 1/8$ relatives. The $r = 1/8$ curves are closer to the average curve because of the greater genetic distance between deceiver and the deceived compared to $r = 1/2$ kin.

Deception should also be influenced by the cost of detection to a person being deceived a point developed in Figure 6.5.

We can all concede that most persons are deceived some of the time. This may occur for a number of reasons: Persons are not consistently vigilant; their detection capacities differ; or the costs of detection may exceed the probable consequences of being deceived. From an evolutionary perspective, the abilities of deceivers and detectors should be about equal in a population; otherwise, either deception or nondeception would become the norm. A strategy one commonly uses to reduce the probability of deception is preferentially interacting with persons who, in one's experience, infrequently deceive (e.g., known altruists). However, limiting one's interactions to such persons is often inconvenient, especially in social environments in which important interactions are carried out between unfamiliar persons. In such circumstances, the probability of deception increases. Figure 6.5 illustrates this point by showing that the probability of deception increases when the cost of detecting deception by a person being deceived increases.

The list of disorders in which deception is thought to be a prominent feature is long and includes factitious condition with psychological symptoms, Ganser's syndrome,

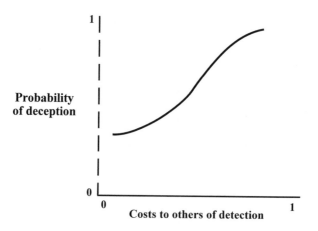

Figure 6.5 Probability of deception versus the costs to another of detection. Adapted from McGuire and Troisi (1990).

pseudopsychosis, factitious posttraumatic stress disorder, somatoform disorders, factitious bereavement, factitious child abuse or neglect, chronic factitious disorder with physical symptoms (Munchausen syndrome), atypical factitious disorder with physical symptoms, malingering,antisocial and histrionic personality disorders, substance abuse disorders, eating disorders, compensation neurosis, pseudologia fantastica, pathological lying, and imposture (Cunnien, 1988). For each of these disorders, deception probably qualifies as a trait, although not necessarily a well-honed trait because detection does occur. While there have been no comprehensive epidemiological studies of the frequency of most of these disorders, clinical experience is consistent with the view that minor forms are present in a larger percentage of persons than is usually suspected.

The preceding long list of disorders raises an obvious question: Is deception a far more fundamental feature of conditions than is usually assumed? Our prediction is yes. If pursued, this prediction could lead to significant reformulations of a variety of *DSM*-classified disorders. Aside from the deception in somatoform, personality, and factitious disorders, deception also seems highly probable in substance abuse disorders, impulse disorders, elimination disorders, sexual dysfunction disorders, and many communication disorders.

A number of predictions follow from the preceding discussion. For example, to the degree that deception is a trait, variations from the relationships shown in Figures 6.4 and 6.5 are expected. In Figure 6.4, the curves would shift to the left. In Figure 6.5, a shift left is also likely, and it would be accompanied by reduced expectations of the costs to others of detecting deception. Further, disorder simulation should coexist with strategies of social manipulation. Clinical data support this prediction. For example, in histrionic personality, somatic complaints are associated with sexual seductiveness. Further, somatic signs and symptoms that are manifestations of deceptive traits should be resistant to change in the absence of therapeutic interventions which provide individuals with alternative strategies to succeed at a reduced cost in their social environment. Providing such alternatives is far from easy, a point suggested by the fact that persons who chronically deceive are difficult to treat effectively. From another perspective, deception should pose a particular challenge to clinicians because it violates the basic assumption of trust underlying clinician-patient relationships. Because strong negative emotional reactions (e.g., moral indignation) have evolved as counterstrategies for limiting the successful use of deception in social interactions (Trivers, 1971), clinicians generally have great difficulty maintaining neutrality and a spirit of supportive empathy when confronted with obvious or presumed deceit.

Detection of Deception

Detection of deception can be thought of as a form of recognition. But detection of cheaters is difficult for some of the same reasons that recognizing others' behavior rules is difficult. Most persons who deceive do so infrequently, and they deceive in different ways across social contexts and goals. Traits may have evolved to counter deception, for example, the capacity to be suspicious or an enhanced memory for the faces of cheaters (Mealey, 1995). In addition, preliminary studies suggest that individuals without conditions are more adept at detecting cheaters than are persons with conditions (Janicki, 1995). And as noted, a number of studies suggest that deception is accompanied by signals such as distinctive hand movements, gestures, voice pitch, and facial expressions (Ekman et al., 1976; Cody and O'Hair, 1983). But clearly, what one finds in the research laboratory does not always generalize to day-to-day events; otherwise, we would be better detectors.

The preceding points notwithstanding, it is important to be realistic about the costs of detection: Excessive efforts to detect cheaters are not only time-consuming but often also unproductive. Thus, a balance between reasonable efforts to detect cheaters, yet without constant concern over the possible strategies of others, may be the most cost-efficient strategy in familiar social settings. The costs involved in the use of different strategies to detect nonreciprocators illustrate the preceding points. For example, from a potential altruist's perspective, a key factor in deciding whether to help another is the estimate of the likelihood that the recipient will reciprocate. In effect, an altruist will make an assessment of a potential recipient's likelihood of paying back help, which in turn influences the potential costs and benefits of providing help (Trivers, 1971, 1985; Cosmides, 1989). The information required for such assessments is not always available, however. Figure 6.6 examines these points through the use of diminishing-returns curves.

The two curves shown in Figure 6.6 represent two extremes in the relationship between interaction effort and information acquisition. For Curve A, information is obtained rapidly with little effort (at low cost), but the maximum amount of information that is likely to be obtained is low relative to that for Curve B. Thus, a saturation point is reached rapidly. Curve A depicts situations in which people identify themselves primarily by the use of language, the earlier mentioned language-symbol system. Language allows one to identify oneself quickly, but not necessarily with sufficient detail, clarity, or accuracy. In effect, as long as one presents a plausible story about one's past and one's motives, others are likely to believe the story. In contrast, for Curve B, information is obtained more slowly and at a greater cost because of greater time and effort requirements. Curve B is the kind of curve one expects when information is gained through repeated experiences with another, the earlier mentioned

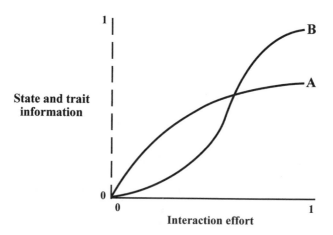

Figure 6.6 State-trait information and interaction effect. Adapted from McGuire et al. (1994).

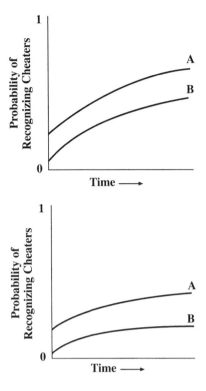

Figure 6.7 Probability of recognizing cheaters across two social environments. Adapted from McGuire and Troisi (1990).

experience system. While relatively inefficient at first, experience-based knowledge is likely to provide the most accurate information from which to judge whether another person is a good reciprocator. Within limits, the shapes of the curves are expected to vary as a function of cultural, situational, recognitional, and personality factors (McGuire et al., 1994).

How the behavior of persons with conditions differs from the behavior shown in Figure 6.6 is explored in Figure 6.7, which takes into account suboptimal and dysfunctional recognition capacities and their fate in two social environments.

Figure 6.7 plots the probability of detecting cheaters against the time spent in two social environments. The top graph depicts a stable social environment where expected interaction outcomes and reciprocation rules are well known by the participants. The bottom graph depicts a rapidly changing social environment in which interaction outcomes and reciprocation rules are less predictable. In both parts of the figure, Curve A depicts persons with optimal detection capacities and Curve B depicts persons with suboptimal or dysfunctional recognition capacities.

In the top graph, the probability of detecting cheaters is relatively high for persons on both Curve A and Curve B because the behavioral expectations are specific to the social group, and knowledge of others increases over time. However, detection errors continue, with the number of errors for Curve B exceeding those for Curve A. In a rapidly changing environment (the bottom graph), the number of detection errors increases for both curves because of the unpredictability of the social environment, the changing social membership in the environment, and the difficulty in predicting interaction outcomes.

In Figure 6.7, Curve B models persons with conditions, who, on average, are less capable of detecting cheaters than persons without conditions. Moderate instances of poor detection are observed in most personality and state-related conditions (e.g., brief periods of anxiety). Extreme instances are observed in persons with borderline and narcissistic personality disorders, hypomania, and paranoid schizophrenia.

In what ways do the preceding points apply to conditions? There should be an inverse relationship between the degree to which an individual's recognition and signaling capacities are optimal (Figures 6.1A and 6.2A) and detection. For recognition, Figure 6.1A depicts a situation in which individuals are capable of making the most accurate interpretations of others' signals. The states and traits depicted in Figures 6.1B to 6.1E reduce the chances of accurate recognition. Signaling competence is important because detection is often hastened if others respond in particular ways to a signal. Thus, knowing what one is signaling and having the ability to control one's signals can be advantageous.

Finally, it is tempting to view some conditions as examples of excessive detection efforts. Paranoia is an obvious example, but a closer look at the features of this and related conditions suggests that suboptimal or dysfunctional information processing is the primary contributing factor, not skewed tendencies to detect others' deceptions.

Concluding Points

This chapter has focused on information-processing algorithms and trait and state recognition and signaling. Considerable variation in these capacities is observed in persons without conditions. Far greater variation is associated with conditions. Some conditions are classified primarily on the basis of atypical recognition or signaling features. Some are characterized by temporary changes in capacities (state conditions), and some are associated with an amplification of traits. Evolutionary explanations of deception, detection of deception, and self-deception were reviewed, and their possible adaptive functions and condition-related features were explored.

PART III

DISORDERS AND CONDITIONS IN AN EVOLUTIONARY CONTEXT

7

Evolutionary Models
of Depression

Evolutionary models of mental conditions differ from psychiatry's prevailing models in a number of ways. Most critically, they are based on a theory of behavior that includes ultimate causes, biological motivations-goals, sexual selection, infrastructures, traits and trait variation, and the social environment. Proximate events, such as genetic mistakes, predispositions, adverse environments, psychosocial stress, wrong or inadequate learning, dysfunctional physiological systems, and intrapsychic conflicts, are sometimes, but not always, part of explanations. Viewed this way, many conditions turn out to be minimally adaptive; some turn out to be adaptive, and some features of conditions represent attempts to act adaptively.

A word about the phrase "attempts to act adaptively." With the exception of extreme situations (e.g., obvious life-and-death decisions), what actually constitutes acting adaptively is often unclear, primarily because one cannot foresee the future. Although miscalculations are possible, people nonetheless act, and when they do, they seldom act indiscriminately. Thus, actions reflect in part capacities to develop workable and efficient scenarios and to translate them into behavioral strategies. As noted, scenarios are biased in several ways: Other things being equal, they reflect better-safe-than-sorry thinking, overestimation of the value of one's time and effort, a tendency not to invest heavily in persons who are unknown, and a tendency to preferentially invest in kin. Most of the time, these biases work in one's favor. But there are exceptions, as when one invests in kin who abuse the investment. It is within these constraints that the term *act adaptively* is used.

Evolutionary models of mental conditions often include more variables than the prevailing models, and because they do, they are usually at odds with prevailing wisdom. Attempts to explain behavior by using a large number of variables often muddle more than they clarify, and many a nascent theory has lost its bearings for just this reason. Yet, if disorders such as acute schizophrenia, depression, substance abuse, bipolar illness, and panic can occur with and without known precipitants, can sometimes run in families and sometimes not, can show different concordance rates among monozygotic twins, can sometimes appear together (comorbidity), can sometimes last a lifetime and at other times remit spontaneously, then it is probable that multiple-variable models offer the hope of achieving greater explanatory validity than models based on one or a few variables. In short, there are good reasons not to shy away from complex explanations.

This and the next seven chapters are about evolutionary models of conditions. This chapter begins with a brief discussion of three topics: models that include more than one variable (the 15% principle), conditions grouped by behavior systems, and sexual selection. These topics set the context for the review of models of depression that appears in the second half of the chapter. Regulation-dysregulation theory, which can account for the onset and the often-differing course of many conditions, is discussed in chapter 8. Other conditions are discussed in chapters 9 to 14.

The 15% Principle

The *15% principle* is a term we have coined to underscore the fact that conditions have multiple contributing factors; rarely are there one or even a few factors. We are not the first to make this point. Nor are most clinicians and investigators unaware that the majority of conditions result from multiple rather than single causes. Nevertheless, the full implications of multiple causality are only occasionally explored (e.g., Wake, 1990).

The 15% principle decreases the likelihood of reductionistic explanations; that is, it decreases the chances that important condition-contributing features will be overlooked. Compared to biomedical explanations, it increases the distance between phenotypes, genes, and atypical physiological states by allowing for the independent or semi-independent function of infrastructures and the possibility that one or more infrastructures will offset the negative function of another infrastructure (e.g., Post and Weiss, 1992). The principle also facilitates recognizing attempts to adapt that often go hand in hand with dysfunctional and suboptimal infrastructures. Distinguishing between these possibilities is critical for both understanding and treating conditions. Compared to psychiatry's prevailing models, the 15% principle increases the number of possible condition-contributing causes without necessarily rejecting other explanations, for example, physiological contributions to conditions. When applied to most conditions, the principle leads to formulations like the following: Currently, 10% of the phenotypic features are explained by trait variation, 20% by dysfunctional automatic systems, 15% by dysfunctional algorithms, 20% by adverse environmental features, and 35% by attempts to act adaptively. These percentages represent the postulated contributions to clinical states at a given moment in time, and they will differ both within and across conditions. For a few disorders, such as Down syndrome due to chromosome breakage, a single causal factor may have a high percentage. Yet, more often than not, causal hypotheses dealing with multiple low-percentage contributions are required (a point that also applies to nondisordered states).

Conditions Grouped by Behavior System
and Functional Consequences

Insights into both ultimate and proximate condition-contributing causes can be gained from grouping conditions that have similar functional consequences within the same behavior system. This is an obvious extension of the functional approach to classifying disorders discussed in chapter 4. Table 7.1 presents a provisional grouping of some two dozen *DSM-IV* (APA, 1994) classified disorders when behavior systems (reproduction, survival, kin investment, and reciprocation) are used as the basis for grouping. The following caveat applies to the table: Depending on their functional consequences, disorders in which affiliative behavior is compromised can be associated with any one of the four behavior systems.

In what ways are the groupings in Table 7.1 informative? There are at least two answers to this question. First, disorders within each behavior system have similar

Table 7.1   *DSM-IV* disorders grouped by behavior systems

Reproduction behavior system

　Anorexia Nervosa (307.1)
　Dysthymia (300.40)
　Sexual and Gender Identity Disorders (302.xx)
　Hypoactive Sexual Desire Disorder (302.71)
　Sexual Abuse of Child (V61.21)
　Histrionic Personality Disorder (301.50)

Survival behavior system

　Separation Anxiety Disorder (309.21)
　Agoraphobia without History of Panic Disorder (300.22)
　Dependent Personality Disorder (301.6)
　Autistic Disorder (299.00)

Kin investment behavior system

　Adjustment Disorders of Childhood (309.xx)
　Conduct Disorders (321.8)
　Oppositional Defiant Disorder (313.81)
　Physical Abuse of Child (V61.23)

Reciprocation behavior system

　Antisocial Personality Disorder (301.70)
　Malingering (V65.20)
　Factitious Disorders (300.xx and 301.xx)
　Narcissistic Personality Disorder (301.81)

Mixed behavior systems

　Mood Disorders
　Schizophrenia (295.xx)
　Obsessive-Compulsive Disorder (300.3)
　Schizotypal Personality Disorder (301.22)

*Note.*The numbers in parentheses refer to *DSM-IV* codes (APA, 1994).

features. For example, feelings of personal danger and exaggerated responses to fearful situations are prominent features of the disorders grouped in the survival system (Separation Anxiety Disorder, Agoraphobia without History of Panic Disorder, Dependent Personality Disorder, and Autistic Disorder). Anorexia Nervosa, Dysthymia (see chapter 14), Sexual and Gender Identity Disorders, Hypoactive Sexual Desire, Sexual Abuse of Child and Histrionic Personality Disorder are grouped in the reproductive behavior system because confused reproduction-related feelings and actions are associated with each. The mixed-behavior-system category is introduced for those disorders in which functional analysis suggests that two or more behavior systems are compromised. Second, conditions within the same behavior system are likely to have similar ultimate but dissimilar proximate causes. For disorders in the reproductive behavior system, reproduction-related motivations-goals (ultimate cause) are postulated along with dysfunctional processing of somatic information dealing with one's sexuality and possibly also environmental cues, such as information about one's social attractiveness (a proximate cause). Said another way, the disorders listed for the reproductive behavior system in Table 7.1 are unlikely to have developed, or they would have taken different forms, if there were no such thing as reproductive motivations-goals. Disorders grouped within the other three behavior systems can be viewed in a similar way. (An analogous approach has been used for grouping addictive, impulsive, and compulsive conditions, which are postulated to have similar functional consequences and possibly a common genetic basis; Blum et al., 1996.)

There are of course other ways to group conditions. Physiology-based, genetic-profile-based, and *DSM*-based systems have been discussed (chapters 2 to 4). Among other things, these discussions reveal that the theoretical orientation one brings to classification tasks can significantly influence how conditions are grouped. Evolutionary classification is a taxonomy of function in which specific functions or functional failures are closely tied to causal hypotheses, while *DSM*-type classification is a mixture of signs, symptoms, and functions that are not tied to causal hypotheses. More in line with an evolutionary approach are Klein's and Gilbert's suggestions that conditions reflect failed adaptive processes (Klein, 1978), or that they can be grouped into functional categories such as reciprocal altruistic (cooperative) conditions, care-eliciting conditions, and intraspecific competitive conditions (Gilbert, 1992).

Sexual Selection

Evolutionary concepts permit the development of predictions about sex-related strategies, behavior, condition probabilities, and condition manifestations. Sexual selection theory is the basis of these predictions, and the theory is usually distinguished from survival-related selection as follows: Survival selection addresses the evolution of survival-related capacities from birth until reproductive age, while sexual selection addresses reproduction-related behavior. Male and female behavior differ because of the different adaptive problems encountered by females and males in the remote past (Darwin, 1872; Cronin, 1992). Differences in male and female reproductive strategies illustrate these points.

From afar, males and females engage in many of the same behaviors. Both invest in spouses, offspring, and kin; both engage in sexual behavior inside and outside marriage; both are jealous and possessive of their mates; and both are competitive in their attempts to acquire and retain resources. However, a closer look reveals that the time, effort, and frequency of these activities differ. For example, compared to males, females allocate more time and energy to having and caring for offspring and to developing and maintaining social support networks. Males allocate more time and energy to resource acquisition and to status-related competition.

Given that females invest more time and energy on their offspring than males, it may seem odd that they spend half of their reproductive energy producing males. A likely explanation for the near 1-to-1 sex ratio in offspring is that males are essential because their genes contribute to immunological competence (defenses against parasites; see chapter 3). Division of labor may also be relevant: Males are probably better adapted to physical-resource-acquisition tasks, and they are reported to have better spatial abilities; females are reported to be better adapted to bonding (e.g., Nyborg, 1994). And it may seem odd that reproductive senescence (menopause) has evolved in human females, but not in many other primate species. But if menopause facilitates greater investment in offspring and, in turn, improved offspring survival and fertility, then menopause may represent an adaptation largely peculiar to *Homo sapiens* (Peccei, in press; Turke, in press).

The effects of sexual selection extend to the details of mate selection preferences and strategies (Buss, 1994). A sampling of findings from studies shows that females more than males value cues among potential mates that reveal resource acquisition capacities, such as earning potential (Buss, 1985, 1987, 1988a, 1989; Wiederman and Allgeier, 1992; Buss and Schmitt, 1993); high social status (Sloman and Sloman, 1988; Baenninger, Baenninger, and Houle, 1993; Cashdan, 1993; Townsend, 1993; Walsh, 1993); and industriousness. Females also prefer males who are somewhat older in age and with broader jaws (Grammer, 1995). Males value females who are younger and submissive; who have full lips, smooth and clear skin, and high cheekbones, and who are symmetrical (Gangestad, Thornhill, and Yeo, 1994; Brown and Kenrick, 1995; Gangestad and Thornhill, 1995; Grammer, 1995; Møller, Soler, and Thornhill, 1995; Singh, in press; Singh and Young, in press). Males are aroused by unknown females, while females tend to view unknown males as a threat (B. P. Lewis, Linder, and Kenrick, 1995). Many of these differences are reflected in male and female fantasies (Ellis and Symons, 1991; Kenrick and Sheets, 1993). In both sexes, these preferences are thought to reflect the search for good genetic quality in potential mates and, by implication, in potential offspring. That subtle appearance differences have reproductive implications is suggested by findings showing that the degree of female physical asymmetry is reduced when females are fertile compared to the beginning and end of the menstrual cycle, when they are not fertile (Manning et al., 1996), and that females copulate more frequently and have more sexual orgasms per copulation with symmetrical compared to asymmetrical males (Thornhill, Gangestad, and Comer, 1995). Further, physical symmetry—similarity in the size of ankles, wrists, and ears—positively correlates with measures of phenotypic quality (Gangestad, 1995).

Females signal immediate sexual access when their objective is a short-term relationship, but they signal sexual restraint when their objective is long term (Buss, 1988b). Characteristic female mate-attraction signals, such as posture, intensity of eye contact, and frequency of hair touching, have been reported (Feierman and Feierman,

1992). Males signal their willingness to invest in offspring when their objective is long term (Cashdan, 1993; Hirsch and Paul, 1996; Paul and Hirsch, 1996). The possibility that females have greater voluntary control over their mate selection behavior than do males has recently been suggested (Hamida, 1996), and male behavior may be more easily influenced by female seductive signals than the reverse. The fact that the findings cited above are observed across a wide range of cultures favors the view that both male and female mating strategies and preferences reflect strongly predisposed traits, partially refined by culture and experience.

Not unexpectedly, differences in reproductive strategies lead to male-male and male-female conflict. Competition among males for females explains in part the observation that males more than females attempt to conceal mates. In addition, because males are uncertain of their paternity, they are more inclined to try to possess and physically dominate females (Daly and Wilson, 1978). For their part, females can control male sexual access and tend to engage in infidelity threats rather than physical coercion during disagreements (Buss, 1988b). In effect:

> In species with internal female fertilization, males risk both lowered paternity and investment in rival gametes if their mates have sexual contact with other males. Females of such species do not risk lowered maternity probability through partner infidelity, but they do risk the diversion of their mates' commitment and resources to rival females. (Buss et al., 1992, p. 251)

For both sexes, assessment of the physical and mental health of potential mates is also important. For example, in high-pathogen areas, physical attractiveness, which is one measure of health and genetic quality, takes on greater importance than in low-pathogen areas (Gangestad and Buss, 1993). Males who are mentally healthy while they are young are more likely to be mentally healthy and alive 20 years later (Vaillant, 1976, 1979; Tsuang and Woolson, 1977, 1978; Tsuang, Woolson, and Fleming, 1980). And studies indicate that physical asymmetry, as measured by absolute finger ridge count, is greater in persons with mental conditions than in persons without (e.g., Markow and Gottesman, 1989).

Condition manifestations should also differ between the sexes. For example, the superior capacities for language (e.g., Dunbar, 1993), the greater importance of social bonding, and the greater capacity to empathize among females are very likely consequences of selection for functions in which females make greater investments, such as raising offspring, assisting kin, and developing and maintaining social support networks. It follows that females should be more articulate about the features and consequences of conditions, more sensitive to distressing emotional states, and more concerned about conditions that compromise empathy. Clinical experience is consistent with these views.

Given the preceding points, condition frequency can be expected to reflect sex-related differences (Archer, 1996). For example, conditions associated with failed attempts to attract desired mates, with an inability to reproduce, with failed attempts to develop social support networks, or with intrusions into one's "somatic territory" (one's sense of the boundaries of one's body) should be more prevalent among females. Conditions associated with losses in male-male competition, with failure to gain sexual access to females, with failure to acquire resources, and with declining social status should occur more frequently among males. Findings support these predictions. Histrionic personality and erotomania have mate attraction components; anorexia nervosa and depression due to infertility have reproductive components; and agoraphobia and hypochondriasis have somatic territory components. All of these disorders are reported to occur significantly more frequently among females than among males (APA, 1994; Kessler et al., 1994).(Differential female-to-male psychological disruptions associated with marital infertility—37% to 1%—were mentioned earlier.) In males, impulse-related conditions, depression and suicide following competitive losses, as well as alcoholism, have components of both male-male competition and resource acquisition, and sexual deviance has female access components. These disorders are more common among males (APA, 1994; Kessler et al., 1994). An obvious implication is that the consequences of sexual selection, which include male-female sexual dimorphism in neuroanatomic, neurochemical, and hormonal function, should be a part of condition explanations.

When we apply some of the preceding points to depression, the finding that males tend to prefer submissive females should increase the likelihood that females who are excessively submissive will be victims of male domination. Greater male possessiveness and the hiding of females may contribute to the increased incidence of depression among females because of constraints on female social options. Further, male proclivities for short-term relationships, when combined with female preferences for more enduring relationships, should result in a higher percentage of disappointing and costly relationships for females. Failure to develop social support networks should also be a factor among females (see chapters 8 and 14). Males of course are not immune to depression, and as noted, they should be particularly vulnerable to status declines and female tactics that suggest infidelity or involve cuckoldry (Trivers, 1985; Buss, 1994). Nevertheless, when the condition-contributing factors are summed, females turn out to be far more at risk for depression than males, and this difference may explain much of the reported sex-related prevalence differences.

Evolutionary Models of Depression

The review of evolutionary models of depression that follows illustrates several important points: (1) how these models can integrate multiple condition-influencing factors (e.g., sexual selection, infrastructural suboptimality); (2) how prevailing model explanations can be integrated into evolutionary models; and (3) how the search for condition-related adaptive features is contingent on functional assessments. First, some background.

In evaluating evolutionary models it is important to keep two points in mind. First, as noted in chapter 2, there is no generally accepted definition of mental conditions. The definition we have adopted has been taken from the work of a number of authors (e.g., Scadding, 1967; Klein, 1978), and its essence is that something has gone wrong with the evolved capacities that allow for adequate functioning. Relative to the prevailing models, it is both the "evolved capacities" and the "adequate functioning" parts of the definition that are expanded in evolutionary models. Second, evolutionary models postulate that certain types as well as certain features of depression are adaptive. When conditions are adaptive, the various nonevolutionary definitions of mental disorders do not apply.

Statistically, depression is clearly a common human affliction. For major depression (severe, debilitating depression) and dysthymia (chronic mild to moderate depression) the combined male-female life-prevalence estimate in the United States is 24.5%, with a 29.3%-to-17.5% female-to-male differential. The 12-month prevalence estimate is 13.8%, with a 15.9%-to-9.8% female-to-male differential (Kessler et al., 1994; cf. Wilhelm and Parker, 1989; Harris et al., 1991). Major depression and dysthymia are the two most frequently diagnosed mood disorders in which depression is a primary feature. Conditions in which mania is either a primary or a secondary feature are also classified as mood disorders. Mania will not be discussed in detail in this chapter. Our focus is on depression.

Clearly, disorders that appear with the frequencies cited here deserve consideration as possible adaptations. It is relatively easy to construct an adaptive scenario for externally elicited, mild to moderate, time-limited, spontaneously resolving depressions: The symptoms of depression may warn a person that past or ongoing strategies have failed; physiological slowing and social withdrawal may remove a person from high-cost, low-benefit social interactions; and signaling one's state to others may initiate others' help without requiring long-term payback. Variants of this type of depression are seen in response to personal losses (e.g., the death of a relative), destruction of personal property (e.g., by an earthquake), and a discontinuation of employment associated with financial hardship. Although good epidemiological estimates are unavailable, it is likely that 15% to 20% of depressions (some of which will not find their way into life prevalence estimates) fit into this category.

At another extreme, there are individuals who become severely depressed and exhibit signs and symptoms associated with what is classified as major depression (APA, 1994). Markedly diminished interest or pleasure, weight loss, psychomotor agitation or retardation, chronic insomnia or hypersomnia, fatigue, the inability to concentrate, social withdrawal, feelings of worthlessness, and striking reductions in functional capacities are indices of severity. Often, external precipitants cannot be identified. Some individuals experience only a single episode. Others have recurrent episodes. And for yet others, the condition may become chronic despite all available medical and social interventions. Severe depressions are less easily interpreted as adaptations than mild to moderate, time-limited depressions.

Between these extremes are a number of conditions in which the signs and symptoms of depression are readily apparent but not necessarily the only clinical manifestations. Dysthymia, or dysthymic disorder, is an example. Chronic mild to moderate depression, usually coupled with functional capacity limitations, failed social strategies, and varying degrees of social isolation are typical features (Essock-Vitale and McGuire, 1990; APA, 1994; see chapter 14). Both dimensional and quantitative differences distinguish dysthymia from time-limited depression and major depression. Although symptom intensity may vary across social environments, persons are seldom symptom-free, and limitations of functional capacity are more often enduring than not (McGuire and Essock-Vitale, 1982; Essock-Vitale and McGuire, 1990). Dysthymia may or may not qualify as an adaptation.

As we noted as early as chapter 1, there are many theories of depression, and each of psychiatry's four prevailing models covets its own explanation (e.g., Brown and Harris, 1978; Beck, 1976, 1983). Epidemiologists (e.g., Dohrenwend and Dohrenwend, 1978), psychologists (e.g., Oatley and Bolton, 1985), and chaos theorists (e.g.,

Gottschalk et al., 1995) have offered their own causal theories. A careful review of these many theories would lead us far afield. Thus, while acknowledging they exist, we will limit our focus to evolutionary models of depression and their ability to accommodate features of the prevailing models.

What are the ways in which evolutionary theory can inform our understanding of depression? An answer to this question would be less difficult if clinical cases of depression were confined to only those signs and symptoms that are most readily explained by adaptive models. But such cases are rare. By far the more common clinical findings associate depression with a host of signs and symptoms that are also present in other conditions (e.g., memory dysfunction, somatic pain, anxiety, paresthesia) and a history of precondition functional limitations, as well as some inflexible personality features. Answers are further complicated by the fact that many medical findings don't easily fit into existing adaptationist interpretations. For example, over a thousand medical diseases and syndromes are thought to reflect atypical genetic information (Nesse and Williams, 1994). There is little about the majority of these illnesses to suggest that they are adaptations. Moreover, given the broad range of these illnesses, there is no a priori reason to discount the possibility that instances of depression not only reflect atypical genetic information but also are minimally adaptive.

Yet another complicating factor has to do with common final pathway phenomena, which, as we have noted, is a shorthand term referring to conditions in which different causes lead to similar phenotypes because of constraints on phenotypic expression (e.g., delirium due to substance withdrawal and elevated temperature). In a significant percentage of depressions, historical and physiological data are consistent with this explanation. For example, some persons with depression grow up and live in adverse social environments, while others do not; some come from families in which depression is common, while others do not; and significant individual differences in putative depression-causing physiological systems (e.g., norepinephrine, serotonin) have been reported. Further, some instances of depression remit spontaneously; some respond to one type of antidepression medication but not to another; some do not respond to any type of medication but do respond to electroconvulsive treatment; and some do not respond to any known intervention. This array of outcomes would be expected if similar phenotypes have different causes.

While multiple causes and constraints on phenotypic expression do not obviate adaptive interpretations (e.g., adaptive systems often function separately from the causes of conditions), their relevance to explaining conditions requires a further look at what is implied in the term *constraint*. As noted, the usual view of common final pathway phenomena is that different underlying dysfunctional systems have similar phenotypic outcomes because the options for expression are limited: In effect, one can have a broken toe because of a falling object, bone weakness, or kicking a wall. But there is another possibility: There are a limited number of ways to act adaptively. This possibility is suggested by the observation that many conditions have common features. Distinguishing between traditional signs, symptoms, and behaviors that represent attempts to adapt thus becomes essential.

The possibility that many instances of depression do not easily fit adaptationist models leads to different evolutionary approaches to explaining depression. One is to disregard such instances, to identify the *core* features of depression, and to assess their potential adaptiveness. Over the last three decades, hypotheses tying specific environmental contingencies (e.g., others' competitive behavior) to depression have been given evolutionary interpretations, primarily by Price (Price et al., 1994), Sloman (Sloman et al., 1994), Gardner (1982), Gilbert (1995), and Birtchnell (1993). Here again, clinical findings both enter and complicate the discussion. Even in the best controlled studies (e.g., in which controls and experimental subjects are the same age and sex), significant cross-person response differences to the same stimulus are observed. The prominence of such differences suggests that both control and depressed individuals bring very different psychological and physiological states to their environment, and that these states significantly influence person-environment interaction outcomes.

Nonetheless, the *core* approach does have advantages: It minimizes the need to explain the diversity of clinical findings, and it facilitates the development of unitary causal hypotheses. Insights into possible core features usually follow, but on balance, they remain distant from what one encounters clinically. An alternative approach is to take the clinical data as they are and use more than one evolutionary concept to explain core features and individual differences.

We employ both approaches. *We view individuals as mosaics of independent or semi-independent traits, many of which vary independently of each other, within themselves, and across persons.* To cite only one finding from a cross-person trait variation study, when normal healthy male volunteers were exposed to stress tests (e.g., a quiz, an arithmetic task), and peripheral cortisol concentrations were used to measure stress responses, the result was a continuum between "complete reactors and nonreactors" (Berger et al., 1987). *We also view individuals as differing in the degree of baseline optimality of specific traits* (McGuire and Essock-Vitale, 1982). *Further, trait clusters and multiple causes, rather than unitary traits or core causes, are postulated to be responsible for the majority of depressions.* The views in italics are consistent with reports from pedigree studies, personality assessments, and behavioral genetics studies, which show, respectively, familial trends toward depression (e.g., Kendler, Heath, et al., 1993); different cross-person, enduring behavioral and physiological measures (e.g., Silberman, Weingartner, et al., 1983a; Silberman, Weingartner, and Post, 1983b; Essock-Vitale and McGuire, 1990); and heritabilities of specific depression-related traits (e.g., McGuffin et al., 1994; Plomin et al., 1994). The net effect of the preceding points is that there is a significant degree of cross-person variation (cf. Nierenberg et al., 1996). When subsets of traits have different baseline optimalities, and when individuals differ in their degree of risk for depression, cross-person manifestations of depression differ, and theories of depression should reflect these points.

The preceding points conflict with hypotheses developed by evolutionary psychologists (e.g., Tooby and Cosmides, 1990a), which emphasize phenotypic plasticity, cross-person similarity in adaptive capacities, and selection favoring the development of psychological mechanisms or rules (traits) that mediate behavior largely in response to environmental contingencies. Although this view may characterize some persons, individuals with conditions provide an exception: *If they are nothing else, most conditions are examples of compromised plasticity, rule use, and functionality.* Further, the plasticity view is not easily reconciled with findings showing that a significant percentage of depressed persons have chronic compromised information-processing capacities (McGuire and Essock-Vitale, 1982; Silberman et al., 1983a, 1983b; Silberg et al., 1990; McGuire, Troisi, and Raleigh, in press); that there are significant sex differences in life prevalence estimates for depression; and that when all mental disorders are lumped together, the life prevalence estimates for both males and females exceed 48% (Kessler et al., 1994).

From an evolutionary perspective, one might argue that the preceding discussion confounds more than clarifies, and that evolutionary interpretations of conditions should focus primarily on *core* adaptations and disregard the wide array of clinical findings. But to take this position is to miss the implications of the preceding discussion in several ways:

1. Different types of depression may develop in response to different adaptive problems; depression associated with the loss of an important other may differ from depression associated with the loss of access to resources (e.g., employment) or to social group membership (Gilbert, 1992).
2. Suboptimal traits may contribute to the onset of depression (as often appears to be the case in repeated instances of personal failure) and may influence both the immediate clinical manifestations and their subsequent course.
3. Some traits may not be influenced during condition exacerbation, as when depressed persons can accurately signal their mood state to others. Further, noninfluenced or minimally influenced traits may improve clinical outcome.
4. Traits that are sometimes viewed as adaptive, such as a mother's 24-hour-a-day care for an ill child, may increase the chance of depression.
5. Most important, competing evolutionary systems may explain much of the clinical "noise" of depression that is often disregarded.

In short, viewing depression as a unitary adaptive strategy may serve the aesthetics of theorizing but may also delay attempts to develop comprehensive evolutionary explanations as opposed to core explanations.

The models of depression that follow incorporate many of the ideas and findings from the work of Price (1967, 1969a, 1969b; Price and Sloman, 1987; Price et al., 1994; Price and Gardner, 1995), Sloman (1976; Sloman and Price, 1987; Sloman et al., 1994), Henderson (Henderson et al., 1980), Gardner (1982), Klein (1974), Nesse (1990a, 1990b, 1991a, 1991b), Gilbert (1989, 1993, 1995; Gilbert and Allan, 1994), Plutchik (1991), Gut (1989), Pezard (Pezard et al., 1996), Salzen (1991), and their colleagues. Individual differences and personality influences on type and degree of depression (e.g., Boyce et al., 1990) are not emphasized, but they return to the discussion in chapter 9, where personality disorders are addressed. Consistent with these authors, we assume that persons are goal-directed and that their capacity to signal that they are depressed has evolved in the same way that the capacities to spot danger and to take action have evolved.

Evolutionary models of depression have both ultimate and proximate cause features, although the degree of their contribution differs across models. For convenience, the models in the following discussion have been subdivided into three, not necessarily mutually exclusive, groups, which emphasize ultimate causes, developmental disruptions, and ultimate-proximate cause interactions.

Models of Depression Emphasizing Ultimate Causes

Three models are considered.

Depression as an Adaptive Trait

The possibility that depression is an adaptive response to adverse external conditions has at least a six-decade history. In 1936, Lewis suggested that depression is a way of eliciting help from others. The idea that depression is a response to the intolerability of low social status has been a central theme in the work of Price (1967), Sloman (1976), and their colleagues. Engle (1980) postulated that depression conserves energy and functions as a homeostatic regulatory process. Klerman (1974) identified several possible adaptive functions of depression, including its often positive social communication effects. And Gut (1989) argued that coping with depression often results in individuals' becoming more psychologically healthy as well as self-aware. Much of this history has been reviewed by Gilbert (1989).

The depression-as-an-adaptive-trait model is primarily an ultimate cause explanation that requires a proximate trigger. The model builds on the idea that depression has evolved as a strategy to respond to an actual or potential reduction in goal achievement (a negative cost-benefit balance), such as a fall in status or a loss in resource-holding power (Price, 1967; Price and Sloman, 1987). Pathogenic events are usually external, although internal precipitants are not precluded. The emotion provides information about one's negative cost-benefit state, and physiological slowing constrains further costly behavior. Affects inform others of one's condition. Coevolution is assumed to have favored the capacity to recognize and respond to persons who are depressed, although care provided by others is contingent upon the presence of established social support networks. If the precipitating factors abate, depression may also abate. The model is not limited to depression, and it does not assume or require depression to be shaped by common final pathway constraints, although this possibility is not precluded.

In this model, information-processing and signaling capacities are assumed to be functional, and condition-triggering events are expected to differ between males and females, for example, loss of status versus infertility.

When precipitating events can be identified (e.g., failing to achieve an important goal), and where there is *no* evidence of previous periods of depression, this model is an obvious explanatory candidate; for example, persons who have experienced a loss or an important competitive defeat frequently become depressed, and members of kin and nonkin social networks often provide help without requiring paybacks, thereby easing the requirements of continuing group membership for depressed individuals. While exact percentages are not known, a reasonable estimate is that approximately half of the mild to moderate, time-limited depressions that are triggered by adverse events are adaptive and resolve satisfactorily without professional intervention. For those that don't, it is essential to consider the possibility that in certain environments (e.g., urban environments), the requisite help is not available.

In what ways can prevailing model hypotheses be integrated into the depression-as-an-adaptive-trait model? Depression-triggering environmental events, such as a significant loss or living in a stressful and depriving social environment, are consistent with those sociocultural and psychoanalytic views that emphasize that the social environment can have pathogenic properties (see Oatley and Bolton, 1985, for an explanation of the effects of extreme disruptions on self-definition). These hypotheses can be integrated, but integration is more difficult with the other prevailing model hypotheses. For example, neurochemical and hormonal explanations of depression are usually framed so that physiological contributions are discussed independently of possible adverse social events. If physiological changes are viewed as secondary, they may be integrated; if they are viewed as primary, they don't easily qualify. Likewise, developmental disruption explanations are not easily integrated, at least in the narrow interpretation of the model. As they are usually understood, disruptive events increase vulnerability, which is not a requirement of this model, although undoubtedly, combinations of developmental disruptions and aversive environments are implicated in many instances of depression. Similar reasoning applies to pedigree and inadequate learning models, both of which place importance on the presence of compromised capacities prior to the onset of conditions. The model does not require precondition infrastructural suboptimality.

Ms. E

After having spent three years in a nearly full-time effort to write a book, Ms. E was unable to obtain a publisher. Following her 11th rejection by publishers, she became depressed, refused to leave her home, and avoided social interactions with friends and family. Friends and family continued to provide support, and a close friend took it upon herself to contact other publishers, one of which took an interest in Ms. E's book. Book negotiations followed, and eventually, a contract for the book was signed. Within a month, the signs and symptoms of Ms. E's depression began to resolve. After three months, she was functioning normally and was actively involved in the final editing of her book. In this vignette, a key contributing factor to Ms. E's depression was most likely her inability to achieve a high-priority goal.

To be sure, the causes and ramifications of Ms. E's depression were not as simple as the vignette might imply. This and other vignettes in this chapter are introduced not to develop complete formulations, but to call attention to possible causal factors, which are often overlooked in clinical formulations.

The Pleiotropy Model of Depression

Pleiotropy is another ultimate cause explanation, and it refers to a type of selection in which a gene (or set of genes) controls one or more phenotypes. It is the possibility of more than one phenotype that makes this model attractive, especially if the non-condition-related phenotype is highly adaptive—high fecundity, social attractiveness, and so on. The model is not limited to depression, and it provides an explanation of how condition-related genes can remain in a population and avoid strong selection effects. It does not require that persons become depressed or that common final pathway constraints be present, although constraints are not precluded. The possibility that persons who become depressed will attempt to act adaptively, or that others will provide help, is also not precluded. Further, the model does not specify at what stage of life depression is most probable, although a reasonable assumption is that severe depression would be less likely among females during their prime reproductive years because of the possible negative effects on mate choice and offspring rearing. Other causes of depression are thus the probable bases of depressions that appear during puberty and early adulthood.

The model has been used to explain senescence (Williams, 1957), and it may apply to the first-time occurrence of depression in postmenopausal women. External precipitating events are not required in this explanation: Changes in CNS anatomy or physiology associated with aging may be the pleiotropic trait, and thus a major contributing factor to condition onset. It is also a potentially attractive explanation for both premenopausal depression and bipolar illness, which are often associated with superior intelligence and creative capacities (Goodwin and Jamison, 1990). Superior intelligence and creative capacities may be the adaptive traits and, for example, foster efficient social navigation and increase mate choice options. Information-processing capacities need not be suboptimal in this model, although periods of dysfunctionality are assumed to occur during exacerbation.

In the pleiotropy model, female and male differences in the prevalence of depression should be influenced by the advantages conferred by the associated adaptive trait(s). For example, if the associated trait is an above-average capacity to read others' behavior rules (i.e., others' minds), the cross-sex frequencies of depression should be about equal because reading others' rules is important to both sexes. However, if the associated trait is above-average reproduction, a greater prevalence of depression among females would be predicted because the adaptive trait may be sex-linked.

Several nonevolutionary hypotheses can be integrated into this model. Those emphasizing neurochemical or physiological dysfunctionality qualify if the pleiotropic trait is neurochemical (e.g., dysfunctionality of the norepinephrine system). Chronic low self-esteem, intrapsychic conflicts, and constrained functional capacities may represent phenotypic expressions of the pleiotropic trait. However, they are not required by the model. Hypotheses dealing with maturational disruptions and environmental perturbations are not required.

Mrs. N

Mrs. N was a 64-year-old female in good physical health who lived with her husband. She was an active member of a close and supportive family, and her three daughters and two sons lived nearby, as did her 14 grandchildren. She had no prior history of depression. Without any evidence of precipitating incidents, Mrs. N developed signs and symptoms of depression. Weight loss, as well as withdrawal from her family and friends, followed. An array of antidepression medications minimally altered her condition. Eventually, she received electroconvulsive treatment, which resulted in a return to her predepression state. In this vignette, a first-time bout of depression in a woman with above-average fecundity, and without obvious external or internal provocations, is consistent with the pleiotropy model.

Mr. P

Mr. P was a successful, physically attractive 42-year-old male writer who had been married numerous times. Beginning in his early 20s, he had experienced intermittent periods of hypomania and depression. Females found him most attractive during his hypomanic periods, and it was at these times that he married. Subsequent periods of depression usually ended in divorce. Mr. P had seven children. In this vignette, the presence of superior creative capacities may have increased Mr. P's mate choice options, which would qualify as the adaptive trait. The biopolar illness would be the result of the expression of the associated phenotype.

In the two preceding models, as well as the trait variation model below, we are not implying that precipitated depressions should necessarily remit in the face of positive social signals. There are many reasons that such signals have limited effects; for example, lack of social skills may maintain a depression and may even invite an intensification of depression once it has begun.

The Trait Variation Model of Depression

This model assumes that cross-person differences in trait clusters and differential within-cluster trait features influence the probability of depression. Trait profiles may be due to the chance effects of genetic mixing at conception or biased genetic information or may reflect incomplete trait refinement. Pedigree data strongly suggest that genetic information influences the probability of depression; and assortative mating (individuals of similar phenotype mate more often than would be expected by chance) among persons with conditions is a well-documented finding. Different trait clusters should result in different types, intensities, and clinical courses of depression, for example, short-term versus chronic depression. Common final pathway constraints are assumed to apply in this model (Mikhailova et al., 1996; Nierenberg et al., 1996). A large percentage of chronic, treatment-refractive depressions are consistent with the trait variation model, while once-only bouts of depression associated with clear precipitating incidents and rapid resolution are more parsimoniously explained by other models.

The model requires evidence of either functional capacity limitations or condition vulnerability prior to condition onset. In the narrow interpretation of the model, biomedical hypotheses are potentially applicable; suboptimal traits may be physiological. Sexual selection may also be a factor. For example, because (on average) the requirement of social interaction skills is greater among females than among males, equivalent male and female suboptimal social skills should contribute to a higher incidence of depression in females than in males. The model does not assume that depression is adaptive, but like the two preceding models, it does not preclude the possibility that help will be provided by others or that persons who are depressed will attempt to act adaptively. Moreover, it does not assume that this form of depression is the "inappropriate expression of evolved propensities concerned with adaptive behavior in the domains of group membership" (Stevens and Price, 1996, p. 29). More likely, depression is caused by deficits that result in decreased capacities for social maintenance and social exchange.

Mr. A and Mr. B

Mr. A and Mr. B were monozygotic twins who were separated at birth and who did not know of each other's existence until they met in their late 20s. One had grown up in a warm and supportive family, the other in a stern and often verbally abusive family. Both had graduated from college; both had jobs in which they were successful; and both were married and had children. Each had suffered periods of moderate to severe depression, beginning when they were teenagers. Psychotherapy (Mr. A) and multiple trials of antidepressants (Mr. B) did not alter their clinical conditions significantly. In this vignette, the similarities in the clinical histories of two persons with the same genotypes, but with different upbringing, favor a genetically influenced trait variation interpretation.

Developmental Disruption Model of Depression

One model of this type is considered here, the disrupted-maturation-programs model. The narrow interpretation of this model assumes that infants have normal genetic information for maturation programs; maturation programs are disrupted; and disruptions lead either to depression or to increased vulnerability to depression. Put another way, an individual who otherwise would have grown up normally becomes vulnerable to depression because of disruptive events during development.

This model is similar to the psychoanalytic, behavioral, and psychosocial models, which postulate that developmental insults or atypical upbringing environments are associated with an increased probability of mental conditions. Ultimate cause explanations provide the framework that informs these hypotheses. In condensed form, one such explanation runs as follows: Mothers have been selected to bond with and care for their infants (e.g., to provide nutrients and protection), and infants have been selected to bond with caretakers and to engage in behavior that facilitates caretaker bonding. However, selection has not favored a high degree of self-sufficiency in infants during the early months of life, presumably because, in the past, successful bonding between mother and infant occurred frequently enough so that alternative selection paths were not favored. One outcome of these events is infant dependency, that is, slowly unfolding maturational programs that are vulnerable to adverse social conditions.

The causes of maturational disruptions vary in timing, type, intensity, and consequence; for example, the effects of excessive maternal alcohol use differ from the effects of maternal rejection. As infants grow, they become increasingly capable of managing adverse events (e.g., social tension, periods of caretaker absence), although periods of increased susceptibility may occur during critical transition periods. While the model does not assume that depression is adaptive, it does not preclude attempts to act adaptively by those who become depressed or the provision of help by others.

In this model, suboptimal information processing is implicated. Further, males and females are likely to be differently affected by disruption type. A number of prevailing model hypotheses can be incorporated into the narrow interpretation of the model, explanations dealing with the effects of toxins on DNA encoding and, in turn, infrastructural suboptimality, among others. If the constraints on the hypothesis are relaxed so that almost any developmental disruption is admissible (e.g., excessive sensitivity to rejection due to condition predispositions), the hypothesis loses power, as it does when studies point to the importance of innate preferences and nonshared environmental factors and their interaction with condition probabilities (e.g., Silberg et al., 1990).

Mr. P

For reasons unrelated to his health or behavior, Mr. P was placed in four different foster homes before he was 15 months old. Reports by stepparents indicated that his pleasant, outgoing nature had changed when he was switched from the third to the fourth home. There was no history of depression among his first-degree relatives. Signs of depression first appeared when Mr. P was 8 years old, and chronic symptoms of depression, coupled with intermittent bouts of more debilitating depression, continued into adulthood. Despite numerous interventions, no satisfactory treatment was found. In this vignette, it is the disruptive effect on bonding due to multiple placements that is a key postulated contributing factor leading to maturational disruption and depression.

Models of Depression Emphasizing
Ultimate-Proximate-Cause Interactions

Three variations of ultimate-proximate-cause-interaction models of depression will be discussed. Each requires the occurrence of an external event (proximate cause) that compromises goal achievement (ultimate cause). Ultimate and proximate mechanism explanations interact in these models. Ultimate causation explains why persons are *locked in* to their social environment (even an adverse environment) to achieve goals, as well as why they create response strategies when they fail to achieve goals. Proximate events explain how conditions are triggered. Because depressed persons often identify themselves as having failed in their efforts to achieve goals, and because they often believe that environmental events are the basis of their failures, clinical histories compatible with these models are frequently provided by the patients. Depression may be adaptive in these models, and the models do not preclude others' providing help when an individual is depressed.

Variation 1: The Competitive-Loss or Decline-in-Social-Status Model

In this variation, one's perception of a fall in status or communication by others that such a fall has occurred results in infrastructural change. The model builds on the evolutionary concepts of competitive interactions and hierarchical relationships. It assumes that both the symptoms and the signaling features of depression are ultimately caused, and it is consistent with the findings from regulation-dysregulation theory, discussed in chapter 8. The clinical data are compatible with this model in that social-status decline or competitive losses often go hand in hand with depression. In this model, males more than females would be expected to develop depression.

Variation 2: The Failure-to-Resolve-Interpersonal-Conflict Model

The key feature of this variation is that interpersonal conflict can result in either a dominant or a ritualized submissive response toward the person with whom one is in conflict, the ritualized submissive response manifesting as depression (Price et al., 1994):

> It is postulated that the depressive state evolved in relationship to social competition, is an unconscious, involuntary losing strategy, enabling the individual to accept defeat in ritual agonistic encounters and to accommodate to what would otherwise be unacceptably low social rank. (p. 309)

The model offers a proximate triggering explanation of depression. Capacities to engage in ritualized agonistic behavior are ultimately caused. A reasonable assumption is that many persons who are unable to resolve interpersonal conflicts also have compromised functional capacities that increase the probability of unresolved conflicts. However, this assumption is not required by the model.

The model has clinical utility in that depression is often associated with unresolved interpersonal conflicts, particularly when the interacting parties are interdependent.

Cost-benefit interpretations are also relevant. For example, ritualized submissive behavior may be more costly than discontinuing a relationship unless depression reduces both conflict intensity and conflict duration. If depression results in a decline in costs and resolves a conflict, the model may qualify as a variant of the depression-as-an-adaptive-trait model applied to situations of interpersonal conflict. The relationship type should determine whether males or females are more likely to be depressed, females having a greater probability of depression than males because of male tendencies to dominate and guard females.

Variation 3: The Response-to-Loss Model

This is the most familiar of the nonbiomedical models of depression, and it is similar to models developed by psychoanalytic and psychosocial theorists. Losses can be real, anticipated, or imagined. This model differs from the competitive loss model because competitive losses may be reversed, while interpersonal losses (e.g., the death of a crucial other) often cannot. A relaxation of the model's constraints allows for other types of losses, such as the loss of a capacity (e.g., a decline in technical skills associated with aging). The model has clinical utility in that loss frequently correlates with depression. An associated factor is the potential cost involved in attempting to replace a critical other. Such costs can be considerable.

In the prevailing models of depression, the sociocultural (e.g., excessive stress, poverty) and the psychoanalytic (e.g., response to loss) are most easily integrated into ultimate-proximate-cause-interaction models. Biomedical models dealing with neurochemical dysregulation remain secondary. Variations in responses to triggering conditions point to the presence of common final pathway constraints.

This chapter has outlined evolutionary models that are applicable to depression, has tied different types of depression to specific functional outcomes, has incorporated the potential condition-related influences of sexual selection, has aligned the prevailing model hypotheses with a theory of behavior, and has set forth the conditions under which features of the prevailing models can be integrated with evolutionary explanations. In turn, some of the explanatory power of the prevailing models has been detailed. For example, if some cases of depression are adaptive, and if dysfunctional physiological systems accompany such depressions, not only will explanations differ, but intervention strategies will also differ. From the same perspective, evolutionary models provide a novel framework within which to view traditional classifications of depression (e.g., endogenous depression, ontogenic depression, exogenous depression), underscoring the importance of collecting new data, a point already discussed in chapter 4. For example, information about which traits increase the risk for depression and which traits do not change during periods of depression becomes critically important. It becomes apparent that evolutionary models are not necessarily mutually exclusive and that multiple-cause explanations are more likely to be valid than single-cause explanations. A consequence of the multicause view is that behavioral clusters reflecting pure forms of the preceding models will be observed infrequently (Gilbert, 1989).

Concluding Comments

In this chapter, the 15% principle and the concepts of sexual selection and grouping conditions by behavior system were introduced to facilitate the identification of ultimate and proximate cause contributions to and influence on conditions. The second part of the chapter focused on evolutionary models of depression, and where applicable, the prevailing model explanations were incorporated into these models. The review of evolutionary models established that there is not just one model of depression in evolutionary theory, but several; that different models are likely to be applicable to different types of circumstances and to be associated with different types of depression; and that the prevailing models of depression are compromised by their failure to embrace an evolutionary perspective.

8

Regulation-Dysregulation Theory and Condition Triggering

Regulation-dysregulation theory (RDT) addresses the effects of social interactions on infrastructural functionality. The theory integrates findings from several disciplines, as well as points discussed in earlier chapters. It permits insights into (1) how specific types of social interactions influence CNS physiology, infrastructural functionality, cognition, and behavior; (2) how the behavior of others in seemingly normal social environments trigger the onset of conditions in individuals who are either condition-predisposed or condition-vulnerable; and (3) how the behavior of others in extremely adverse environments explains the onset of conditions in individuals who are neither predisposed nor vulnerable to conditions (McGuire and Troisi, 1987a; McGuire et al., 1994). Because most of the relevant research has been conducted by investigators in the fields of physiology, behavior, and psychology, the vocabulary of these disciplines dominates our discussion of theory. Recall, however, that infrastructures are integrated anatomic-physiological-psychological systems and that changes in one part can lead to changes in other parts.

The term *regulation* refers to the state in which infrastructures are functioning optimally. Optimal functioning is synonymous with the term *homeostasis* as it is used in the medical literature. One feels well, has the energy to do what one wants to do, thinks clearly, goes about achieving goals efficiently, and is asymptomatic. For many reasons (e.g., trait variation, compromised infrastructures), individuals differ with respect to their modal level of functioning, that is, their "set point" (see Figure 4.4). Modal set points range from those in which individuals are symptom-free and function efficiently to those in which individuals are continually symptomatic and function inefficiently. Persons also differ in the amount of infrastructural variation they can tolerate without either the onset or worsening of symptoms already present or a decrease in function. On average, the more optimal the set point, the greater the tolerance for variation.

The term *dysregulation* refers to compromised infrastructure function. Moderately compromised infrastructures are associated with a reduction in goal achievement and mild symptoms and signs, including depression, anxiety, anger, boredom, difficulty concentrating, and repetitive but inefficient actions. Severely compromised infrastructures are associated with a more severe reduction in goal achievement, as well as intense and debilitating signs and symptoms. Suboptimal infrastructures are usually implicated.

There is of course nothing new about the idea that social interactions can have psychological and somatic effects. Only the details remained to be clarified, and only in the last few decades has research focused on the physiological and genetic (perhaps) consequences of socialization. Social interactions commence at birth and continue until death. They are important every moment of the way. Key points in the ontogeny of individual socialization were discussed in chapter 3, and three points from that chapter are particularly relevant here: (1) For Freud, and later for Bowlby (1958, 1969, 1973) and others, specific types of interpersonal interactions were essential for the optimal unfolding of maturational programs; (2) infants require frequent holding, touching, and vocal input to maintain physiological and psychological homeostasis (Hofer, 1984); and (3) social deprivation or chronic aversive environments lead to a blunting of maturational programs (Spitz, 1945).

Social interactions are no less important to adults, although they sometimes like to think otherwise. Adults need to talk, touch, and receive others' recognition; otherwise, dysregulating physiological and psychological changes occur (McGuire and Troisi, 1987a; McGuire, 1988). The nearly total elimination of auditory and visual stimuli that characterized the sensory deprivation studies conducted during the 1960s (which used normal adult subjects) led to such a high prevalence of adverse psychological and physiological consequences in the subjects (dysregulated infrastructures) that the studies were discontinued (Schultz, 1965). And unusually well-controlled studies have shown that significant increases in depression, psychosis, and attempted suicide (behavior associated with extremely dysregulated infrastructures) occur in persons in penal institutions who are placed in solitary confinement (Volkart et al., 1983). When one is alone, pleasurable fantasies and meditation can partially regulate dysregulated states, but there are limits (McGuire and Troisi, 1987b). In effect, humans, like other primates, live in a world of conspecifics with whom they frequently interact, a world of social noise, visual stimuli, physical contact, thoughts, and feelings. It is not surprising that humans seek out and defend those who, because of the ways they interact, increase infrastructural regulation. Such interactions are as important to CNS homeostasis as glucose is to cell life. It is equally understandable that persons avoid social interactions that have dysregulating effects.

The complexity and the potential negative consequences of certain types of social interactions are important enough to pursue further. Let us consider complexity first: Persons who socialize with ease and who decide to turn a serious eye to the study of social interactions are often surprised by what they find. Persons meet and recognize one another; they allow context, relationship history, and nonverbal signals to influence how they interpret each other's signals; they discern hidden motives; they respond emotionally at different moments; and so forth. In all, to interact socially is to engage in a highly complex and subtle process that, during any extended interaction, continually undergoes modification—a process in which the average individual engages during approximately 40% of his or her waking day, and a process whose success is contingent on the efficient operation of multiple infrastructures. Given this complexity, it is no wonder that minor system alterations (e.g., missing connections, reduced neurotransmitter levels) have disruptive effects.

With regard to potentially negative consequences, a spiraling sequence of events is often observed among persons who interact socially in atypical ways: Atypical signals by A (e.g., due to minor infrastructural dysregulation) leads to B's avoidance of A because of the adverse effects of A's signals on B's physiological (emotional) and psychological states which causes greater dysregulation in A because of the absence of social signals essential to homeostasis and consequently the onset of a condition in A.

Despite the preceding points, the social environment is often dismissed as a critical condition-initiating factor, particularly if persons are thought to suffer from conditions in which predispositions are suspected. Yet, as we have stressed, discounting the social environment is an invitation to misunderstanding the vast majority of conditions. The social environment is often highly competitive, excessively demanding, rejecting of certain behaviors or individuals, depriving of certain information, and constraining of attempts to achieve goals. In such environments, maintaining a regulated state is next to impossible.

All the above is not to suggest that the social environment is the primary cause of all conditions or that if social environments were somehow ideal, conditions would disappear. Individuals bring their predispositions, vulnerabilities, personalities, infrastructural states, and motivational priorities to environments. What we are saying is that (1) individuals differ in the ways they interact with their environment; (2) social environments differ in the way they interact with individuals; and (3) different social environments can significantly affect infrastructural functionality. In this chapter, we focus on two key features of person-environment interactions: the impact of others' signals on one's infrastructural functionality, and one's capacities to select and manage environments so as to remain regulated (i.e., to avoid dysregulation). The concept of self-other separateness provides a convenient paradigm within which to look at these two features.

Physical self-other separateness is a biological fact, and most of the time, persons recognize that they are physically distinct from others. When they do, others are perceived as leading their own lives and having their own aspirations, values, priorities, short-term strategies, and so forth. Yet, there are moments in which the boundaries of separateness blur, such as during sexual orgasm, periods of intense infatuation, moments of empathy and dependence, intoxication, or periods of group excitement. Blurring is associated with an increased responsivity to others' signals and their physiological and psychological influence (McGuire et al., in press). Conversely, when separateness is intensified, as when one is preoccupied with one's own thoughts, one's responsivity to others' signals declines, as does the influence of others' signals on one's physiological and psychological states. Between these extremes is the modal state that most persons experience most of the time: basic separateness, with greater or lesser degrees of blurring during interactions with different individuals. The range within which one moves back and forth between greater and lesser degrees of separateness can be referred to as a *window of intimacy*. Windows differ from person to person.

Both precondition states and conditions are associated with windows that are either too-narrow-and-too-rigid or too-wide-and-too-labile. In both instances, infrastructures are compromised. When windows are too-narrow-and-too-rigid, as in persons with schizoid, paranoid, and obsessive-compulsive personality disorders, chronic self-other separateness is present. This state was modeled in chapter 6 (Figures 6.1B and 6.1C), showing the consequences of missing or inhibited connections which have the effect of reducing the amount of available and accurate social information. Chronic dysregulation is a likely consequence, primarily because of insufficient signals and inaccurate interpretation of the signals essential to ensure homeostasis. When windows are too-wide-and-too-labile, as in persons with borderline and dependent personality disorders and impulse-related disorders, dysregulation is also present, but for other reasons: Others' signals are insufficiently screened and the available information is distorted. Figure 6.1D (hyperactive connections) and Figure 6.1E (unstable connections) model these conditions. In clinical settings, both extremes are seen, and social contact has a different effect on each. For example, talking with individuals with paranoid or obsessive-compulsive personality disorders usually has minimal effects on their behavior. Conversely, talking with persons with borderline or dependent personality disorders usually does have an impact on their behavior, although often only briefly.

From a research perspective, our knowledge about the moment-to-moment details of social interactions is still in its infancy, partly because of methodological problems. Studies of person-person interactions are difficult to design and to execute correctly. Cross-subject differences in recognition capacities and motivational priorities, as well as subtle environmental contingencies and difficult-to-measure features of social interactions (e.g., subtle response delays), contribute to these problems. Nevertheless, the evidence that is available is impressive, and much of it comes from three sources: clinical observations, nonhuman primate studies, and PET studies.

Numerous individual case and research reports document how alterations in the social environment correlate with changes in both condition-related behavior and internal states. Figures 4.2 to 4.6 provided clear illustrations of this point, and so also do a variety of other studies (e.g., Kiritz and Moos, 1974; Fairbanks et al., 1977; McGuire et al., 1977; chapters 5, 6, and 7). Social environment features can have positive or negative effects. On the negative side, the consequences of sensory deprivation and solitary confinement have already been mentioned. Other studies show that individuals with chronic forms of schizophrenia who enter "emotionally charged" nonhospital social environments have an increased chance of sign and symptom recurrence compared to when they enter "emotionally sensitive" environments (Vaughn et al., 1984; cf. Parker, Johnston, and Hayward, 1988). The impact of emotionally sensitive environments is illustrated in the clinical situation with which literally every person trained in psychiatry is familiar: An individual enters the hospital in a psychotic state, only to appear significantly less psychotic following several hours of sensitive and empathic inpatient care. Enduring dysregulated states due to chronic adverse environmental conditions are also known. Conditions that are continually experienced as unpleasant and depriving and from which individuals believe they are unable to escape correlate positively with stress-related physiological measures (e.g., Cohen and Williamson, 1991). On the positive side, we have already mentioned findings showing that demonstrated disorder-characteristic glucose metabolism in the caudate nucleus of persons with obsessive-compulsive disorder can be modified in response to behavior therapy (a form of environmental input) (Baxter et al., 1992). In each of the preceding examples, social interaction type and frequency are implicated in infrastructural change.

Because clinical findings can often be given more than one interpretation, animal studies, in which experimental conditions can be controlled, are frequently helpful in resolving research questions. Nonhuman primate studies are a case in point. For example, consider the physiological effects of a one-time introduction of monkeys previously unknown to each other into new groups from which they cannot escape. Such studies have resulted in significantly elevated cortisol levels (McGuire et al., 1986). When group membership changes were repeated, severe cardiovascular damage could be a consequence (Kaplan et al., 1982, 1983). In these studies, animals seemed incapable of adjusting to rapid changes in group membership without undergoing physiological and anatomical alterations that compromised their longevity. As a rule, nonhuman primates (like humans) live in groups in which membership is stable and from which animals depart infrequently or, in some instances, not at all. There are good reasons for this behavior. Stable social groups have predictable social interactions, while just the opposite is true for up to 12 months in groups that are formed of animals unknown to each other. Within wide ranges, predictable interactions are not only preferable to uncertain interactions but also regulating, and they increase affiliative behavior while reducing aggression (Kaplan et al., 1982, 1983; McGuire et al., 1986).

Findings from studies of vervet monkeys, first discussed in chapter 5, are even more revealing. High-status or dominant males have peripheral serotonin levels averaging almost twice those of low-status or subordinate males, and they have significantly higher measures of CNS serotonin sensitivity. When CNS serotonin sensitivity is high, the frequency of initiated aggressive behavior is low, animals are relaxed socially, they are tolerant of the behavior of other animals, and they frequently initiate and respond to affiliative gestures by other group members (Raleigh et al., 1984). Essentially the opposite relationships apply to animals with low CNS serotonin sensitivity. This condition is associated with low social status, fewer initiated and received affiliative behaviors, a high frequency of received and initiated threats from other low-status animals, high levels of interanimal vigilance, and frequent dominance displacement by high-status animals (McGuire et al., 1983). *While the advantages of high status (e.g., preferential access to females) undoubtedly contribute to male-male competition for dominant social status, findings also permit the interpretation that animals compete for high status because of the somatic effects associated with elevated CNS serotonin sensitivity.*

Figure 8.1 illustrates some of the preceding points through an analysis of interactions between the frequency of dominance displays by dominant males, the frequency of submissive displays by subordinate males, and peripheral serotonin levels.

In Figure 8.1, the vertical axis measures peripheral serotonin levels in a socially dominant male, while the horizontal axis measures time. At the top of the figure, the downward arrows depict the frequency per unit time of both dominance displays by a dominant male directed at a subordinate male and submissive displays by a subordinate male directed toward a dominant male (i.e., submissive response to dominance displays). Among vervets, high-status males initiate dominance displays toward low-

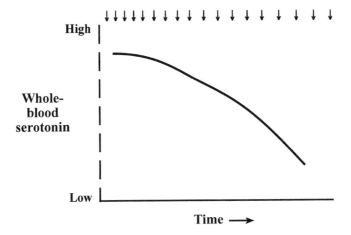

Figure 8.1 Changes in peripheral serotonin levels in high-status male vervet monkeys as a function of the frequency of submissive displays by low-status males.

status males approximately 30 times per day. In the majority of instances, subordinate males respond submissively by lowering their hindquarters, backing off, shifting their weight, angling their head to the side, or positioning themselves to flee. In effect, they send submissive signals. On occasion, subordinate males respond with a counterdisplay or a threat. High-status males are tolerant of such displays provided they are infrequent and not intense. If they become frequent or intense, dominant males may respond aggressively.

If the frequency of submissive displays received by a dominant male declines, his whole-blood serotonin levels also decline. One can demonstrate this relationship experimentally by manipulating the frequency of submissive displays that high-status animals receive (e.g., temporarily separating dominant and subordinate animals), in which case dominant males stop displaying, or by placing dominant animals behind one-way mirrors, where they will threaten subordinate animals but will not receive submissive displays in return. One-way mirror studies are used to determine if serotonin measures decline in the same way as they do when dominant animals are separated from their groups. If the decline in peripheral serotonin levels is essentially the same under both conditions (which it is), submissive displays by subordinate males, not a reduction in dominance displays, are implicated as a key factor in the decline in serotonin levels.

In both natural and captive settings, the behavior of high-status males can be characterized by their movement back and forth along the top left-hand half of Figure 8.1, where the following sequence of behavioral and physiological events is postulated to occur:

1. High-status males with high CNS serotonin sensitivity and high levels of peripheral serotonin lead to

2. Reduced frequency of dominance displays by high-status males toward low-status males causing

3. Reduced submissive displays by low-status males toward high-status males which then causes
4. Decline in CNS serotonin sensitivity in high-status males (a physiological state associated with unpleasurable feelings) which leads to
5. Increased frequency of dominance displays by high-status males,
6. Increased frequency of submissive displays by low-status males, and
7. Increased CNS serotonin sensitivity and peripheral serotonin levels among high-status males.

The preceding sequence can be translated into the vocabulary of signaling, recognition, and infrastructures as follows:

1. A high-status male recognizes an unpleasant somatic state (automatic system) associated with decreased CNS serotonin sensitivity.
2. He initiates an algorithm-generated behavior strategy that involves seeking out and displaying to low-status males.
3. Low-status males recognize the displays and respond with algorithm-initiated submissive displays.
4. The high-status male's recognition of the submissive displays initiates automatic system alterations by a variety of routes (e.g., visual and auditory, which are thought to influence both pineal and pituitary gland activity and to transform selected stimuli from the social environment into endocrine information; Makara, 1985; Fuchs and Schumacher, 1990), hence causing
5. An increase in CNS serotonin sensitivity and the attainment of a desired physiological-emotional state (automatic system) in the high-status male,
6. Discontinuation of algorithm-generated behavior strategies that increase CNS serotonin sensitivity, and
7. A reduction in the frequency of dominance displays by the high-status male.

In this model, subordinate males remain moderately dysregulated physiologically, while high-status males seek out and engage subordinate males in ways that facilitate achieving a desired somatic state and infrastructural regulation (McGuire, Raleigh, and Brammer, 1984). Behavioral evidence is consistent with the idea that subordinate males are dysregulated. For example, in their response to novel environments and maze exploration tests, subordinate males are hyperactive, are reluctant to explore novel environments, and perform poorly in maze exploration tests, while dominant males' reactions are the opposite (McGuire et al., 1983). When a subordinate animal becomes dominant, these behaviors change to those characteristic of dominant males.

Humans, of course, present a more complex picture than nonhuman primates. Although submissive displays by others are important (e.g., in the armed services, business hierarchies, religions, sports teams, clubs), equally, if not more important, are the positive signals one receives indicating that one is liked, socially attractive, important to others, and valued. Human social structure also differs from that of nonhuman primates, a point illustrated by the following list of minimal functional requirements for successful membership within most human hierarchies:

- Know how and when to respond to social signals in order to affirm status relationships and to maintain one's membership within a hierarchy.
- Devise ways of engaging individuals with whom one is competitive without generating excessive antagonism.

- Tolerate differences in cost-benefit outcomes; not all privileges are equal within hierarchies.
- Engage in short- and intermediate-term reciprocal and affiliative interactions, often with others who are disliked.
- Accurately read others' behavior rules.
- Recall the history and state of social relationships and act in ways consistent with such histories.
- And, recognize the appropriate times and settings for challenging status.

Although the preceding list of functional requirements is far from complete (cf. Barkow, 1989; Salter, 1995), it is sufficient to suggest why persons with compromised information-processing capacities often find it difficult to maintain social group membership. It also suggests why minimally competitive and strongly supportive environments, such as psychiatric inpatient settings, which are characterized by sensitive caretaking (win-win) in contrast to competitive (win-lose) interactions, are often beneficial to persons with suboptimal capacities: Such environments are associated with increased physiological regulation.

Among humans, there appears to be at least a three-factor relationship between social status, personality, and serotonin activity. For high-status, aggressive competitors (sometimes called *Machiavellians*), the relationship between whole-blood serotonin and social rank is strongly positive. For more deferent high-status individuals (sometimes called *moralists*), the relationship is negative (Madsen, 1986). Among groups studied thus far, the ratio between Machiavellians and moralists is 7 to 1 (Madsen, personal communication, 1996). In terms of the self-other separateness discussed earlier, moralists usually stand back, evaluate, and judge; that is, *windows of intimacy* are narrow and self-other distinctions are intensified. Conversely, Machiavellians are more likely to manipulate others in face-to-face interactions, a type of behavior that presupposes an openness to others' signals not characteristic of moralists. Postulated relationships between different neurotransmitter profiles, personality types, and personality disorders were noted in chapter 5 (Cloninger, 1986), and recent studies which for example, show an inverse correlation between harm avoidance scores and platelet serotonin 2A receptors (Nelson et al., 1996) are in line with the neurotransmitter–personality type formulation. Thus, among humans, social status is likely to be only one of several determining factors of serotonin activity.

To be sure, there are other sides to status relationships. High-status persons receive more valuing signals from those in their environment. They are often sought out and popular and are often objects of adoration. These behaviors are likely to be psychologically and physiologically reinforcing to the high-status person. But in a way, this is just our point: It is easy to move from the preceding findings on both vervet monkeys and humans to postulate that one of the "driving forces" to attain high social status is the desire for its physiological benefits (e.g., desired feeling states such as feelings of power, elation, and control).

Findings from PET studies of both normal populations and persons with depression inform us on a number of the preceding points. For example, studies demonstrate that both regional cerebral blood flow and glucose metabolism differ in normal controls and persons with depression (e.g., Baxter et al., 1985; Bench et al., 1992; George et al., 1993; Biver et al., 1994). Other studies show that depression-characteristic glucose metabolism findings tend to normalize following successful treatment (Martinot et al.,

1990). However, in a subset of depressed persons, both frontal cortex and whole-cortex hypometabolism persist following successful treatment (Martinot et al., 1990). Further, regional blood flow studies show an anatomical dissociation between the effects of depressed mood and depression-related cognitive impairment (Bench et al., 1992). These findings may partly explain differences in disorder vulnerability and perhaps also the instances of chronic cognitive suboptimality observed in some persons with depression (Essock-Vitale and McGuire, 1990). Another finding shows regional blood flow differences in normal males and normal females when they experience dysphoric thoughts (Pardo, Pardo, and Raichle, 1993; George et al., 1995). This finding opens the door to the possibility that different parts of the male and female brain are active during depression. If so, such differences may partially account for the sex-related prevalence findings discussed in chapter 7. Further, should females have wider windows of intimacy than males, females would be more vulnerable to the influences of negative external information.

The PET findings suggest the following points: (1) Different parts of the brain are activated in persons with and without conditions; (2) males and females differ in the ways they process some external or internal information; and (3) apparently normal individuals may harbor infrastructural vulnerabilities that become apparent only in particular social environments.

Figure 8.2 generalizes the findings from Figure 8.1 to humans in a way that is consistent with both the clinical and the PET findings and that provides an explanation of the triggering of conditions in some environments but not others.

The figure shows changing infrastructural states for Persons A and B in social environments X and Y. The horizontal axis depicts time. The vertical axis shows areas depicting regulated and dysregulated states. Zero (0) on the horizontal axis references physiological regulation. In Environment X, others are supportive, moderately competitive, and moderately demanding, and there are numerous social options. Envi-

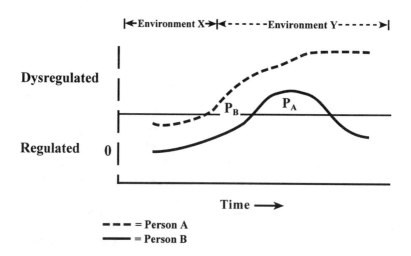

Figure 8.2 Relationships between physiological dysregulation, two social environments, and the probability of conditions.

ronment X represents a moderate-stress environment in which social interactions are usually regulating. In Environment Y, others are not supportive and are highly competitive and socially demanding, and there are fewer social options. Environment Y is a high-stress environment with few regulatory social interactions, that is, one that can easily contribute to dysregulation. Person A has optimal infrastructural capacities. Person B has suboptimal infrastructural capacities.

In Environment X, Persons A and B are physiologically and psychologically regulated (not suffering from symptoms), although Person B is less regulated than Person A. Person A is not symptomatic, and on occasion, Person B is mildly symptomatic. When Persons A and B are in Environment Y, they become dysregulated and symptomatic (e.g., tense, anxious, depressed, preoccupied, bored, angry) because of increased competition and social demands, negative or off-putting responses, and limited social options.

Critical differences in Person A's and Person B's responses to Environment Y occur at points P_A and P_B, the points at which Persons A and B first experience unpleasant emotions. Person B becomes dysregulated and symptomatic (Point P_B) sooner than Person A. At Point P_A, Person A has two options to avoid further dysregulation: she or he may change behavioral strategies and social signals and engage persons in Environment Y in ways that lead to regulation (e.g., become socially dominant), or she or he may leave Environment Y and return to Environment X, where less competitive and less stressful social interactions will facilitate regulation. The first alternative is shown in the figure: Person A changes behavior strategies and regulates. Hypothetically, the same options are available to Person B. However, Person B lacks capacities to change behavior to the degree required to attain dominance status and to regulate in Environment Y. Should Person B remain in Environment Y for an extended period—dysregulation doesn't occur instantly—she or he will become dysregulated, symptoms will appear, and a disorder may be triggered (Post, 1992). In principle, Person B could regulate by returning to Environment X, but clinical experience suggests that a large percentage of persons with suboptimal capacities fail to locate themselves in environments that are optimal for regulation (see chapter 14). The opposite picture may also appear. Person B may avoid entering Environment Y as a self-protective strategy.

One point concerning Person B in Figure 8.2 deserves further emphasis: the often-observed inability of B-type individuals to remove themselves from environments that are dysregulating. Almost no research addresses this point. However, clinical findings suggest that the following work against changes: (1) Familiar social environments, despite their often negative effects, are preferred because the social interactions are predictable; (2) possible negative features of new environments are overestimated; (3) decreased capacities to generate novel scenarios result in fears that one will be unable to manage new environments; and (4) exaggerated estimations of one's importance in one's social environment make one reluctant to leave it (Gilbert, Allan, Ball, and Bradshaw, 1996a; see chapter 14).

RDT predicts that prolonged periods of infrastructural dysregulation will trigger the onset of conditions or make existing conditions worse. Dysregulation can affect more than one automatic system, and if trait variation is disregarded, the degree of dysregulation is contingent primarily on two factors: which automatic systems are influenced by which type of social interactions and how much influence they have.

The same point applies to algorithms, with the caveat that certain dysfunctional algorithms, such as those responsible for cost-benefit assessments and for reading others' behavior rules, are likely to have far greater consequences than other dysfunctional algorithms.

Mr. L

Mr. L was a 47-year-old married male with three grown children when he sought therapy for depression.

His history revealed that he was physically healthy, and he had not previously suffered a mental condition. He had been president of his high school class and captain of his high school football team. He was a member of a close-knit and supportive family, he enjoyed recreational activities, and he was an active participant in community projects. His history also revealed that he was very selective in his choice of friends and that he was unusually competitive with other males. At work, Mr. L frequently spent extra hours to finish tasks before others were able to do so, and when others were prematurely or unfairly promoted, he suffered periods of intense anger. He lacked psychological insight. (Lack of psychological insight is revealed when persons are oblivious to the possibility that their motives or behavior may contribute to social interaction outcomes. Clinically, one simply asks patients if they believe they contribute to situations that distress them. Most of the time they believe they are "innocent".)

Five months prior to seeking therapy, Mr. L was informed that he had failed in his attempts to gain the presidency of a large manufacturing firm for which he had worked for 15 years. A younger and, in his view a less competent, male had been appointed president. Mr. L's anger was so intense that he resigned from the firm and took a lower-status job with a competing company. Within weeks, symptoms of depression appeared. Over the ensuing months, his symptoms became worse, although he was able to continue work.

Interventions involved both medications and psychotherapy. Medications were moderately effective in relieving his symptoms. However, undesirable side effects reduced his compliance. Psychotherapy was unproductive. Then, the sudden departure of the president of the company for which Mr. L was working led to his appointment as acting president. Two months later, he was appointed president. These changes were followed by a dramatic decline in his symptoms.

This case illustrates how important social status can be among males, as are interactions between status level, infrastructural functionality, and symptoms.

Concluding Comments

This chapter reviewed key points of regulation-dysregulation theory: Specific types of social interactions are essential to maintaining infrastructural regulation; persons differ in their capacities to utilize the social environment to regulate themselves physiologically; and persons seek out specific social environments because of their desired physiological effects. RDT provides a bridge between the physiological, the psychological, and the social and begins the process of more precisely identifying factors in the social environment that may contribute to mental conditions.

9

Personality Conditions

The prevailing psychiatric orientation to personality disorders is that they are enduring, relatively inflexible, minimally adaptive patterns of behaving and information processing. If these disorders are looked at in this way, an evolutionary analysis might seem to promise few insights into either their causes or their possible functions. However, a closer look not only invites a different conclusion but also suggests that such an analysis can be highly informative.

The chapter begins with discussions of adaptive genetic variation and phenotypic plasticity, life history strategies, short-term strategies, prevailing models of personality disorders, and traits. These topics introduce new points and briefly review key points discussed earlier in order to set the context for the second part of the chapter, where personality disorders are interpreted in an evolutionary context. Once again, our analysis raises questions first addressed in chapter 3, which dealt with how conditions are identified and defined.

Setting the Context

Adaptive Genetic Variation and Phenotypic Plasticity

Adaptive genetic variation and phenotypic plasticity are two of the ways evolutionary biologists explain phenotypic differences. In the adaptive variation view, genes and genetic expression are major contributing factors to both phenotypic differences and phenotypic change. Internal and external factors influence the expression of genes, and changes in genotype and phenotype are thought to occur over a *relatively few generations* (Wilson, 1994). This view builds on two assumptions: Because of the presence of a within-species "genetic reserve," the full array of genetically influenced phenotypes is never realized, and the genetic reserve can account for many instances of rapid phenotypic change (e.g., facultative or contingent adaptations). Phenotypes are closely tied to genes and sensitive to the effects of natural selection. The rate of selection for a gene is contingent on the degree to which an allele confers an adaptive advantage. And through the different reproductive outcomes of different phenotypes, selection favors those phenotypes that are adaptive and eliminates those that are minimally adaptive.

The phenotypic plasticity view is relatively recent in origin (Cosmides and Tooby, 1987; Tooby and Cosmides, 1989; Symons, 1990; Barkow et al., 1992). Its key arguments are that the last *major* period of intense natural selection affecting *Homo sapiens* was at some point in the past (e.g., the EEA, see chapter 3); that during this period, selection favored the development and refinement of psychological systems (what others and we have called *algorithms* or *universal psychological rules*) that mediate behavior; and that subsequent behavioral changes to environmental changes have been due largely to algorithm-mediated behavior.

Genes and selection are also critical in the phenotypic plasticity view:

> The concept of universal human nature, based on a species-typical collection of complex psychological adaptations, is defended as valid, despite the existence of substantial genetic variation that makes each human genetically and biochemically unique. These apparently contradictory facts can be reconciled by considering that (a) complex adaptations necessarily require many genes to regulate their development, and (b) sexual recombination makes it improbable that all the necessary genes for a complex adaptation would be together at once in the same individual, if genes coding for complex adaptations varied substantially between individuals . . . An evolutionary approach to psychological variation reconceptualizes traits as either the output of species-typical, adaptively designed developmental and psychological mechanisms, or as the result of genetic noise creating perturbations in these mechanisms. (Tooby and Cosmides, 1990a, p. 17)

In effect, genes are closely tied not to specific behaviors, but to psychological systems that mediate behavior. Said another way, in the phenotypic plasticity view, the distance between genes and behavior is far greater than in the adaptive genetic variation view.

The phenotypic plasticity view acknowledges male and female differences, but it is attributes that humans have in common, not their differences, that merit study. There are some allowances for atypical behavior. Such behavior may be due to tensions that can develop between previously selected capacities and the current environment (e.g., genome lag), and algorithm dysfunctionality (e.g., impaired memory consolidation) may be a contributing factor to some conditions (Cosmides and Tooby, 1995). Nevertheless, the focus remains on attributes that humans hold in common and universal psychological rules. (One consequence of this view is that the search for genetic "markers" might better focus on markers associated with evolved psychological structures than on markers for signs, symptoms, or conditions; e.g., Belmaker and Biederman, 1994).

The two preceding views—the one that favors phenotypic plasticity and the one that favors adaptive genetic variation—have been at the center of considerable recent controversy (much of which is best understood as a matter of territorial disputes among academics). The available evidence does not provide a resolution, nor is it our intention to enter the controversy. Rather, our aim is to point out what features of both views are present in the models developed throughout this book. For example, there are readily identifiable, socially important traits, such as infant-parent bonding, preferential investment in kin, male and female sexual strategies and preferences, sexual orgasm, and anger at being cheated, that are most parsimoniously explained as ultimately caused by and closely tied to genes. Phenotypic plasticity may explain some of the variations in these traits, but not the universal presence of the traits themselves. There are also readily identifiable behaviors that reflect interactions between behavior and environmental contingencies; contrasting the daily lives of Eskimos and Wall Street stockbrokers illustrates this point. Many of these behaviors are most readily explained from the phenotypic plasticity perspective. Thus, the importance of both views should be kept in mind.

One point in the preceding quotation deserves a closer look, however: "sexual recombination makes it improbable that all the necessary genes would be together at once in the same individual" (Tooby and Cosmides, 1990a, p. 17). We agree, but in our analysis, it does not follow that cross-person behavioral differences are best explained in terms of "adaptively designed psychological mechanisms." We find the following findings more compelling: *(1) Enduring features of individual variation and behavior are far more prevalent than would be predicted by the phenotypic plasticity view (see chapters 2 to 4, and below); (2) genetic recombination and genetic influence on traits are the strongest candidates for explaining much of this behavioral variation (see below); and (3) humans are more accurately viewed as mosaics of traits, some of which are independent, some semi-independent, and some more plastic and optimal than others.* Conceptualized this way, trait predispositions and trait strengths make significant contributions to individual differences, although these contributions fall short of fully explaining all phenotypic features. Two points already mentioned are relevant here: *Evolution has thrived on both genetic and phenotypic variation, one consequence of which is the expectation of a high degree of cross-person variation, and selection favoring the refinement of infrastructures may be secondary to selection for immunological capacities (see chapter 3), in which case a high degree of cross-person variation in "psychological mechanisms" is likely.*

Where do the preceding points leave us? From one perspective, and contrary to some characterizations, some plasticity can be observed among all persons—with or without conditions (see Figure 4.2). Infrastructures, therefore, remain responsive to changing circumstances. Still, evidence favoring genetic influences on behavior systems, infrastructures, and particular traits is equally compelling: *A cardinal sign of the vast majority of conditions, and particularly those classified as personality disorders, is constrained plasticity.* Thus, when considered alone, neither the adaptive genetic variation nor the phenotypic plasticity views can be expected to provide adequate explanations of personality conditions. If behavior is as malleable as those advocating the phenotypic plasticity view suggest, we should see fewer mental conditions. In particular, there should be fewer conditions in which certain traits are enduring, minimally adaptive, and relatively unchanging across environments. Similar points apply to the adaptive genetic variation view: If conditions are minimally adaptive, in which case they should be selected against, and genetic change can occur rapidly, fewer conditions in which genes play a critical role would be expected. On this point, recall that the lifetime prevalence estimates for disorders classified by *DSM-III-R* (APA, 1987) are reported to be approximately 50%, a percentage that, among other things, suggests that clinically observable phenotypic variance is as prevalent as phenotypic similarity.

Life History Strategies

Life history strategies were first discussed in chapter 3, where we concluded that these strategies are ultimately caused, genetically influenced programs for allocating resources that enable individuals to achieve biological goals. These allocation programs are subject to trait variation, although, like motivations-goals, cross-person variation is not thought to be a major causal factor in conditions, and environmental contingencies can lead to changes in allocation priorities and their associated behaviors.

When the life-history-strategy concept is applied to personality conditions, there are several important implications:

1. Age- and sex-related program changes provide insights into the prevalence profiles of some conditions. For example, the behavior characteristics of antisocial personality disorder often first appear during adolescence and disappear during the fifth decade.
2. Normal allocation programs, coupled with compromised infrastructural capacities, frequently distinguish persons with and without conditions, a point we pursue below.
3. Resource allocations can be characterized in terms of their upper and lower limits, that is, the limits on the resources that persons will allocate to achieving a goal. Most of the time, neither upper nor lower limits are reached. In turn, both within-individual and cross-individual behavioral fluctuations tend to disguise the fact that there are limits. Some exceptions are likely, however. For example, a normal adult male spends approximately one hour per day in personal hygiene, while the same male suffering from obsessive-compulsive disorder may spend up to six hours. Or a female with anorexia nervosa may spend up to four hours per day exercising, while a same-age normal female might spend half an hour. In such instances, resource allocations may approach their upper limits; for example, a person with obsessive-compulsive disorder could spend 15 hours per day in personal grooming but does not.

Short-Term Strategies

Short-term strategies differ from life history strategies in several critical ways: (1) They are developed to achieve short-term goals; for example, a large percentage of persons who suffer from low-intensity depression or anxiety resolve their depression or anxiety (the short-term goal) by devising strategies to reduce the unpleasant features of their emotions; (2) they may or may not be responsive to self-monitoring information; (3) they are often jettisoned if they fail; and (4) strategies may take multiple forms; for example, phobias may be understood as strategies to achieve separation, and amnesia may be understood as a strategy to isolate information. Persons with severely debilitating conditions also act to achieve short-term goals, although many of their strategies fail. Failures do not make the assessment of these strategies less important, however. Not only are failures key factors in explaining conditions, but knowing the reasons for failure is critical to optimizing interventions designed to improve capacities and to successfully execute strategies.

At a more detailed level, strategies can be somatic, functional, or informational. Somatic strategies involve the use of somatic information to achieve goals. Anorexia nervosa is perhaps the clearest example, but somatic strategies are also seen in bulimia nervosa, somatoform disorders (e.g., conversion disorder, hypochondriasis, pain disorder), factitious disorders, and mood disorders. Functional strategies involve the use of specific behaviors to achieve goals. Adjustment disorder, impulse disorders, antisocial personality disorder, malingering, and attention-deficit/hyperactivity disorder (ADHD) are examples. Informational strategies involve the use of information to achieve specific goals. Paranoid and narcissistic personality disorders, phobias, and dissociative disorders are examples. None of these strategies are assumed to be volitional.

Which strategy will be utilized depends on several factors, but the most important are an individual's motivation-goal priorities and his or her degree and type of infrastructural functionality; that is, in enacting strategies, persons tend to favor those infrastructures that are least compromised. For example, an adolescent male who is highly dissatisfied with his home environment might become depressed, develop an adjustment disorder, or reject his parents' teaching and psychologically decouple himself from his family. Depression would reflect a somatic strategy, the rejection of parental teaching would reflect a behavioral strategy, and psychological decoupling would reflect an informational strategy.

Prevailing Models of Personality and Personality Conditions

The extensive literature dealing with psychological, psychiatric, and genetic models of personality (the term *temperament* is sometimes used) and personality disorders has been reviewed elsewhere (e.g., Wiggins, 1968; Millon, 1981; MacDonald, 1995). Here, we mention only a few points with the aim of establishing how these models compare with evolutionary models.

Psychological Models

A fundamental premise of personality research in psychology, and to a lesser degree in psychiatry, is that persons with both normal personalities and personality disorders have enduring traits that can be characterized dimensionally. Since Eysenck (1947) first reported on the dimensional features of personality in the 1940s, a mass of empirical evidence has appeared that supports the trait dimensionality concept (e.g., Bouchard, 1993, 1994; Lykken et al., 1993; Svrakic et al., 1993). There are varying degrees of correspondence among the traits that interest investigators; for example, both psychological (e.g., Eysenck, 1947) and psychiatric models (e.g., Cloninger, 1986; N. L. Pedersen et al., 1988; Cloninger, Svrakic, and Przybeck, 1993) have focused on dimensions such as novelty seeking and harm avoidance. Personality characterizations based on dimensional characterizations approximate many clinical findings, and to the degree that they do, they tend to take on an aura of validity. We remain cautious, however, primarily because dimensional measurements are seldom closely tied to detailed functional assessments, and there are a number of reports that contest the stability of dimensional measures; for example, studies show that the cross-situational generality of traits is moderate when situations are similar, but generality is not present when the functional requirements of situations change (Moskowitz, 1994; see Rowan, 1990, for another perspective).

Psychiatric Models

Early assessments of personality by psychiatrists (e.g., Freud, Otto Rank) relied on clinical findings that were interpreted within psychoanalytic models. While these efforts are important historically, they are largely tangential to empirically oriented psychiatry (biomedical and behavioral models), primarily because of the difficulties involved in testing psychoanalytic hypotheses. Currently, psychiatry's most influential model of personality disorders is the work of Cloninger and his colleagues (e.g., Cloninger 1986, 1987a; Cloninger et al., 1993), who have attempted to integrate psychometric measurements, clinical assessments of personality, and physiological measures:

> Evidence suggests that variation in each dimension is strongly correlated with activity in a specific central monoaminergic pathway: novelty seeking with low basal dopaminergic activity, harm avoidance with high serotonergic activity, and reward dependence with low basal noradrenergic activity. These neurobiological dimensions interact to give rise to integrated patterns of differential responses to punishment, reward, and novelty. The combination of high novelty seeking, high reward dependence, and low harm avoidance (histrionic personality) or the combination of high harm avoidance, low reward dependence, and low novelty seeking (obsessional personality) are [sic] each associated with information processing patterns that lead to unreliable discrimination of safe and dangerous situations and hence chronic anxiety. (Cloninger, 1986, p. 167)

Findings from a host of studies are consistent with the idea that there are interactions between physiological measures, psychometric measurements, and clinical findings. To cite only a few examples, inhibited children differ from normal children in their hypothalamic-pituitary-adrenal and sympathetic nervous system activity (Coll et al., 1984; Kagan et al., 1987, 1988). Higher levels of CNS monoamine activity are observed in persons who are repetitive sensation seekers and risk takers (Zukerman, Buchsbaum, and Murphy, 1980; af Klinteberg et al., 1987). Atypical CNS serotonin function is thought to be a feature of obsessive-compulsive personality disorder (Zohar et al., 1987). And testosterone levels and female personality features are reported to interact (Baucom, Besch, and Callahan, 1985). As would be expected from such findings, a number of reports show positive correlations (comorbidity) between personality disorders and other disorders (e.g., Alnaes and Torgersen, 1990a; Bronish and Hecht, 1990; Duggan, Lee, and Murray, 1990; Jackson et al., 1991; Andreoli et al., 1992).

Genetic Influence Models

Although studies tying DNA to specific traits are in their infancy, there are a sufficient number of replicated studies of putative gene-trait relationships to strongly implicate genetic influences. For example, when monozygotic twins are reared apart, and personality features such as well-being, social potency, achievement, social closeness, stress reaction, alienation, aggression, self-control, harm avoidance, novelty seeking, interests, and intelligence are measured, estimates of genetic influence (heritability) range from .20 to .60 (e.g., Tellegen et al., 1988; Bouchard, 1993, 1994; Cloninger et al., 1993; Lykken et al., 1993; Plomin et al., 1994). Further, as noted, one of the key implications of behavioral genetics research deals with the part played by nonshared environments in the development of personality (Plomin and Daniels, 1987; Plomin et al., 1994): In effect, children in the same household create and experience very different microenvironments, and both parents and children contribute to these differences through their own genetically influenced preferences and the influence of these preferences on their interactions (Scarr, 1992). While these models are highly consistent with much of evolutionary theory, as well as with the emphasis we have placed on trait-condition interactions, for the most part they have been developed outside an evolutionary framework.

Evolutionary Models

A number of investigators have studied personality in evolutionary context. Plutchik's characterizations of both normal and atypical personalities were reviewed in Chapter 5. Cognitive and evolutionary theory have been combined to interpret personality disorders (e.g., Beck, Freedman, and Associates, 1990), and a number of psychological models (which often include dimensional assessments) are based on evolutionary concepts (e.g., Buss, 1991; Schroeder, Wormworth, and Livesley, 1994; MacDonald, 1995). When sexual selection is introduced into personality evaluations, differences in male and female trait strengths and age-relative resource allocations can all influence personality. A number of the sex-related strategy differences have been noted; for example, males allocate more resources to achieving social status, acquiring resources, and competing with other males over mates, while females allocate more resources to bonding and nurturing (Chapter 7). Additional differences are probable. *Window of intimacy*, duration of pleasurable states (e.g., multiple orgasms in females), impulse control, attention to detail, and persistence are likely candidates.

Evolutionary theorists have also attempted to explain how certain traits and strategies enter and remain in populations. Evolutionary stable strategy models provide, perhaps, the most familiar examples. Different, often competing,strategies, such as those associated with hoarding and sharing resources, can evolve within a population. When both strategies are present, an equilibrium is reached (frequency-dependent selection), and a population is composed of different percentages of individuals exercising each strategy (Maynard Smith, 1982). Temporary fluctuations in percentages can occur because of environmental changes that are more advantageous to individuals using one of the strategies, but eventually, a stable equilibrium is again achieved. For example, assume that 20% percent of a society is composed of persons who hoard and the remaining 80% percent is composed of persons who share. If the percentage of those who share declines, there will be less to hoard, the number of hoarders will decrease, and the proportions will be reestablished. Evolutionary stable strategy theory is a population genetics explanation; thus, it deals with trait features of large populations. Applied to mental conditions, the theory has the potential to explain how certain traits or personality types that are viewed as socially undesirable can persist over time; for example, if most populations have a high prevalence of caretakers, a certain percentage of persons seeking excessive amounts of care can remain in the population and reproduce.

A recent addition to the personality literature is found in the work of Birtchnell (1993) (see also Soldz et al., 1993). He viewed personality and personality disorders using a framework with two central concepts: social status and closeness. The two main axes of his model are upperness ←→ lowerness (respectively, relating to positions of social strength and weakness), and closeness ←→ distance (respectively, remaining socially involved or becoming socially separate) (Birtchnell, 1993, in press-b). The main axes (e.g., upperness and closeness) represent different interpersonal objectives. The four points of the two main axes, plus the four intermediate positions, make up an octagon in which different personality types can be located in two-dimensional space. Emotions are responses to either successful or failed efforts to achieve objectives (Birtchnell, in press-a), and personality disorders are viewed primarily as consequences of repeated but unsuccessful attempts to achieve closeness (Birtchnell, in press-a, in press-b).

The importance of the Birtchnell model lies in several areas: (1) It achieves a close fit with two important evolutionary concepts: status and affiliation; (2) it assigns a key place to social interaction type in its explanations of both normal personalities and personality disorders; (3) it deviates from existing dimensional approaches to personality (e.g., harm avoidance) by creating new conceptual and measurement categories that are applicable to social interactions; and (4) it is the source of some of the interpretations developed below.

The Influence of Genetic Information on Traits, Within-Trait Variation, Trait Clusters, and Functional Consequences

Although there is overlap between the personality models discussed above and those developed below, none of the models discussed thus far take into account what we view as a critical point in the explanation of personality disorders: *the degree of individual variation observed among persons currently classified as having the same disorder*. To deal with this point, it is necessary to return to the topics of genetic influences on traits, within-trait variation, trait clusters, and their functional consequences.

Genetic Influence on Traits

Figure 9.1 summarizes findings from studies of genetic influences on traits of monozygotic and dizygotic twins (Plomin et al., 1994).

For genetically influenced traits, higher intraclass correlations are expected among monozygotic twins than among dizygotic twins. Findings in the figure are in agreement with this expectation: Each of the measured items shows clear intraclass correlations between the two types of twins.

Although intraclass correlations like those shown in Figure 9.1 have been replicated across numerous studies involving different populations (e.g., Floderus-Myrhed, Pedersen, and Rasmuson, 1980; Fulker, Eysenck, and Zuckerman, 1980; Rose et al., 1988), debates continue over whether it is actually traits or something else that is being measured. Any attempt to address these debates would take us far afield, with little guarantee of a satisfactory answer. What is important to emphasize here are the following points: (1) Individuals can be characterized as mosaics of traits, each trait

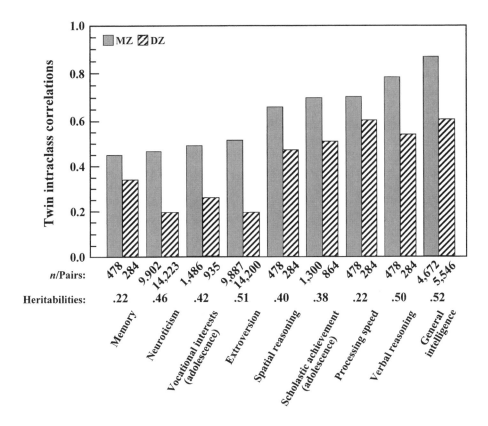

Figure 9.1 Twin intraclass correlations for selected personality measures. Monozygotic and dizygotic twin intraclass correlations for personality (neuroticism and extroversion), vocational interests in adolescence, scholastic achievement in adolescence (combined results for English usage, mathematics, social studies, and natural science), specific cognitive abilities in adolescence (memory, spatial reasoning, processing speed, and verbal reasoning), and general intelligence. Adapted from Plomin et al. (1994).

varying in its degree of optimality; (2) assessments of traits, such as verbal reasoning, provide rough estimates of functional capacities; (3) heritability estimates (shown at the bottom of Figure 9.1) are more consistent than not across studies and provide estimates of genetic contributions to traits; and (4) the traits measured in the figure (e.g., neuroticism, verbal reasoning, memory, general intelligence) are frequently used to characterize conditions and to assess outcomes.

Within-Trait Variation

There is of course nothing new either inside or outside psychiatry about the observation that people differ. What may be new is that these differences are more important than is usually appreciated, and particularly so in attempts to adapt. Within-trait variation is a way of referencing cross-person differences in traits. With few exceptions, the traits that are of interest to psychiatry are present in persons with and without conditions. Figure 4.4 is particularly relevant to this point because it demonstrates three critical trait features: (1) Persons with disorders have limited functional capacities; (2) however, none of the functional capacities in the figure was absent among persons with disorders even during periods of disorder exacerbation; *thus, within-trait variation rather than all-or-none characterizations of traits is a descriptively accurate way of characterizing traits*; and (3) persons without disorders vary in the degree to which their traits are optimal.

Figure 9.2 further develops the implications of Figure 9.1 by addressing within-trait variation in a hypothetical but not necessarily random-breeding population (e.g., assortative mating may be present). The figure should *not* be interpreted as reflecting only the effects of genetic information. Rather, it depicts phenotypes, the expression of which is influenced by both genetic makeup and events that occur between the time of conception and the time phenotypes show signs of becoming enduring.

In Figure 9.2, A represents high-frequency traits for which there is minimal cross-person variance, and for which the vast majority of persons have the same genes. B-type traits represent high-frequency traits that have a high degree of cross-person within-trait variance. C represents low-frequency traits. A-type traits include anatomical features (e.g., number of eyes), basic physiological systems (e.g., cardiovascular system), and reflexes (e.g., cough). These traits are shared by literally all humans. While there are variations in A-type traits (e.g., the nerves for pain or the genes responsible for certain enzymes may be absent; Goedde et al., 1984, 1985), thus far, only a few examples of conditions associated with the absence of A-type traits have been reported (e.g., Brunner et al., 1993). C-type traits represent unusual attributes, such as perfect pitch, double-jointedness, and tongue curling. For many of these traits,

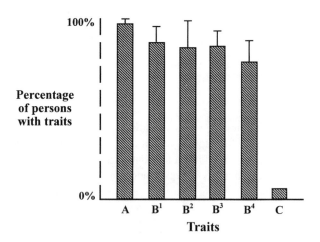

Figure 9.2 Postulated distribution of traits and trait variance in a population. A = physical characteristics; B = algorithms; C = unusual traits; B^1 = capacity to read others behavior rules; B^2 = casual modeling; B^3 = self-monitoring; B^4 = capacity to utilize monitoring information.

there is discontinuous variation: one is either double-jointed or not. C-type traits also are infrequently implicated as causes of conditions.

This leaves B-type traits. Variant forms of B-type traits are apparent in the majority of conditions (see Figures 4.4, 4.5, and 9.1). Moreover, an indisputable fact of clinical life is that individuals vary significantly when any one of their condition-related traits is studied in detail. The same point applies to psychometric and psychological data, such as measurements of neuroticism or CSF 5-HIAA, which vary across persons diagnosed with the same disorder. And, a fact of therapeutic life is that the same intervention with persons of the same age and sex, diagnosed with the same disorder, sometimes helps, sometimes doesn't help, and sometimes makes things worse. Each of these points is consistent with the cross-person within-trait variation concept.

In our view, cross-person within-trait variation is an obvious outcome of evolution, and models of conditions would be unwise to disregard this point. Variation is as much a part of clinical phenotypes as "core" features of conditions. Also in our view, the lack of appreciation of the cross-person within-trait variance in B-type traits, of their independence or semi-independence, and of the possibility that B-type traits do not closely tie to reproductive success is one of the major obstacles to developing novel insights into personality disorders. Nesse and Williams (1994) emphasized two points that set the framework for the preceding points: Traits that may have been valuable in the past lead to a directional bias in subsequent selection, and selection is characterized by trade-offs. Biases and trade-offs not only render some current traits far from optimal (e.g., bone strength) but may also increase susceptibility to conditions (Nesse and Williams, 1994). (Note: The concepts of directional bias and trade-offs differ from concepts underlying the genome lag hypothesis.)

Trait Clusters

Trait clusters are bundles of independent and semi-independent traits that make up the phenotypic features of personality disorders (and many conditions). The concept of trait clusters does not imply that all of the traits in a cluster are suboptimal or dysfunctional. For example, persons meeting the diagnostic criteria for dependent personality disorder may differ significantly in their intelligence and thus in their management of their disorder.

Although a number of known and postulated reasons for genetic influences on both within-trait variation and trait clusters have been discussed, assortative mating deserves to be singled out. Assortative mating, or selecting mates with similar traits, is known to influence the probability that certain traits will appear in offspring (Merikangas, 1982; Merikangas, Bromet, and Spiker, 1983). Studies suggest that individuals prefer mates who are genetically similar (Buss, 1985), and that assortative mating effects are more prominent in traditionally defined personality traits than in physical and sociodemographic traits, attitudes, or values (Merikangas, 1982). An obvious implication of these findings is that assortative mating can increase homozygosity and, in turn, influence trait strength. Although many details of assortative mating remain to be clarified (e.g., factors accounting for mate preferences; Thiessen and Gregg, 1980; Burley, 1983), evidence is consistent with the view that persons with conditions, as well as those who are at risk for conditions, mate more frequently with each other than would be expected by chance (Coryell, 1980; Guze et al., 1970); if this were not the case, pedigree data would be less compelling. Thus, a bias due to assortative mating is a likely contributing factor to personality disorders and their within-kin continuation across generations.

Summary

To summarize this section:

1. The phenotypic features most often used to characterize personality disorders are rarely present all the time (Clark, Watson, and Mineka, 1994; McGuire et al., 1994). Thus, an improved understanding of these disorders is likely to come from closely observing interactions between specific environmental and phenotypic features.
2. In the assessment of enduring phenotypic features, it is important to distinguish between these features and the effects of constructs. Numerous studies have shown that persons who are "primed" for certain attitudes (i.e., their knowledge structures dealing with social behaviors, such as "Others are aggressive, rude, or pleasant," are activated) alter their subsequent judgments and behavior (e.g., Bargh, Chen, and Burrows, 1996).
3. Within limits, the social effects of trait clusters are a function of the characteristics of the environment in which a person locates himself or herself; an individual who is a member of a large, closely knit, and supportive family may go unnoticed as a dependent personality, while the same person living in a depersonalized urban environment may be quickly recognized.
4. Reports continue to appear in the literature that implicate adverse early experiences as a major disorder-contributing factor. For example, high frequencies of child abuse are thought to increase the chances of multiple personality (Boon and Draijer, 1993). Such reports perpetuate the assumption that personality disorders are *primarily* the outcome of social influences. The available evidence simply does not support this view unless environments are extremely adverse.
5. In the interpretations offered below, persons with personality disorders can be seen as having suboptimal, and at times dysfunctional, infrastructures. They are locked into their social environment because of their attempts to achieve goals, and they attempt to achieve goals by utilizing those strategies and capacities that are least compromised.

Personality Disorders

Two points further set the context for this section. First, a general point: From an evolutionary perspective, there is no reason to suspect that selection has favored an ideal personality. Many personality types are far more likely. Moreover, there is good reason to suppose that a number of traits associated with disorders can be adaptive. For example, above-average capacities for deceptiveness may be adaptive in selected environments (e.g., rapidly changing urban environments, wartime environments). Yet, deception is unlikely to be adaptive across all environments. Thus, both the strength of traits and their discretionary use can be expected to influence adaptiveness. Second, a specific point: Our view of *DSM-IV* (APA, 1994) personality disorders is that these disorders are overlapping groupings of enduring trait clusters in which some but not all traits are suboptimal. Within these groupings, two general subgroups are discussed: trait clusters that can be adaptive when judged by evolutionary criteria and trait clusters that reflect attempts to act adaptively. Antisocial and histrionic personality disorders, malingering, and ADHD are examples of the first subgroup. Paranoid, narcissistic, borderline, and dependent personality disorders are examples of the second subgroup.

Personality Disorders That Can Be Adaptive

Antisocial Personality Disorder (Sociopathy)

According to *DSM-IV* (APA, 1994), "The essential feature of Antisocial Personality Disorder is a pervasive pattern of disregard for, and violation of, the rights of others that begins in childhood or early adolescence and continues into adulthood" (p. 645). Individuals with this disorder frequently lack empathy and tend to be callous, cynical, and contemptuous of the feelings, rights, and suffering of others (p. 647). Moreover, "glibness, superficial charm, self-assurance, sexual exploitation, the absence of guilt when exploiting others, and substance abuse are frequently associated behaviors" (p. 647). The disorder is thought to affect approximately 3% of adult males and to occur more frequently than expected by chance among the first-degree relatives of those with the disorder, a finding suggesting genetic influence. It is less frequently reported among females. A recent in-depth review of this disorder considers two possible etiologies, one emphasizing ultimate causation (i.e., selection favoring specific traits), and the other emphasizing developmental factors (Mealey, 1995). Here, we will focus on the former possibility.

A number of investigators have discussed antisocial personality disorder in an evolutionary context. A central theme in their analyses is that nonreciprocation (or cheating) is observed consistently. The disorder closely fits predictions from a model of nonreciprocators (Harpending and Sobus, 1987; Dugatkin, 1992; Mealey, 1995), a model in which persons frequently use what they calculate as the least costly way of obtaining goals. In effect, individuals carry out a strategy of social defection (McGuire et al., 1994), but awareness of deceptiveness is not required. The earlier discussion of evolutionary stable strategies provides an explanation for how disorder-related traits may be favored by selection: In a society made up primarily of reciprocators, genes for cheaters can enter the population and remain, provided persons with such genes reproduce. (If, as we have suggested, the traits associated with the vast majority of conditions are not strongly selected against, then not only will these types of traits remain in populations, but there is no reason to assume that they will become less prevalent or less varied.) The greater prevalence of this disorder among males may be due to selection favoring stronger migratory tendencies among males than among females; that is, migration and detection should correlate inversely.

Reasonable assumptions are that approximately 50% of the persons who could meet the criteria for this disorder go through life undiagnosed and undetected, and that they are successful by evolutionary criteria (McGuire et al., 1994). They acquire mates and resources, have offspring and social influence, invest in kin, and so forth. The 50% figure is conservative when compared, for example, to statistics on crime and lawbreaking from which assaults are excluded. Criminologists estimate that less than 20% of persons who engage in repeated forms of deceptive behavior, such as scams, fraud, bunko, and bigamy, are identified and apprehended (Jeffers, personal communication, 1993). The probable low percentage of detections makes age of onset and remission estimates speculative. Nevertheless, clinical experience suggests that those instances of the disorder that first appear in adolescence and remit during the fifth decade represent a high-risk strategy associated with resource acquisition and reproduction. Those instances that appear during childhood are less likely to represent a high-risk strategy (Mealy, 1995). What is being suggested is shown in Figure 9.3.

Figure 9.3 graphs the probability of detection as a function of the tendency (frequent versus infrequent) of persons with antisocial personality disorder to engage in deceptive behavior. The figure assumes that the distribution of persons with different tendencies has an inverted U distribution. As expected, the probability of detection is greatest among persons who engage in deceptive behavior most frequently. A clear implication of the figure is that selection against antisocial behavior is likely to remain weak because of the low probability of detection of a large percentage of persons with the disorder.

Persons with this disorder who avoid detection are capable of developing novel behavior strategies, accurately assessing the costs and benefits of short-term social interactions, accurately reading others' behavior rules, utilizing self-monitoring information to alter their strategies, and successfully disguising their intentions (MacMillan and Kofoed, 1984; McGuire et al., 1994). Thus, many algorithms appear to operate efficiently. However, limited capacities to experience guilt may be present. Experimental studies indicate that there is a significant reduction in psychophysiological responses to guilt-eliciting stimuli (e.g., Hare, 1983). Findings of reduced guilt point to specific types of automatic system suboptimality for the nonkin- and kin-investment systems. Missing connections (Figure 6.1B), which limit the availability of emotional information, are implicated. Thus, self-deception may not be a major feature of this disorder. Further, the absence of guilt may increase the chances that these individuals will achieve their biological goals; for example, persons who do not experience guilt

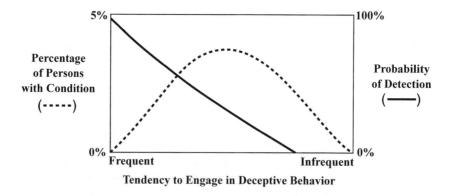

Figure 9.3 Percentage of persons with antisocial personality disorder and the probability of detection. Broken line = persons with antisocial personality disorder. Solid line = probability of detection.

may act where others are constrained from doing so. (In principle, what is suggested here differs little from what has been postulated for a variety of other conditions that can be explained as a consequence of missing connections or structural defects; e.g., dyslexia is a condition in which many persons with mild forms are undetected.)

Histrionic Personality Disorder

An explanation similar to that proposed for antisocial personality disorder—the excessive use of social defection strategies—closely fits histrionic personality disorder. *DSM-IV* (APA, 1994) describes this disorder as one in which there is a "pervasive and excessive emotionality and attention-seeking behavior" (p. 655), coupled with a tendency to be dramatic and engage in fantasy. The disorder is often coupled with inappropriate sexually provocative or seductive behavior, and clinical assessments are consistent with the view that a significant percentage of time and effort are spent in defecting (e.g., by feigned illness) from promised or implied reciprocations or intimacy. In these instances, guilt is usually absent. The prevalence of the disorder is estimated at between 2% and 3% among adult females, and individuals differ in their identifying features. It is seldom diagnosed among males. Pedigree studies indicate that the disorder runs in families (Coryell, 1980).

As with antisocial personality disorder, clinical experiences suggest that a large percentage of females who meet the criteria for this disorder turn out to be successful by evolutionary criteria. They acquire mates, marry, have children, command resources, and invest in kin, although generally, they experience difficulties in achieving emotional intimacy. These individuals appear to have normal motivational and resource allocation programs. As in antisocial personality disorder, self-deception may not be a major causal factor. Experience also suggests that cultural stereotypes may mitigate against the early diagnosis of this disorder both inside and outside the clinician's office; for example, a married woman who has children is likely to be treated for symptoms associated with a personality disorder rather than for a personality disorder.

With advancing age, a percentage of women with this disorder are diagnosed as suffering from a somatoform disorder (usually hypochondriasis), while others show "burnout" features similar to those observed in some males with antisocial personality disorder. Thus, the disorder may have several causes, and the possibility that a percentage of persons are enacting a high-risk strategy (the "burnout" group) while others (somatoform group) attempt to adapt within enduring constraints remains an attractive hypothesis.

Malingering

While malingering is not classified as a personality disorder in *DSM-IV*, it fits within the category of conditions characterized by the excessive use of social deception strategies. According to *DSM-IV* (APA, 1994), "The essential feature of Malingering is the intentional production of false or grossly exaggerated physical or psychological symptoms, motivated by external incentives such as avoiding work or military duty, obtaining unearned financial compensation, evading criminal prosecution, and/or obtaining drugs. Under some circumstances, Malingering may represent adaptive behavior—for example, feigning illness while a captive of the enemy during wartime" (p. 683).

In the models developed here, malingering can be understood as a variant of antisocial personality disorder. There is a similar lack of guilt feelings about the consequences of one's behavior for others, and social defection strategies are prominent. What tends to differentiate the two disorders is their customary strategies: Persons with antisocial personality disorder are more likely to involve others in their schemes, while malingerers generally do the opposite. Although false productions are often thought to be intentional (*DSM-IV*), even if they are there is no evidence to suggest that these productions are associated with feelings of guilt or that the same behavior would not occur if the person was unaware of his or her intentions. In our experience, false productions seldom appear to be intentional. Rather, they represent strong perceptual biases. As in antisocial personality disorder, resource allocation programs appear to be normal, and the behavioral strategies used are those that are most beneficial and least costly to the person who malingers. Evolutionary predictions applicable to malingering suggest that the condition will be characterized by average or above-average capacities to assess others' behavior rules, and thus to manipulate others' responses, and that malingerers will migrate frequently to social environments in which the possibility of detection is low. Clinical assessments are in agreement with these predictions: Persons who malinger generally have elaborate reasons for failing to participate socially or to reciprocate; for example, they may complain about recent or chronic illnesses, aversions to certain types of work, and dissatisfactions with others.

Attention-Deficit/Hyperactivity Disorder

Like malingering, ADHD is also not a personality disorder in *DSM-IV*. On initial evaluation, ADHD would seem to present a counterintuitive example of an adaptive strategy. Yet, because a percentage of persons with this disorder turn out to be successful by both social and evolutionary criteria, the possibility that they are engaging in a high-risk strategy requires consideration. *DSM-IV* (APA, 1994) describes the disorder, in part, as follows: "The essential feature of Attention-Deficit/Hyperactivity Disorder is a persistent pattern of inattention and/or hyperactivity-impulsivity that is more frequent and severe than is typically observed in individuals at a comparable level of development" (p. 78). Studies of second-degree relatives show a greater incidence of ADHD in grandfathers than in grandmothers, and in uncles than in aunts (Faraone, Biederman, and Milberger, 1994), and thus suggest possible male influence in the transmission of this disorder.

An evolutionary interpretation of this disorder might be thought of in this way. (We thank Kevin MacDonald for this example.) Clinically, the behaviors that are characteristic of this disorder include sensation seeking, impulsivity, reward seeking, and aggressive interaction styles (e.g., pushy or assertive). Clusters of these behaviors may range from minimally adaptive to adaptive. On the minimally adaptive side, persons diagnosed as having ADHD are often unable to constrain certain behaviors (e.g., impulsivity), and they are usually viewed as lacking conscientiousness and as being undisciplined, unplanful, and unreliable. They engage in asocial and antisocial behaviors (e.g., risk taking, aggression, petty criminality) with greater than average frequency, and they reciprocate helping behavior less often than the average. In part, this clinical profile can be explained by a bias toward the use of the behaviors designed to achieve goals when persons have limited capacities to accurately read others' behavior rules, develop efficient behavioral scenarios and strategies, and accurately self-monitor the consequences of their behavior. Automatic system suboptimality may extend to all four automatic systems.

As in the three preceding disorders, there is considerable variation across trait clusters. For example, children who are moderately inquisitive and aggressive often eagerly engage in and master the environment. These children may be curious and exploratory, may be good athletes, and may enjoy being in highly stimulating environments—fast-paced events where the focus of activity and reward shifts rapidly. In addition, such children often show enormous energy and enthusiasm, are easily motivated by rewards, and behave assertively. Rough-and-tumble play may be paradigmatic of the type of activity that such children find most attractive. Conversely, the attention of children who exhibit many of the same behaviors may become so diffuse, and their impulsivity so out of control, that they have difficulty organizing and putting to use important information, as well as coordinating their activities with those of others. Such children are often overly impulsive and aggressive, socially rejecting and rejected, and diagnosed as suffering from ADHD.

Among a percentage of ADHD children, minimally adaptive behaviors continue into adulthood, and in this group, antisocial behavior is common (Mannuzza et al., 1991). However, a subset of previously diagnosed ADHD children, particularly those who do not have associated mental disorders, are remarkably similar to control populations without a history of mental disorders (Mannuzza et al., 1989). In this group, studies suggest that there are no differences in occupational adjustment, social functioning outside school, angry behavior, alcohol or drug abuse, or antisocial behavior. There are also studies showing that some adults who are diagnosed with ADHD as children go on to achieve high levels of success as entrepreneurs, salespeople, and entertainers (Cantwell and Hanna, 1990; Cantwell and Baker, 1992). These individuals often exhibit an extraordinary amount of drive in pursuit of their goals, and typically, their jobs are free of excessive routine and excessive attention to detail. Although exceptionally successful individuals are not typical of mature ADHD children, they are of theoretical importance in that ADHD-related behavior can be viewed as a product of a strategy that is highly adaptive in some environmental circumstances, but minimally adaptive in others.

Summary

To summarize this section, what characterizes persons with the four disorders discussed thus far is that their behavior can be understood as a high-risk strategy, and despite evidence of partially suboptimal infrastructures, they can be successful by evolutionary criteria. Their resource allocation programs appear to be normal. Evidence of genetic influence is present for all of the disorders except malingering, and studies may eventually demonstrate that malingering is also influenced genetically.

The possibility that a number of currently classified disorders are adaptive by evolutionary criteria does not lead to the conclusion that these disorders should not be identified and treated. Clearly, a frequent outcome of these disorders is that others are victims. What the preceding points do suggest, however, is that the four disorders are highly likely to be resistant to treatment. This prediction closely fits clinical experience.

Personality Disorders as Attempts to Act Adaptively

Attempts to act adaptively do not require conscious intent, and the independence or semi-independence of infrastructural systems permits the hypothesis that some infrastructural systems operate to offset or regulate those systems or traits that are compromised. Both points are critical to understanding disorders in this category.

Personality disorders that can be interpreted as attempts to adapt include borderline, narcissistic, paranoid, avoidant, and dependent personality disorders. The first three will be discussed here, with an emphasis on functional interpretations. These disorders can be characterized by the presence of suboptimal automatic systems and algorithms essential for successful social navigation, in particular, limited capacities to read others' behavior rules, to develop novel behavior strategies, and to efficiently utilize self-monitoring information. Descriptively, these disorders can be likened to the circumstances of a person who has a cast on one leg and is trying to cross the street rapidly in the face of oncoming vehicles: This person hops, jumps, and even crawls if necessary, and it is this behavior that is adaptive.

Borderline Personality Disorder

In *DSM-IV* (APA, 1994), borderline personality disorder is essentially characterized by "a pervasive pattern of instability of interpersonal relationships, self-image, and affects, and marked impulsivity that begins by early adulthood and is present in a variety of contexts" (p. 650). The prevalence rate is estimated at 2% of the general population, and the frequency of the disorder among females exceeds that among males (*DSM-IV*). Compared to control populations, a fivefold greater prevalence of the disorder is reported in first-degree relatives (*DSM-IV*). *DSM-IV* diagnostic criteria for the disorder were listed in chapter 2, but for convenience, they are repeated here: (1) A frantic effort to avoid real or imagined abandonment; (2) a pattern of unstable and intense interpersonal relationships characterized by alternating between extremes of idealization and devaluation; (3) identity disturbance, with a markedly and persistently unstable self-image or sense of self; (4) impulsivity in at least two areas that are potentially self-damaging; (5) recurrent suicidal behavior, gestures, or threats, or self-mutilating behavior; (6) affective instability due to a marked reactivity of mood; (7) chronic feelings of emptiness; (8) inappropriate, intense anger or difficulty controlling anger; and (9) transient, stress-related paranoid ideation or severe dissociative symptoms (p. 654). Five of the nine criteria are required for diagnosis.

Persons who have this disorder severely enough to meet the diagnostic criteria have a minimally adaptive type of personality organization. They switch between engaging but clinging personal styles to angry and hostile affect, characterized by manipulative and self-destructive acts. A wide variety of signs and symptoms are observed, disorganized thinking, depression, and anxiety being among the most common. Symptom type and intensity fluctuate with relationship state. There is often a pattern of undermining one's efforts at the moment a goal is about to be realized (e.g., destroying good relationships when it is clear that the relationships could last). Behavior is often coupled with ideas of reference and undue suspiciousness toward the person with whom one is involved. Altering the social environment to meet these persons' needs may temporarily, but not permanently, reduce the intensity of their signs and symptoms. Social interaction failures may result from their inability to activate others' reward systems. High frequencies of sexual and physical abuse during childhood continue to be reported (Ogata et al., 1990; Boon and Draijer, 1993), and possible responses to these experiences include increased external vigilance and heightened fears of being hurt. These responses may explain rapid shifts in mood and affect (Gilbert, 1995).

As noted, the most frequently encountered features of this disorder can be understood as manifestations of a suboptimal reproductive automatic system and compromised capacities to read others' behavior rules, to develop novel behavior strategies, and to utilize self-monitoring information to guide and stabilize behavior. Unstable connections for both information processing and signaling model the disorder (Figures 6.1E and 6.2E). When functionality is severely compromised, individuals are seldom attractive to others, although even those persons who clearly meet *DSM-IV* diagnostic criteria have periods in which many disorder features are not apparent. Persons with less severe forms of the disorder may be successful by both evolutionary and social criteria. During periods of extreme stress, otherwise normal persons often show borderline-like signs and symptoms.

When this disorder is viewed in an evolutionary context, the majority of the DSM-IV *diagnostic criteria can be interpreted as indices of failed efforts to achieve goals that require others' participation.* Examples include a frantic effort to avoid real or imagined abandonment; transient, stress-related paranoid ideation; and a pattern of unstable and intense interpersonal relationships characterized by alternations between extremes of idealization and devaluation. These behaviors are best understood as strategies used by persons who are *active participants* in the social arena, attempting, although often unsuccessfully, to engage and interact with others.

Narcissistic Personality Disorder

DSM-IV (APA, 1994) describes the essential feature of narcissistic personality disorder as "a pervasive pattern of grandiosity, need for admiration, and lack of empathy that begins by early adulthood and is present in a variety of contexts" (p. 658). The prevalence rate is estimated at 1% in the general population, with the frequency in males exceeding that females (*DSM-IV*). Five or more of the following nine criteria are required for diagnosis: (1) has a grandiose sense of self-importance; (2) is preoccupied with fantasies of unlimited success, power, brilliance, beauty, or ideal love; (3) believes that he or she is "special" and unique and can be understood only by, or should associate only with, other special or high-status people; (4) requires excessive admiration; (5) has a sense of entitlement; (6) is interpersonally exploitive; (7) lacks empathy; (8) is often envious of others or believes that others are envious of him or her; and (9) shows arrogant, haughty behaviors or attitudes (p. 659).

Clinically, there is little doubt that many persons with this disorder have difficulty navigating the social environment and that they are undesirable social partners. Clinical histories reveal poor social adjustment lifelong, with frequent loss of friends and spouses. Their frequently noted vulnerability in self-esteem is thought to be closely linked with the observation that these individuals are highly reactive to criticism and defeat (e.g., an unusually wide and paradoxical *window of intimacy*), against which they must defend themselves. This feature implicates the automatic survival system. Symptom type and intensity fluctuate with relationship state and no doubt in part reflect the interaction styles of others. Socialization capacities (e.g., lack of empathy, difficulty in recognizing the desires of others) are limited. Compromised and distorted recognition features are present in persons with this disorder, particularly in the frequently encountered belief that others are totally interested in what they say, think, and feel. Missing and hyperactive connections (Figures 6.1B, 6.1D, 6.2B, and 6.2D) model the main features of this disorder.

In an evolutionary context, fluctuating emotions generally implicate suboptimal infrastructures punctuated by periods of dysfunctionality. However, like some persons with ADHD and antisocial, histrionic, and borderline personality disorders, a percentage of persons meeting *DSM-IV* criteria for narcissistic personality disorder are unusually talented, are highly successful socially, and acquire resources and mates. Thus, it is likely that many persons with minor forms of this trait cluster are undetected or, if detected, tolerated. Further, both the criteria for characterizing this disorder and the clinical manifestations are consistent with the view that these persons are actively involved in their social environments; for example, belief in one's uniqueness requires excessive admiration, a sense of entitlement, and interpersonal exploitation.

Paranoid Personality Disorder

DSM-IV (APA, 1994) describes paranoid personality disorder as follows: "The essential feature of Paranoid Personality Disorder is a pattern of pervasive distrust and suspiciousness of others such that their motives are interpreted as malevolent. This pattern begins in early adulthood and is present in a variety of contexts" (p. 634). Prevalence estimates range from 0.5% to 2.5% in the general population, and frequency estimates for males exceed those for females (*DSM-IV*). The disorder is usually associated with solitariness, poor peer relationships, social anxiety, excessive suspiciousness, hypervigilance, hostility, and underachievement in school, all of which point to suboptimal infrastructures. Others' behavior is often viewed as a threat or a form of malevolent deception. There is some evidence of an increased prevalence of this disorder among relatives of probands with chronic forms of schizophrenia (Lenzenweger and Loranger, 1989). The persistence of the behavioral features, as well as the fact that alterations of the social environment have a minimal effect on clinical states, points to severe automatic system suboptimality, particularly in the survival system. Goal achievement is usually compromised. A *DSM-IV* diagnosis requires that at least four of the following criteria be met: (1) suspects, without sufficient basis, that others are exploiting, harming, or deceiving him or her; (2) is preoccupied with unjustified doubts about the loyalty or trustworthiness of friends or associates; (3) is reluctant to confide in others because of an unwarranted fear that the information will be used maliciously against him or her; (4) reads hidden demeaning or threatening meanings into benign remarks or events; (5) persistently bears grudges, that is, is unforgiving of insults, injuries, or slights; (6) perceives attacks on his or her character or reputation that are not apparent to others and is quick to react angrily or to counter-attack; and (7) has recurrent suspicions, without justification, regarding the fidelity of spouse or sexual partner (p. 637).

On initial consideration, it seems counterintuitive to view this disorder as an example of a failed strategy to achieve social goals. Yet, as with the two preceding disorders, if the diagnostic criteria are carefully interpreted, literally all of the criteria are consistent with the idea that persons with this disorder are actively involved in their social environments. Put another way, and removed from judgments about their social acceptability, the criteria of excessive fear and suspicion of others, doubts about others' loyalty, and unforgiveness of insults point to social involvement.

Summary

In each of the three disorders in the attempt-to-adapt category, the resource allocation programs appear to be normal. Moreover, allocations are directed in ways that are often typical of persons without disorders (e.g., in achieving biological goals). However, the capacity of individuals to translate allocations into behaviors that consistently achieve goals is another matter. The strong possibility that algorithms for self-monitoring and developing novel behavior strategies are suboptimal is suggested by the finding that persons with these disorders continue to repeat the same strategies even though these strategies are often costly and unsuccessful.

An obvious question is: Might borderline, narcissistic, and paranoid personality disorders be selected in the sense that we have suggested antisocial and histrionic personality disorders might be selected? This seems unlikely, if only because others frequently reject such persons. A more plausible explanation is that the severe forms of these disorders are the result of extreme degrees of within-trait variation applicable to key infrastructures; for example, infrastructural features that offset borderline, narcissistic, and paranoid tendencies, particularly mood regulation and affective expression, are underdeveloped or underrefined.

Concluding Comments

To summarize the key points in this chapter:

1. Personality disorders can be understood as composites of suboptimal, independent, and semi-independent traits (trait clusters) that persist across time and environments.
2. Persons with personality disorders have essentially normal motivations-goals and resource allocation programs.
3. Phenotypic manifestations of personality disorders that can be adaptive change with age because of age-related changes in motivations-goals and allocation programs.
4. Persons with and without personality disorders enact many of the same behaviors (e.g., deception). The place and frequency of their use are what usually differentiate persons with and without disorders.
5. Persons with personality disorders have limited behavioral plasticity.
6. Compared to persons without disorders, persons with personality disorders are either less (e.g., antisocial personality disorder) or more (e.g., borderline personality disorder) responsive to environmental information (i.e., extremes on a *window-of-intimacy* scale).

7. Personality disorders are not uniformly associated with reduced reproductive success.

8. For minimally adaptive disorders, reduced goal achievement increases the probability of dysregulation and associated signs, symptoms, and disorder comorbidity.

9. Much of the behavior of persons with personality disorders can be understood as attempts to change others' behavior in order to facilitate goal achievement, that is, to structure the environment to their own advantage.

10. Tying functional outcomes to explanatory hypotheses leads to novel characterizations of disorders (e.g., the importance of social participation in paranoid personality disorder despite the presence of often severely constrained functional capacities), as well as novel causal hypotheses (e.g., disorders viewed as the consequences of ultimately caused adaptive strategies).

11. Trait variation and trait clusters explain a significant number of the features normally associated with disorders.

Two related points deserve special emphasis. The first deals with whether statistically atypical physiological states should be expected in the disorders discussed in this chapter. In those persons who would meet the criteria for any one of the disorders, yet who remain undetected, there is no reason to suspect atypical physiological measures. This point applies particularly to those disorders that may be adaptive, and it probably also applies to a percentage of persons with disorders who have been identified clinically; for example, even the striking psychophysiological findings that are often reported for antisocial personality disorder (reduced anxiety and guilt) apply to only a percentage of persons receiving this diagnosis. On the other hand, when infrastructures are severely suboptimal or dysfunctional, dysregulation and atypical physiological and psychological states are likely.

Second, it is worth asking if the interpretations offered here can be integrated with recent prevailing model reconceptualizations of personality and personality disorders. For example, Cloninger and his colleagues (1993) examined self-concepts associated with temperament and character to determine how persons with personality disorders view themselves as autonomous individuals, as integral participants in human society, and as an integral part of the universe as a whole. These categories overlap in part with the ways in which we have conceptualized disorders and thus reflect an increasing interest in the importance of social functioning as a major condition-related factor.

10

Anorexia Nervosa

Evolutionary models promise insights into a large number of conditions in addition to those already discussed. This and the following four chapters explore this promise in discussions of anorexia nervosa; schizophrenia (chapter 11); phobias (chapter 12); somatoform, adjustment, dissociative, and other disorders (chapter 13); and dysthymic disorder (chapter 14).

"The essential features of Anorexia Nervosa are that an individual refuses to maintain a minimally normal body weight, is intensely afraid of gaining weight, and exhibits a significant disturbance in the perception of the shape or size of his or her body" (APA, 1994, p. 539). Self-imposed dietary limitations and atypical patterns of handling food, coupled with an intense fear of obesity and weight gain, are usually present. Loss of appetite tends to occur during the later phases. If weight loss is extreme, physical and laboratory signs of starvation may be present. The disorder has been recognized since antiquity (Ploog and Pirke, 1987).

Features frequently associated with the disorder include delayed ovulation, amenorrhea in postmenarche females, diminished concentrations of several reproduction-related hormones (e.g., estrogen and follicle-stimulating, luteinizing, growth, and thyroid hormones; e.g., Vaccarino et al., 1994), depressed mood, social withdrawal, irritability, anxiety, insomnia, diminished interest in sex, and obsessive-compulsive behavior. Many of the features are also present in other disorders, including major depressive disorder, schizophrenia, and social phobia (*DSM-IV*, APA, 1994). PET findings suggest that the disorder is associated with glucose hypometabolism in selected cortical regions (Delvenne et al., 1995).

Anorexia is limited almost entirely to females, who account for more than 90% of the diagnosed cases, 85% of which occur between the ages of 13 and 20, with peaks bracketing ages 14 and 17. The clinical course varies greatly, *although the most common form is a single period with resolution.* The earlier the age of onset, the better the prognosis. Prevalence estimates range between 0.51% and 3.70% (Walters and Kendler, 1995). The peak times of onset implicate the hormone changes (or a lack of them) associated with different phases of puberty as an important disorder-contributing factor. Monozygotic twins are reported to have a higher concordance rate than dizygotic twins, and co-twins of twins with anorexia are at significantly higher risk of developing anorexia and related disorders, such as bulimia nervosa and major depressive disorder (Walters and Kendler, 1995; APA, 1994). Genetic contributions are thus a likely second disorder-contributing factor. A third possible factor has been identified by psychoanalysts, who have emphasized the importance of family dynamics (particularly excessively controlling parents), food phobias, and avoidance responses to the sexual tensions associated with puberty. The fact that the disorder is largely limited to industrial countries implicates cultural influences (e.g., age-specific social expectations of female appearance and behavior) or culturally typical patterns of child rearing (e.g., separation of mother and infant for extended periods, bottle feeding). Culture is thus a fourth potential contributing factor. Possible costs and benefits of adolescent weight control have been modeled (Anderson and Crawford, 1992), and a number of authors have discussed treatment options (e.g., Nygaard, 1990).

Evolutionary Analysis

Evidence pointing to genetic (e.g., monozygotic and cotwin data) and cultural influences (e.g., the preponderance of cases in industrial countries), clinical findings implicating developmental influences (e.g., excessively controlling parents), the narrow time frame within which the disorder manifests itself, and several distinct clinical courses—mild-brief-remitting, moderate-severe-but-usually-remitting, and severe-and-sometimes-unremitting—is typical of this as well as other disorders. Disorders with these characteristics implicate different clusters of contributing factors (the 15% principle), within-trait variation, different degrees of compromised infrastructural functionality, and the use of disorder-related strategies.

Motivations-goals and resource allocations associated with the reproductive behavior system are assumed to be normal. *Indeed, it is the persistence of reproduction-related motivations-goals that appears to be both a critical and a necessary disorder-contributing factor, although it is not a direct cause of the disorder.* The observation that females with anorexia are often competitive with other females supports this interpretation. The unusually high prevalence rate among females (>90%), as well as the effect of the disorder on physical attractiveness, points to yet a fifth contributing factor: the use of strategies to avoid sexual maturation. Females compete among themselves for mates, and physical attractiveness, because it so predictably commands male attention, is a high-priority element in such competition. Adopting a strategy that renders one unattractive (e.g., excessive weight loss) contributes to a reduction in one's competitiveness. Further, males (on average) are more sexually exploitive of females than the reverse, and particularly between adolescence and the early 30s, the period during which more than 90% of anorexia cases are reported to occur. Thus, for females who fear the consequences of adult womanhood (Mahowald, 1992), including the possibility that their physical attractiveness will trigger male exploitation or that they may fail to attract a desirable mate, strategies leading to weight loss and unattractiveness offer a way of avoiding these consequences. In terms of the three basic strategy types (somatic, functional, and informational; see chapter 9), anorexia represents a combination of somatic and informational strategies, with a primary emphasis on the former type.

In mild and brief forms, anorexia deviates minimally from socially defined norms for weight management. Many females, young and old, seriously pursue a course of weight reduction; witness, for example, the number of highly successful weight-oriented support groups and businesses (special diets, sports centers) that cater to weight regulation and female attractiveness in the United States. Women living in industrial societies are strongly influenced by these norms, and for potential mates, reduced weight may signal capacities for restraining impulses as well as suggest sensitivity to cultural values. The fact that many individuals pursue weight reduction strategies similar to those used by persons with mild-brief forms of the disorder contributes to the porous boundaries separating normal (and possibly healthy) weight control and mild anorexia. Thus, a percentage of persons with anorexia are likely to remain undetected. Distortions due to automatic system and algorithm dysfunction (not suboptimality in the brief-mild form of the disorder) are a sixth contributing factor. Those distortions that are present in the mild-brief form are time-limited and self-correcting. Self-correction suggests that nonaffected infrastructures limit the influence of, as well as offset the effects of, dysfunctional structures and eventually normalize behavior.

When anorexia is moderate-severe-but-usually-remitting, greater infrastructural dysfunctionality is implicated, and both medical and psychological treatment are often required. Recognition distortions dealing with the self, intimacy, sexual activity, pregnancy, and male availability are frequently observed even in social environments in which males are plentiful. These distortions implicate relatively severe dysfunctionality of the reproductive automatic system and perhaps also suboptimality, which is one possible inference from the monozygotic and cotwin data. Weight loss not only contributes to a delay in the onset of puberty (delayed menstruation and ovulation) but also signals one's immaturity to others during critical mate-selection and reproduction-related periods. Beyond a certain point, males will interpret weight loss as a sign of illness and decreased genetic quality. Although moderate-severe forms are often associated with others' providing care and attention, these outcomes are not viewed as primary motivations, largely because of the significant reproductive, social, and evolutionary risks associated with the disorder (e.g., physical illness and, in some instances, death). Both dysfunctional causal modeling due to inhibited or missing connections (Figures 6.1B and 6.2C) and compromised scenario development model the moderate-severe form: Individuals develop their own private logic concerning weight, food, and eating, and they experience great difficulty in imagining alternative ways of living. Self-monitoring algorithms appear to be least compromised. In most instances, weight loss is recognized, as is the fact that an increase in weight would improve one's health and attractiveness to males. However, such information minimally influences behavior. Compared to the mild-brief form, in the moderate-severe form the degree to which nonaffected infrastructures compensate for their dysfunctional counterparts is reduced.

Delaying puberty can be viewed as a strategy for conserving reproductive resources in environments in which males are *perceived* as scarce and female competition for males is *perceived to be* intense (Surbey, 1987): In effect, it is "an emergency strategy that is pursued whenever the sovereign handling of one's own reproductive potential is not possible because of socioecological or ontogenetic constraints" (Voland and Voland, 1989, p. 223). Viewed this way, attempting to preserve one's reproductive potential within a context of recognition distortions and algorithm dysfunctionality *is a further example of an attempt to adapt within constraints*. In effect, there is an effort to decouple oneself from a perceived depriving and hostile environment. Said another way, the decreased interest in sex, the frequent social withdrawal, the intensification of self-other distinctions, and the rejection of cultural norms of attractiveness can be understood as a high-risk, often high-cost strategy in which individuals suffer from an inability to tolerate disconcerting information because of recognition distortions. If effective, the strategy allows the individual a temporary withdrawal from reproduction-related competition. And as with other conditions, attempts to adapt may be successful.

In instances in which the disorder is severe-unremitting and leads to death, suboptimal infrastructures due to deficits (missing connections; Figures 6.1B and 6.2B) and hyperactive connections (Figures 6.1D and 6.2D) are implicated. The misinterpretation (e.g., amplification) of reproductive-related social stimuli is consistent with this interpretation. This form of the disorder can be characterized as an outcome of a combination of serious recognition distortions, a failed strategy (which might be adaptive under other circumstances), a significantly compromised ability to utilize information about the consequences of weight loss (Nygaard, 1990), and the absence of capacities essential to discontinuing the weight-reducing strategy once it is initiated.

A number of investigators (e.g., Ploog and Pirke, 1987) have postulated that when weight falls beyond a certain point, the accompanying physiological and mental changes sustain the disorder. Figure 10.1 models this hypothesis, which is applicable to both moderate and severe instances of anorexia.

Three primary causal factors are identified in the figure: predispositions, developmental events, and social influences. In Ploog and Pirke's (1987) formulation, the "hunger drive and its regulating cerebral machinery are perverted," and the addiction-like qualities of the disorder are postulated to interact with CNS reward systems, which contribute to the disorder's repetitive features (p. 854). These authors' postulates can be tied to points already developed. Features of the disorder that can be attributed to predispositions include vulnerability to, or suboptimal contributions to, recognition distortions (e.g., one's body, the unavailability of males), dysfunctional causal modeling (e.g., inaccurately interpreting the consequences of changes in one's body weight), and an inability to discontinue the weight reduction strategy once it is initiated. Brain system rewards (the addictive feature) are thought to be mediated through the an individual's sense of control (however misperceived) over her body and social state. Excessively controlling parents provide the models for the sense of control, and they would constitute developmental contributions. Social influences (e.g., attractiveness equals thinness) were discussed above.

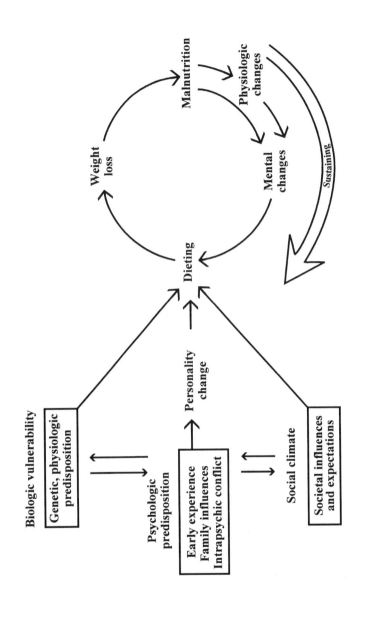

Figure 10.1 A causal theory of anorexia nervosa. From Ploog and Pirke (1987).

Concluding Comments

Of all the disorders that have been and will be discussed, anorexia nervosa provides perhaps the clearest example of interactions between social factors, compromised infrastructural functionality, and attempts to adapt. Hormonal, genetic, and developmental factors are the probable basis of recognition distortions, which are largely confined to the reproductive automatic system. Distortions trigger somatic and informational strategies to exit from a potentially hostile and depriving social environment while concurrently preserving one's reproductive potential. Thus, normal motivations-goals, recognition distortions, and strategies emerge as the key explanatory factors, and the relative contribution of each is thought to determine clinical manifestations, degree of severity, and clinical course.

11

Schizophrenia

The term *schizophrenia* refers to a group of disorders with overlapping signs and symptoms, the causes of which have baffled psychiatry and related disciplines for centuries. At least a dozen taxonomic systems are used to classify this disorder—more accurately, group of disorders—and there are at least as many theories about its causes. An evolutionary analysis offers yet other perspectives.

The diagnostic criteria for schizophrenia include characteristic signs and symptoms associated with cognitive and emotional functioning, such as delusions, hallucinations, and disorganized speech; social and occupational dysfunction; and a period in which the signs and symptoms persist (*DSM-IV*, APA, 1994). Paranoid, disorganized, catatonic, and other forms of schizophrenia are usually recognized. Epidemiological studies have not fully settled incidence questions, in part because there is an absence of agreement on which signs and symptoms are required for diagnosis, and in part because narrow and wide definitions lead to different incidence estimates. Nevertheless, most studies report close to 1% for lifetime prevalence rates, and slightly higher and lower rates are reported for some geographic and cultural areas. It is also unclear if the frequency of the disorder is greater among males or females (*DSM-IV*), although epidemiological data point to an earlier onset in males, and recent studies using magnetic resonance spectroscopy (MRS) suggest possible gender differences in cerebral features of the disorder (Buckley et al., 1994). Sex-linked differences in the transmission of flat affect have also been reported (Goldstein et al., 1995). Compared to different types of depression (chapters 7 and 14), anorexia nervosa (chapter 10), and phobias (see chapter 12), sex differences appear to play a less important etiological role.

Even though a small percentage of persons reveal disorder-predictive signs as early as age 2 (Fish, 1987; Fish et al., 1992; Caspi et al., 1996), most investigators believe that the disorder is rare during the preadolescent years. In the majority of cases, clinical indications first appear during adolescence, although first occurrences have been reported as late as the mid-30s, as have atypical and sporadic cases (Jönsson and Jönsson, 1992). Indices associated with a good clinical prognosis include good premorbid social, sexual, and work histories (e.g., good social-support systems, lasting relationships, employment); an acute and late age of onset; obvious precipitating factors; paranoid features; an undulating course; a family history of mood disorder; the presence of affective symptoms; and positive symptoms (delusions, hallucinations, and disorganized behavior). Essentially, the opposite set of indices, when accompanied by the presence of negative symptoms (affective blunting, poverty of speech and thought, apathy, and poor social functioning), are associated with a poor clinical prognosis (e.g., Brewer et al., 1996). A major distinction between schizophrenia and mood disorders is the failure to return to baseline functioning following remission (*DSM-IV*).

When all forms of the disorder are combined, the World Health Organization reports that 97% of afflicted persons lack insight, 74% experience auditory hallucinations, 70% experience verbal hallucinations and ideas of reference, 66% experience flat affect, and 65% and 64%, respectively, experience voices speaking to them and delusional mood (Kaplan and Sadock, 1989a, p. 768). All of these indices reflect compromised information processing (which may explain why a large percentage of recent research on schizophrenia has focused on the prefrontal cortex). Despite the advent of drugs, estimates are that only 20% to 30% of persons with the disorder are able to lead marginally normal lives, while 20% to 30% continue to experience moderate symptoms and are unable to lead marginally normal lives, and 40% to 60% remain significantly impaired and socially peripheral (Kaplan and Sadock, 1989a). These percentages have changed only slightly over the last several decades (Hegarty et al., 1994). For those occasional cases in which there is an apparent single episode with full remission, the possibility of an alternative condition is likely.

Although causal hypotheses differ, most prevailing models build on some combination of genetic, stress, or multicause hypotheses (D. Rosenthal, 1970). Compared to control populations, first-degree relatives of probands have a 10% higher risk of the disorder, and the reported concordance rates among monozygotic twins far exceeds those for dizygotic twins (*DSM-IV*). Thus, a percentage of the population appears to carry predisposing genes or genes that contribute to disorder vulnerability (e.g., Eaton, 1985; Jablensky, 1987; Torrey, 1987; Allen and Sarich, 1988; Gottesman and Bertelsen, 1989). Whether there is a specific gene, different combinations of genes, events associated with DNA encoding, or interactions among genetic products (e.g., proteins), the key contributing factors remain to be determined (e.g., Fowles, 1992). On the subject of stress, specific vulnerabilities to stress have been postulated, although the nature of these vulnerabilities has successfully defied identification; some investigators have argued that stress is somatic (e.g., physiological change), others favor informational or interactional stress (e.g., a frightening experience, excessive sensitivity to the death of an important other, environmental stimuli), and still others favor functional stress (e.g., failure in an important endeavor). Foster children whose foster mothers are schizophrenic are reported to have a higher incidence of the disorder than foster children with nonschizophrenic mothers (Heston, 1966), although these findings remain to be confirmed. One additional causal candidate is a version of RDT (regulation-dysregulation theory) in which extreme physiological or genetic changes during adolescence interact with compromised infrastructures and changing environmental demands and signals. Hypotheses dealing with viral causes and physical trauma provide an exception to the stress-diathesis hypothesis because in the narrow sense they do not require disorder predispositions or vulnerability. Current reports suggest that, at most, viral causes can account for no more than 2% of cases (Takei et al., 1996).

The possibility that excessive levels of dopamine (dysfunction of the dopaminergic system) may be the source of many disorder-related features remains an active research focus, in large part because of the clinical effectiveness of drugs that modulate CNS dopamine activity. Dysfunctions of other neurotransmitter (e.g., serotonin) systems have been considered, but their causal importance remains to be established. A number of studies have found increases in ventricle size or temporal lobe atrophy in persons with chronic forms of the disorder (e.g., DeLisi et al., 1991, 1992, 1995; Honer et al., 1994; cf. Budinger, 1992; Friston, 1992). Cognitive and motor impairments have been linked to gray matter deficits (Sullivan et al., 1996). And impaired motor skill learning has been tied to corticostriatal dysfunction (Schwartz et al., 1996). Still other studies have found correlations between schizophrenia and minor physical anomalies (O'Callaghan et al., 1995). A recent and interesting line of research has focused on attempts to preclude subvocalizations, an experimental therapeutic technique that is achieved primarily by persuading patients to open their mouths. These studies indicate that auditory hallucinations disappear in a large percentage of patients who undertake this maneuver (Bick and Kinsbourne, 1987). Thus, auditory hallucinations may represent projections of subvocalized verbal thoughts that reach awareness because of deficient cerebral cortical inhibition.

Evolutionary Analysis

Evolutionary interpretations of schizophrenia are plagued by many of the same problems that trouble prevailing model interpretations: Schizophrenia is a loosely defined disorder, and current diagnostic systems produce data of questionable reliability and validity (Costello, 1993b). Nevertheless, evolutionary insights are informative, particularly with respect to information processing, traits, infrastructural function, and strategy features that are common to the multiple forms of the disorder.

Crow (1995) suggested that schizophrenia is a genetic anomaly of relatively late evolutionary origin, and it is the disruptive effects of the anomaly that are postulated to account for many of the disorder-typical information-processing distortions. Although Crow was not specific about the dates at which the anomaly might have occurred, as noted in chapter 3 the relatively uniform distribution of schizophrenia throughout the world suggests that an anomaly would have had to occur prior to 100,000 to 150,000 years ago, when our ancestors are assumed to have been located primarily in Central Africa, and when the last major migratory exodus of *Homo sapiens* from Africa is estimated to have begun. Cosmides and Tooby (1995) postulated that schizophrenia is characterized by compromised capacities to model others' minds. While this hypothesis could in part explain misinterpretations of others' behavior, misreading others' minds is most likely a secondary feature due to recognition distortions. Stevens and Price (1996) viewed schizophrenia as a consequence of selection favoring genes dealing with rapid group splitting, rapid group elimination, selection between groups, and externally mediated sexual selection, which leads to certain individuals' (e.g., charismatic leaders) being more likely to develop features of psychosis and to leave their group. This view would be more compelling if the negative effects of the disorder were less apparent and less destructive of coping capacities.

The preceding theories notwithstanding, in our view the features of the disorder most in need of explanation are those implicated in the enduring suboptimal information processing that is a hallmark of this disorder. A sampling of findings pointing to severely compromised information processing includes slow initial responses to stimuli, attenuated responses to stimuli once they are identified, episodic interruptions in carrying out complex tasks, poor selective attention, slow shifts in attention, and slow and repetitive motor responses (e.g., Cleghorn and Albert, 1990). These are largely measures of suboptimal cognition. Their effects are shown in Table 4.3, which includes the functional categories of social exchange, social manipulation, social maintenance, social understanding, and information processing, each of which is compromised in schizophrenia, and each of which has multiple information, social cognition, and recognition distortion components, in particular, errors in encoding information and signaling. Among persons with chronic forms of the disorder, both cognitive and functional indices tend to endure and significantly reduce plasticity. In the models developed in chapters 4 and 6, these kinds of errors are thought to result from deficit conditions and breakdowns in infrastructural boundaries: Because of missing connections, insufficient information is available for processing, and because of porous infrastructural boundaries, information is not processed primarily by one automatic system but extends to (intrudes into) other infrastructures as well as awareness (much as happens when one is intoxicated). The effects of these events include distortions of external and internal information, confusion, hallucinations, and delusions. In turn, algorithms are less able to make sense of available information or to translate information into goal-related actions.

There is an alternative theory of hallucinations that merits discussion. Glenberg (forthcoming) has postulated that the environment is a primary source of hallucinations unless a selective filtering of environmental information occurs: Selective filtering is essential if one is to avoid being overwhelmed by the nearly infinite number of external stimuli. In effect, focusing attention serves to reduce environmental information and in turn facilitates the organization and prioritization of information. Figure 4.2 is consistent with this hypothesis: Atypical behavior in patients declined as the number of people in a social area increased (more people require more focused attention). Thus, insufficient filtering, which may be a feature of hyperactive, missing, or inhibited connections (Figures 6.1B, 6.1C, and 6.1D), may also be the basis of some of the information-related features of the disorder.

All four of these processes (subvocalization, deficit conditions, porous infrastructural boundaries, and insufficient filtering of environmental stimuli) may be present, although our view is that evidence favoring the latter three is more compelling from a causal perspective. Many persons in stable, nonchanging, low-stimulus environments (individuals alone in their rooms) often actively hallucinate. Deficit conditions and porous infrastructural boundaries best explain these kinds of hallucinations. And prominent features of all forms of the disorder are that individuals isolate themselves socially, resist environmental change, and often deteriorate clinically during periods of environmental alteration. These behaviors implicate overinterpreted environmental stimuli. Depending on the clinical outcome, different types or degrees of compromised infrastructures are predicted. Those instances of the disorder that have the best prognosis point to transient infrastructural disorganization (a period of inhibited and unstable connections), the lingering effects representing a partial continuation of these features. In chronic forms of the disorder with negative symptoms, multiple deficit conditions are probable. These predictions fit with the idea that positive symptoms represent an excess of normal functioning, and negative symptoms represent a diminution or absence of normal functioning (*DSM-IV*).

In our view, the many compromised functional domains that usually make up the disorder strongly point to multicause explanations. Hormonal changes are, of course, implicated in any disorder that begins between the ages of 12 and 20. However, the important events of adolescence are not limited to hormonal changes. There are challenges to both the reproductive and survival systems. Adolescence is a period in which one moves away from parental protection, and for some individuals, moving away may have survival implications. It is also a period in which one begins seriously to seek mates. Thus, in theory, dysfunctionality of one or both of these behavior systems may be a contributing factor. Epigenetic factors, particularly the failure to integrate the functional requirements of infrastructural systems, may also be critical. However, the low percentage of persons developing the disorder during preadolescence argues against this interpretation.

The causal explanation we favor for the majority of instances of the disorder is the multiple-suboptimal-trait hypothesis. Recall Figure 9.2, in which there are A-, B-, and C-type traits, and the A- and C-types are not considered either frequent or significant disorder-contributing traits. This leaves B-type traits, which have the greatest cross-person, within-trait variance. Both chance and genetic loading for specific suboptimal traits may lead to a high percentage of severely suboptimal B-type traits and, in turn, explain disorders such as schizophrenia and schizoid and schizotypal personality disorders, which share features. In this model, genetic loading is not specific for schizophrenia, but for suboptimal traits, and depending on the type of loading and subsequent events, different conditions occur.

The multiple-suboptimal-trait explanation has both weak and strong forms. The weak form states that chance genetic mixing at conception results in suboptimal traits associated primarily with information processing. This explanation is most applicable to situations where pedigree data for the disorder are absent, in other words, sporadic cases. The strong form acknowledges the presence of pedigree data and genetic loading for both suboptimal information processing and other traits. Between these two extremes is a host of possible trait clusters, the expression of which can be influenced by intergenetic interactions, different rates of trait maturation, and developmental events that have different effects on genetic expression.

The possibility that the many forms of schizophrenia reflect clusters of suboptimal traits with overlapping phenotypic features is consistent with findings from two recent studies:

Even though the schizophrenics as a group showed an equivalent level of deficit across all tests composites, *1) the deficits were associated with different aspects of psychiatric symptomatology; 2) the motor deficit was independent of the cognitive deficits; and 3) each neuropsychological domain contributed independently to the deficit pattern.* Thus, what appears to be a generalized functional deficit in Schizophrenia may actually be at least in part, combinations of multiple specific deficits. (Sullivan et al., 1994, p. 641; italics added; see also Woods et al., 1996)

The suggestion of independent contributions from different neuropsychological domains is similar to what is reported in factor analysis assessments of clusters of positive and negative symptoms. For example, Peralta, Cuesta, and deLeon (1994) found that the best fit of clinical findings is with a four-syndrome model that includes positive, disorganized, negative, and relational dimensions, each with different cluster profiles (see also Arndt, Alliger, Andreasen, 1991). Similarly, EEG findings suggest the presence of multiple disorder subgroups (John et al., 1994), and ethological studies of drug-free schizophrenia subjects show deficits in nonverbal behavior among only a percentage of subjects (Troisi et al., 1991). The multiple-suboptimal-trait view is also implicated in two other important findings: (1) Signs and symptoms closely correlate not with functional capacities, but with measures of community functioning (Bellack et al., 1990), and (2) in twin studies, both afflicted and nonafflicted twins may have very different social capacities (Dworkin et al., 1988). In effect, several functional processes appear to be involved in the development of schizophrenia, an idea set forth by Strauss and his colleagues (1978) two decades ago.

Functional interpretations of schizophrenia emphasize that much of the behavior reflects strategies to avoid painful and costly social contacts because of recognition distortions associated primarily with the survival system, and also, but less, with the reproductive system. In principle, this idea is similar to the interpretation of infantile autism discussed earlier: The avoidance behavior characteristic of autism, particularly reduced eye contact, was interpreted as an attempt to avoid what autistic individuals perceive as aggressive behavior by others. Far more is involved in schizophrenia, however. It is easy to understand how individuals would be inclined to try to decouple from a world that they experience as unpredictable and in which they consistently fail to achieve their biological goals. It is equally easy to understand how the world could be viewed as hostile and depriving, and as one in which the expenditure of resources in social activities promises minimal benefits. Thus, social withdrawal may reflect a strategy for exiting from a perceived dangerous or lethal environment (Sorensen and Randrup, 1986), of which a high degree of uncertainty is a critical feature. (Persons without the disorder may view such environments as neutral or supportive.) Moreover, the creation of a private inner world (e.g., hallucinations, which have the effect of drowning out other information) makes sense. One develops a world that one controls, although not necessarily a world that one likes. Differences in the clinical manifestations of the condition may in part reflect attempts to create such worlds in individually different ways (Dittmann and Schuttler, 1990).

There are several difficulties in the multiple-suboptimal-trait hypothesis, the most obvious being the need to explain the postulated capacities to execute strategies despite the presence of severely suboptimal infrastructures. This apparent paradox is partially solved by a recognition that there are different types of strategies and that withdrawal is among the most primitive. It is, for example, the strategy adopted by infant primates when they are separated from their mothers for extended periods of time, and also by many dogs when their owners depart. In both examples, withdrawal appears to be a favored strategy when the pain of existence exceeds the effort required to gain benefits.

There is one additional causal model worth considering. Its best explanation comes from nonhuman primate data, which require that we address some general features of higher nonhuman primate social behavior: (1) The behavior of animals in social groups is *highly responsive* to signals by other animals; that is, the behavior of individual animals is under a high degree of social control; (2) social interactions normalize behavior among participants; (3) animals rarely live alone, and during any given day, they spend relatively few moments by themselves; (4) animals seldom migrate from their groups except at certain periods of development, at which time all animals of the same sex (usually males) migrate; and (5) nonhuman primates rarely exhibit behavior that closely models schizophrenia.

Humans are under far less social control than nonhuman primates: They are often members of several groups; they may change groups frequently; social interactions often do not normalize behavior; and most important, humans, unlike nonhuman primates, withdraw from groups and often lead solitary lives. Two apparent trade-offs in the evolution of *Homo sapiens* are (1) increased requirements for initiating and carrying out novel goal-achieving strategies and (2) increased requirements for managing one's own behavior on the basis of self-monitored information. If persons are minimally capable of developing and maintaining social relationships and achieving goals, as is the case when one is afflicted with multiple suboptimal traits, withdrawal from social systems that are based on these requirements is likely. This model thus postulates that in humans, the release from social control leads to the increased probability that social withdrawal will be utilized when one's world is perceived as depriving, hostile, and uncertain.

The multiple-suboptimal-trait model does not imply that persons with enduring forms of schizophrenia are attempting to adapt to the social environment within the constraints of their condition. Instead, their behavior is more consistent with the view that they attempt to forgo social participation and actively engage in creating a private world. Their signals reflect this interpretation: Unlike those suffering from anxiety, depression, and some personality disorders, whose signals invite the help and participation of others, persons with chronic forms of schizophrenia signal their desire to be left alone and to forgo social interactions. (It is the social withdrawal feature that significantly contributes to the difficulties encountered in the social treatment of schizophrenia.) Further, as noted in chapter 4, this model does not require a reduction in motivations-goals. If motivations-goals were not present and others were not essential to fulfill those goals, the environment would be less depriving.

The preceding discussion suggests several points. When viewed as a whole, the cluster of disorders classified as schizophrenia are minimally adaptive for achieving short-term goals, although, as noted in chapter 2, the impact of the disorder on reproductive success still needs to be determined. In our view, suboptimality of multiple information-processing infrastructures is the explanation most consistent with current findings. The causes of suboptimal traits are likely to be multiple, but the net effect is that there are information-processing deficits, much as described in Figures 6.1B and 6.2B (missing connections). Capacities to develop primitive strategies appear to remain intact, however.

Concluding Comments

This chapter has introduced a number of evolutionary views of schizophrenia. While our analysis is neither exhaustive nor conclusive, it illustrates how evolutionary concepts can alter how existing research data are interpreted and provides new insights into the possible causes and functions of disorders.

12

Phobias

Phobias are intense and persistent fears associated with significant and functionally debilitating anxiety and behavior designed to avoid the perceived source of the fears. They invite an evolutionary analysis.

In *DSM-IV* (APA, 1994), phobias comprise a subgroup of disorders within the general category of anxiety disorders, different types of phobias occurring at different frequencies both within the population as a whole and among males and females. Specific phobias, which are intense fears of animals and special events, have a lifetime population prevalence rate in the range of 10%, with the following male-female differences: 75% to 95% of animal or natural environment phobias are diagnosed in females; similar percentages apply to situational phobias, such as fear of machines; and females account for 55% to 75% of blood or injection phobias (*DSM-IV*). Social phobias, which are persistent fears of embarrassment in social or performance situations, have a lifetime prevalence rate ranging between 3% and 13%, and they occur with approximately the same frequency in males and females (*DSM-IV*). Panic with agoraphobia has a lifetime prevalence rate of 1.5% to 3.5%, and the frequency in females exceeds that in males (*DSM-IV*). Agoraphobia without panic attack, which accounts for approximately 5% of the individuals with agoraphobia, is also observed more frequently in females (*DSM-IV*).

When all types of phobias are lumped together, findings indicate that the incidence is highest during the key reproductive years, beginning with adolescence and extending to the mid-30s. A variety of clinical patterns are observed, including continuous, episodic, and once-only or transient patterns. Naturalistic studies covering periods from 6 to 10 years following initial diagnosis indicate that 30% of persons are well and asymptomatic, 40% to 50% remain somewhat impaired, and 20% to 30% have not changed or are slightly worse (*DSM-IV*). The many physiological and cognitive features characteristic of phobias have been extensively reviewed by Marks (1987; see also Nesse, 1990a). Physiological and cognitive changes are expected, given the often intense somatic, perceptual, and behavioral effects of this group of disorders.

Both genetic and cultural contributions have been implicated as causal factors. Genetic contributions are suggested by findings showing that relatives of probands are four to seven times more likely to develop a phobia than relatives of persons without phobias, and monozygotic twin studies show a higher disorder concordance rate than would be expected by chance. The fact that in some societies persons are phobic about witchcraft and magic, while in other societies they are phobic about other things, implicates culture in the content of phobic fears, perhaps also in prevalence rates: Persistently hostile or frightening environments can lead to excessive fears. Studies of developmental contributions have been inconclusive, except for the earlier-mentioned examples of childhood phobias (chapter 3) concerning the dark and being alone, for which a strong case for ultimate causation can be made.

Evolutionary Analysis

Like anorexia nervosa, phobias are consistent with evolutionary models, and they are most parsimoniously understood as either adaptive responses or exaggerated, unremitting, and minimally adaptive forms of such responses. Within the circle of evolutionary-oriented investigators, Marks and Nesse have said the most about phobias. Excerpts from their works identify and develop key points:

> It is easy to think of fears that enhanced survival in the past (say, fear of animals) or continue to do so in the present (fears of heights, separation, and perhaps strangers). This idea is intrinsic to more recent and related concepts of prepotency . . . and preparedness . . . Prepotency indicates that particular stimuli are salient for a given species, which attends selectively to them rather than to others even at their first encounter. Preparedness is the idea that certain stimuli associate selectively with one another and with particular responses, some connections being more available than others. (Marks, 1987, p. 230)

> The cues that most often elicit panic are those associated with increased risk of attack . . . People who repeatedly experience panic develop agoraphobia, a remarkably consistent syndrome that includes fears of specific cues: wide open spaces, close in spaces, places where intense fear has occurred before, and being far from home, especially if unaccompanied by a trusted relative. These characteristic agoraphobic fears are well suited to avoiding attack in a dangerous environment . . . a person who lacks the tendencies to panic in the face of danger and to experience agoraphobic fears in dangerous situations will, in a natural environment, be at a selective disadvantage. (Nesse, 1990b, p. 271)

Two ideas central to the evolutionary interpretation of phobias are present in the preceding quotations: (1) Individuals enter the world with predisposed tendencies to fear situations that are or may be dangerous ("particular stimuli are salient for a given species"); thus, the presence of transient phobias is expected; and (2) a failure to be fearful and cautious is often associated with a selective disadvantage. Elsewhere, Nesse (personal communication, 1994) has noted that the cost of an exaggerated defensive response to either dangerous or potentially dangerous situations represents a small cost compared to the possible consequences of not responding. This idea is in agreement with the view that humans are preprepared and strongly biased to act defensively to threats to any of the four behavior systems, but its primary focus is on the survival system as the mediator of both the emotional responses and the recognition distortions associated with phobias. Counterbalancing the potential adaptiveness of overresponding is the presence of recognition distortions that are signatures of intense and enduring phobias that often lead to a failure to distinguish between "legitimate fear" and misinterpretations of stimuli. The fact that intense and enduring phobias usually result in significant reductions in behavioral plasticity might well be added to the points above developed by Marks and Nesse: Such phobias are often associated with freezing or fleeing, neither of which may be adaptive, and individuals often go to unusual lengths to avoid the conditions that trigger anxiety and fear.

Viewing phobias as evolved defensive responses to specific contingencies implies that they can be adaptive. While this is an acceptable interpretation of transient phobias, it is not so easily reconciled with enduring and debilitating phobias. Thus, an obvious question is: When might phobias become counteradaptive? In principle, this question is easily answered: They become counteradaptive when they consistently compromise biological goal achievement. For example, in situations such as social phobia, where an individual avoids social encounters and fails to develop social support networks or to engage in potentially beneficial reciprocal behaviors, the potential negative effects are apparent. However, when an individual is phobic about contingencies that are statistically associated with risks, such as heights or predators, the assessment is less straightforward. Generally, if a person goes about mastering risky situations, such as learning the habits of predators or taking adequate precautions with heights, and following such efforts, the anticipation of a phobic response declines or phobias become less intense, the phobia is not counteradaptive. On the other hand, if mastery fails to occur, the condition may be counterproductive.

For all types of phobias, whether transient and mild, or extended and debilitating, resource allocation programs and motivations-goals are assumed to be normal. As we proposed for personality disorders and anorexia nervosa, both the normal allocation of resources and normal biological goals-motivations are thought to be necessary conditions for these disorders. Fears of heights, predators, machines, and social ostracism imply not only that the motivations-goals associated with the survival and other behavior systems are intact, but also that their intactness is essential to explaining fears and anxiety. Cross-person within-trait variation, particularly of recognition capacities, introduces another factor, however. The variety of responses to specific stimuli (e.g., heights, snakes, blood) or situations (e.g., social events), which range from indifference to panic, suggests four important points: (1) Persons differ significantly in the degree to which particular stimuli are salient; (2) the degree of recognition distortions also differs significantly; (3) in enduring and debilitating phobias, suboptimal infrastructures are likely; and (4) for the preceding three points, the 15% principle is applicable to causal explanations.

The consequences of recognition distortions are well illustrated by a comparison of phobias that are transient with those that are enduring. Transient phobias implicate infrastructural dysfunctionality primarily of the survival automatic system. Phobias that grossly exceed the actual danger in the environment and fail to remit implicate suboptimal infrastructures. Both are defensive responses, and both are initiated by external stimuli. However, their clinical courses differ significantly. These differences are explained in part by the observation that transient phobias are seldom associated with the perception that the environment as a whole is hostile, only that the perceived source of one's fears is hostile or dangerous. On the other hand, enduring phobias are often associated with the view of a generally hostile environment that extends well beyond the specific stimuli to which individuals respond with fear. How such views develop may be partly an outcome of upbringing experiences and the information that prevails in one's microculture, much as fears of witchcraft and magic or of all snakes (as distinct from poisonous snakes), are influenced by experience and one's microculture. Further, individuals who experience transient phobias are able to discontinue their anxiety after the stimulus disappears or is understood. In contrast, anxiety-discontinuing capacities are severely compromised in persons with enduring phobias. In transient phobias, discontinuation suggests that the effects of dysfunctional automatic systems (e.g., inhibited connections) are constrained in part by other infrastructures and are eventually normalized by information-organizing and prioritizing revisions, and by changes in causal modeling. The lack of resolution in enduring phobias is best explained by missing and hyperactive connections (Figures 6.1B and 6.1D). Individuals with enduring phobias often adopt strategies of social withdrawal, while those with transient phobias remain active participants in their social world.

Phobias may thus be viewed as informational strategies designed to isolate undesirable external information, that is, to control perceived traumatic situations among persons who are unable to prevent stimulus-related information from reaching awareness and influencing emotions. These strategies may be contrasted with those observed in anorexia nervosa and amnesia (chapter 13), where potentially traumatic information is decoupled from awareness.

A related point concerns the potency of the phobic signals. The intense and usually obvious emotional responses associated with phobias (e.g., sweating, behavioral freezing, anxiety) have a predictable impact on others' behavior and lead to the kinds of responses noted in chapter 5 in the discussion of intense anxiety: Others alter their behavior, provide help, and reduce their expectations concerning certain responsibilities. Such signals are best understood as evolved strategies to obtain help in frightening situations, and their presence implies attempts to adapt within constraints.

Differences in the sex-related prevalence rates of phobias are predicted in evolutionary models. For both males and females, there are obvious advantages in caution and vigilance. For females, these advantages are most closely associated with reproduction and survival. Females need to protect themselves from assault and rape, as well as to remain physically safe during pregnancy and to act in ways that optimize the safety and rearing of their offspring. Thus, females are likely to be especially sensitive to situations that have potentially negative reproductive or survival consequences for both themselves and their offspring. Further, it is likely that females, more than males, have evolved to attend to detail, to be more discriminating in their assessments of the social and physical environments, and thus to be more aware of real or potential dangers. An alternative possibility is that selection has favored a lower threshold for threatening information in females, in effect, a more sensitive flight or avoidance system. In light of either of these possibilities, phobias among females would be expected to be most common during the reproductive years, and this expectation has proved to be true.

For males, there are also advantages in being vigilant and cautious. These behaviors should be most closely associated with remaining safe from possible attacks by predators or competing males, reducing paternity uncertainty, and attaining and defending high social status. However, in males, vigilance for potential environmental dangers (e.g., predators, male competitors) must be counterbalanced with the requirements of mastering and using the environment to access resources and acquire mates. The evolutionary trade-offs required to balance these two needs may be the basis for the greater degree of risk-related self-denial, as well as the higher fear threshold for dangerous situations, that is observed in males, and the extreme risktaking by young males, which applies to approximately 12% of adolescent males, would represent a high-risk variant of such trade-offs. A higher threshold would also partially explain the reported sex differences in the prevalence of phobias.

The preceding interpretations are compatible with models developed in the preceding chapters; for example, female identity is more closely tied to reproduction than is male identity, and (on average) the reproductive consequences for females of environmental dangers exceed the consequences for males (chapter 7). For example, the possibility of rape has both reproductive and survival implications, as does the safety of one's offspring. The potential involvement of two automatic systems leads to several predictions: Phobic responses will be more complex in females than in males, they will be more enduring, and they will be more constraining behaviorally.

Compromised algorithms are also an integral part of phobias, especially in enduring phobias, and the algorithms most frequently involved are causal modeling, scenario development, and self-monitoring. For example, individuals with enduring phobias are often aware that their fears are excessive. Yet, in a ways analogous to being unable to alter how one views visual illusions of which one is aware, recognition of the excessive nature of their response to specific stimuli does not alter either their perceptions of the potential danger of stimuli or their emotional response. When excessive fears turn out to be unfounded and are not coupled with response changes over time, compromised causal modeling is likely, and it may be due either to strongly inhibited or missing connections (Figures 6.1B and 6.1C). Compromised scenario development is suggested by the fact that persons with phobias are usually unable to imagine behavioral strategies other than avoidance and withdrawal, which, as noted in the earlier discussion of schizophrenia, are among the most primitive of responses. Self-monitoring is implicated because, when expected events fail to occur, neither causal modeling nor the emotional responses to specific stimuli are altered.

Concluding Comments

We have reviewed some of the ways in which the interpretation of phobias in an evolutionary context differs from the interpretations developed by prevailing models (e.g., an overlearned response, dysfunctionality of CNS receptors). Many phobias turn out to be normal and adaptive, or if extended and debilitating, they are examples of a normal and adaptive response gone wrong because of reduced capacities to process information accurately, to correct recognition distortions, and to discontinue emotional responses once they are initiated.

13

Other Conditions

Conditions other than those already discussed can also be illuminated by an evolutionary perspective. In this chapter, the perspective serves as the basis for assessing dissociative amnesia, adjustment disorder, two somatoform disorders, alcohol dependence, suicide, spousal abuse, and child abuse.

Dissociative Amnesia

The essential feature of dissociative amnesia (the one dissociative disorder to be discussed) is a disruption in the usually integrated functions of consciousness, memory, identity, or perception of the environment. The disruption may be sudden or gradual, transient or chronic (*DSM-IV*, APA, 1994, p. 477). Diagnosis requires the inability to recall important personal information that is usually of a traumatic or stressful nature, which exceeds ordinary forgetting, and the symptoms must cause clinically significant distress or impairment in social, occupational, or other important areas of functioning (*DSM-IV*). The fact that disruptions of consciousness can be either specific (e.g., loss of memory of a specific past event or period of time) or general (e.g., loss of memory of longer periods, confusion about one's identity, failure to recognize familiar environments) should be added to the preceding description: Type and degree of disruption interact with clinical outcome.

Amnesia has been diagnosed from preadolescence through old age. It is not known if the disorder occurs more frequently among females or males, and increased prevalence rates during specific age periods have not been reported. The greater than chance association of amnesia with depression, trance states, analgesia, and spontaneous age regression (*DSM-IV*) suggests that genetic information influences capacities to tolerate and manage traumatic and stressful events.

While most clinicians believe that the disorder occurs infrequently (the lifetime prevalence rate is estimated to be in the range of 0.5%), during the last decade there has been considerable disagreement over the prevalence of repressed traumatic memories of childhood, which, in *DSM-IV*, are viewed as a type of dissociative amnesia. Whether most reported instances of these memories (usually of sexual abuse) are accurate depictions of past events or products of suggestion remains unclear. The fact that many individuals who recall traumatic memories (usually in psychotherapy) also score high on measures of hypnotizability and dissociative capacity (*DSM-IV*) raises an obvious caution flag about uncritically believing memories of supposedly forgotten past events without corroborating evidence. The preceding points aside, the current interest in repressed memories represents a highly focused area of research and therapy, largely unrelated to the loss of identity or of the memory of recent past events observed in adults. It is these disruptions of consciousness that were the basis for most amnesia-related reports prior to a decade ago, and it is these types of disruptions on which we will focus.

Evolutionary Analysis

There are interesting parallels between dissociative amnesia and adjustment disorder (discussed later in this chapter): (1) Both disorders are thought to represent responses to traumatic or stressful events; (2) neither is associated with high prevalence rates during the key reproductive years; and (3) in the majority of instances, both are time-limited. Because of the importance attributed to stress and trauma in the prevailing model explanations of these disorders, it is worth briefly reviewing what is usually implied by these terms.

Among clinicians, the term *stress* usually refers to repeated, moderately aversive events, the effects of which may be additive. Eventually, additive effects pass a *threshold* and trigger psychological and physiological change and sometimes the development of disorders (e.g., Goldberg and Breznitz, 1982). This formulation of stress is similar to the model of regulation-dysregulation (RDT) developed in chapter 8: Cumulative aversive events eventually result in physiological and psychological dysregulation and, in turn, in condition-related symptoms and signs. The term *trauma* usually refers to a single catastrophic event, such as the unexpected loss of a significant other or of one's employment. Like stress, traumatic events can have significant physiological and psychological effects, but the effects occur much more rapidly and may be of a different nature. There is, however, no formal distinction between the two terms, except perhaps in the extreme, and even then, the terms require clarification. *Stress* is sometimes used to describe single events, such as the loss of a loved one, and *trauma* is sometimes used to describe repeated events. In short, common usage overlaps, and the definitions are porous. The overlap reflects in part the difficulties clinicians and investigators experience in identifying and studying stress and trauma, as well as their effects.

We will use the terms as follows. Stressors are events that are perceived to be moderately adverse, they can have additive somatic and psychological effects, and their effects are modeled by RDT. Thus, the onset of responses to stressful events is gradual, and the effects become increasingly obvious over time. To the degree that stressful events persist, dysregulation is also likely to develop. Traumas are events that are perceived as seriously threatening one's high-priority goals, such as the loss of offspring or the total destruction of one's material assets. Traumas are viewed not as accelerated instances of stress, but as experiences in which important goals are perceived as suddenly becoming improbable. This perception is not necessarily a feature of stress. This distinction provides some clues to why the memory and identity effects differ in the two disorders.

Clinically, events themselves seldom predict how individuals will respond. Clinicians are aware of this fact, as well as of the fact that different individuals respond differently to what are logically equivalent events (e.g., Berger et al., 1987), such as the death of a spouse. They are also aware that many events to which persons respond seem less stressful or traumatic than the responses suggest. *Response types do interact with motivational-goal priorities, however.* For example, the death of an unknown child has only moderate disruptive effects, while the death of one's offspring may be devastating. Said differently, the external events that usually qualify as traumas or stressors are not so much specific events or event features, except perhaps in the extreme (e.g., extended sensory deprivation, repeated physical abuse), as they are events that have a special meaning to the person who responds. Thus, rather than looking to events to predict specific outcomes (a common epidemiological and life-event research strategy), a more instructive approach is recognizing that (1) external events are differentially adverse across individuals; (2) individual differences influence how persons respond to events (e.g., in amnesia, events are temporarily amplified and then erased from memory, while in adjustment disorder, they are amplified and retained in memory); and (3) an individual's infrastructural capacities, as well as his or her degree of dysregulation, are likely to be key factors in determining response type. Thus, a person's response to stressful and traumatic events should reflect interactions between goal priorities, infrastructural functionality, degree of dysregulation, and the environment.

For amnesia, atypical resource allocations and motivations-goals are not predicted. In the models developed here, normal motivations-goals and allocations are required to explain the disorder; that is, amnesia rarely occurs unless high-priority goals are perceived as significantly jeopardized. Further, the absence of evidence pointing to an increased prevalence of the disorder during the key reproductive years suggests that amnesia differs in important ways from a number of disorders that occur most frequently during the years between adolescence and the mid-30s (e.g., anorexia nervosa and phobias). Compromised infrastructures are assumed to be present in most instances of the disorder, and clinical experience suggests that any of the behavioral systems may be affected. Amnesia may follow physically frightening experiences (survival behavior system), the sudden loss of kin (kin-investment behavior system), rape or the death of an offspring (reproductive behavior system), or abrupt social ostracism (non-kin-reciprocity behavior system).

When adults develop disruptions of their consciousness of recent events, suboptimal automatic systems are implicated. The *initial* effects of traumas are most parsimoniously explained as a response to massive information overload affecting one or more automatic systems. Hyperactive connections (Figure 6.1D), which amplify both the features of an event and its perceived consequences, are the postulated basis of such overloads, as well as the sense that information is chaotic. In effect, the usual auto-matic system functions of filtering, organizing, and prioritizing information break down. Once overload begins, there is a rapid and significant increase in inhibited connections (Figure 6.1C), which minimize the amount of information available to awareness and lead to memory and identity loss. Inhibited connections rather than missing connections are likely because in most instances, the disorder remits, and persons return to their preamnestic state (which may be a suboptimal state). It follows that the more optimal an individual's automatic systems are at the time of trauma, the less likely it is that the individual will experience a severe case of amnesia, and the more likely it is that events that occurred prior to and during the period of amnesia will eventually be recalled.

In mild cases of amnesia, algorithms may not be dysfunctional; however, traumatic information may be unavailable because of the effects of inhibited connections. Fur-ther, the gradual recall of traumatic information, which is characteristic of the resolu-tion phase of amnesia, suggests that remembered information gradually becomes available for algorithm processing, rather than algorithms' improving in their process-ing capacities. The postulate that only a subset of infrastructures is compromised is further supported by the observation that individuals with amnesia are often capable of carrying out functions that are usually assumed to require access to specific memo-ries or behavioral strategies that, during amnestic periods, are unavailable to aware-ness; for example, persons with amnesia often don't forget how to dress, drive a car, or make a cheese sandwich.

Can the disorder be adaptive? Mild instances of amnesia can be understood as attempts to adapt within constraints. In terms of the strategies discussed in chapter 9, the disorder is primarily an informational strategy, not a somatic or a functional strat-egy. It is a strategy that leads to the decoupling of both disconcerting (e.g., goal-compromising) information and structures, while potentially preserving the function of nonafflicted structures. Earlier, psychic defenses and compartmentalized delusions were explained in a similar way (chapter 5). The disorder does not appear to be a selected high-risk strategy, as was postulated for ADHD and histrionic and antisocial personality disorders (chapter 9). Further, in mild forms, amnesia elicits help from others. Thus, it may partly reflect a strategy to signal one's disrupted state. Although more severe and disruptive forms of amnesia may represent signals, they are likely to be minimally adaptive because of their significant functional consequences.

Adjustment Disorder

According to *DSM-IV* (APA, 1994), "The essential feature of an Adjustment Disorder is the development of clinically significant emotional or behavioral symptoms in re-sponse to an identifiable psychosocial stressor or stressors" (p. 623). The diagnosis requires a state of distress that exceeds what would be expected from exposure to the stressor and significant impairment in social or occupational functioning. In *DSM-IV*, multiple subtypes of the disorder are identified (e.g., adjustment disorder with anxious mood, adjustment disorder with disturbance of conduct), suggesting once again that the prevailing method of classifying disorders fosters the grouping of multiple over-lapping trait clusters, with different causal profiles (the 15% principle), into single diagnostic categories. According to *DSM-IV,* the disorder may begin up to three months following the stressor(s) and last for six months following the cessation of the stressor(s). Disorder types range from specific to disorganized, and the duration may be brief to extended. Recurrent cases have been reported. Prevalence estimates range from 5% to 20% of the population, percentages that suggests that adjustment disorder is one of the more frequently diagnosed disorders. The available data do not point to different prevalence rates among males and females or to increased frequencies at specific ages. The monozygotic twin data and the pedigree data are inconclusive.

Prevailing causal explanations emphasize both the nature of the stressors and excessive vulnerability to stress, and epidemiologists report a greater prevalence of the disorder among individuals from disadvantaged life circumstances (*DSM-IV*). Psychoanalysts explain vulnerability as being a consequence of disrupted relationships during development, particularly with one's mother. These formulations suggest that the disorder might not occur if one grew up in a certain type of social environment and that the disorder may be attributable largely to environmental perturbations. Both possibilities seem improbable if taken as single-cause explanations, primarily because the effects of adverse environments are not uniform. Thus, the place of genetic information in causal formulations remains to be determined.

Evolutionary Analysis

As noted in earlier chapters, evolutionary analyses permit hypotheses about the types of events that should trigger responses when age, sex, and goal priorities are included in the analyses. Fear of darkness among children, competitive losses among males, reproductive failure and lack of kin in whom to invest among reproductive-age females, and social ostracism—all qualify. The fact that these events often (but far from always) precipitate responses explains in part why individuals follow strategies to locate themselves in environments that they associate with a reduced probability of such events (chapter 8). Yet, as noted in the preceding discussion of trauma and stress, the environment alone is only one condition-contributing factor. It follows that, with the possible exceptions of the period between infancy and adolescence (when one's control over one's environment is often constrained by one's parents) and catastrophic traumatic events, *relationships between individuals and stress are best viewed as bidirectional*: Unless we consider the possibility that individuals contribute to their own responses to external events, most findings from stress and trauma research are of questionable value for improving our understanding of conditions. If we apply these points to adjustment disorders, the kinds of repetitive stressful events that would be predicted to increase the incidence of the disorder are social environments that initiate a three-step process: (1) in vulnerable individuals, they lead to dysregulation and symptoms; (2) in turn, dysregulation reduces the chances of achieving biological goals; and (3) reduced goal achievement intensifies symptoms and hastens disorder onset. If negative environmental features are unremitting, the disorder is likely to endure.

Motivations-goals and resource allocations are assumed to be normal in the disorder; otherwise, it is difficult to explain responses that are best understood as negative reactions to environmental features. Compromised infrastructural functionality is usually present, and different types and degrees of compromise influence the form and intensity of the disorder. Because prevalence estimates for males and females do not correlate with age, compromises of any of the four behavioral systems are likely. The type of the disorder is also important. For example, if a response is focused on a specific stressor, then adjustment disorders with anxious or depressed mood are likely. Infrastructural dysfunctionality, not necessarily suboptimality, is implicated in such responses, and mood states are interpreted as reflecting an inability to favorably alter environmental contingencies. On the other hand, if the adjustment response is not focused, adjustment disorder with mixed disturbance of emotions and conduct is more likely, and infrastructural suboptimality characterized by hyperactive and missing connections, as well as recognition distortions, is probable. Thus, trait differences are implicated in the different forms of the disorder. Unlike in amnesia, decoupling is not a prominent feature, a circumstance suggesting that automatic systems can process (and distort) the available information.

An assessment of algorithm capacities is critical to understanding disorders to which environmental events are contributing factors: Optimal algorithm functionality is associated with an increased probability of avoiding adverse environments or reducing their dysregulating effects, while suboptimal capacities are associated with the opposite outcomes. In adjustment disorder, causal modeling, scenario development, and behavioral strategy execution are the most frequently compromised algorithms, with development being perhaps the most severely compromised. Individuals tend to focus primarily on how their environments should change, while ignoring the possibility that those environments might change, or that they might enter other social environments.

Individuals with different types of adjustment disorder do not withdraw from social participation but remain in their environments and interact—the *locked-in* effect. Thus, most types of the disorder appear to represent attempts to adapt within constraints, although the possibility that the disorder is a selected high-risk strategy remains a possibility. Of the three strategy types discussed in chapter 9 (somatic, informational, and functional), adjustment disorder is primarily a functional strategy. Moreover, disorder manifestations often serve as signals of dissatisfaction, as well as attempts to alter the behavior of those perceived to be the cause of distress. In instances where one's behavior is organized and one's responses are directed at particular contingencies (e.g., parental rejection), the social environment may change. When it does, the disorder can be considered adaptive.

Somatoform Disorders

Somatoform disorders are a group of disorders in which there are physical symptoms suggesting that a person is suffering from a general medical condition but that are not satisfactorily explained by a general medical condition (*DSM-IV,* APA, 1994). Two disorders from this group will be discussed: somatization disorder and conversion disorder.

According to *DSM-IV,* somatization disorder is characterized by "physical complaints beginning before age 30 years that occur over a period of several years and result in treatment being sought or significant impairment in social, occupational, or other important areas of functioning" (p. 449). Diagnosis usually occurs before age

25, but symptoms may begin during adolescence or before. A history of multiple pain and gastrointestinal symptoms, coupled with at least one sexual and one pseudoneurological symptom, is required for diagnosis. Over 30 symptoms are associated with the disorder (e.g., abdominal pain, nausea, bloating, back pain, joint pain, trouble walking). Symptoms are not thought to be intentional. Current estimates are that this disorder occurs among 0.2% to 2% of females (*DSM-IV*) and less frequently among males. Concordance rates among monozygotic twins exceed chance expectations. Pedigree studies indicate that 10% to 20% of first-degree female relatives have similar conditions (*DSM-IV*). An increased incidence of substance abuse and antisocial personality disorder is reported among first-degree male relatives. These findings point to genetic influences and perhaps also to assortative mating. Studies indicating that the condition negatively correlates with social status implicates dysregulation as a mediator of fluctuating symptom intensity. Clinically, the disorder does not appear to remit fully, although definitive longitudinal studies remain to be done.

According to *DSM-IV*, "The essential feature of Conversion Disorder is the presence of symptoms or deficits affecting voluntary motor or sensory function that suggest a neurological or other general medical condition" (p. 452). Psychological factors, such as a lack of concern about one's symptoms (*la belle indifférence*) and suggestibility, are frequently observed. Nonintentionality with respect to symptoms, the absence of an underlying medical condition, and significant impairment in social or occupational areas because of the symptoms are required for diagnosis. The incidence of the disorder is not known. Estimates range from 11 to 300 per 100,000 persons, to reports suggesting that there is a 3% incidence rate among women entering outpatient mental health clinics (*DSM-IV*). The disorder is most frequently diagnosed between the ages of 10 and 35, less frequently in older persons (*DSM-IV*), and it is reported to occur more frequently in females than in males, with ratio estimates ranging from 2 : 1 to 10 : 1. Some studies suggest that conversion symptoms occur at greater than chance frequency among relatives of persons with the disorder (*DSM-IV*), while other studies suggest that the disorder is more common in rural populations, in individuals of lower socioeconomic status, and in individuals less knowledgeable about medical and psychological concepts (*DSM-IV*). As in somatization disorder, an increased incidence of personality disorders is reported. Further, principal diagnoses may change with age; for example, in women, conversion disorder may later manifest as somatization disorder, while in men a relationship between the disorder and subsequent antisocial personality disorder is reported. The disorder has periods of remission and exacerbation, which may continue until the fifth decade, when symptom frequency tends to trail off. Atypical cerebral blood flow patterns during symptom periods have been reported (Tiihonen et al., 1995).

For both somatization and conversion disorders, the importance of developmental influences remain to be clarified, in part because of lack of data and in part because instances of the disorders are observed in persons who have experienced warm and sensitive upbringing environments.

Evolutionary Analysis

The two disorders share a number of features: (1) Both occur more frequently in females than in males; (2) both are most prevalent during the key reproductive years;

and (3) symptoms frequently elicit caretaking behavior by others, as well as a reduction in helpers' expectations of the fulfillment of role-related responsibilities by individuals with these disorders. Atypical motivations or resource allocation programs are not predicted; otherwise, individuals would be unlikely to signal their distress to others and draw them into helping relationships. Different variations of the disorder are very likely the consequence of trait differences and the resulting trait profiles.

When these disorders occur during the prime reproductive years and are enduring, suboptimality of the reproductive behavior system is implicated, an interpretation suggested by the relatively high degree of functional impairment present during periods of disorder remission. In somatization disorder, hyperactive connections are the likely basis of both symptom amplification and recognition distortions, particularly concerning the self. Inhibited connections and tendencies to rapidly decouple somatic and cognitive states are characteristic of conversion disorder.

Algorithm suboptimality is also implicated because, as noted, most persons exhibit functional deficits during periods in which symptoms are minimal. In both disorders, suboptimal causal modeling may lead to primitive views of the body. Further, the clinical observation that scenarios are not easily translated into behavioral strategies suggests that behavioral strategy capacities are compromised: Persons with these conditions are often able to imagine alternative ways of acting yet remain unable to change their behavior.

As much as any of the disorders we have discussed or will discuss, somatization and conversion disorders implicate genetic loading for suboptimal traits rather than for specific disorders. Loading results in trait clusters and clinical profiles with some shared symptoms. The pedigree data, the diversity of signs and symptoms, the predominantly deficient features of infrastructural function, and the lack of capacities to navigate socially—all favor this interpretation. That some capacities are not refined is not precluded by this interpretation, and refinement during later decades may explain instances of gradual clinical improvement.

Neither of these disorders appears to be selected in the sense that we have suggested for some personality disorders (chapter 9); that is, they are not easily interpreted as high-risk behavioral strategies. Rather, they are better understood as attempts to adapt within constraints by individuals who are *locked in* to their social environments and who, because of suboptimal infrastructures, lack capacities to successfully navigate their social environments in ways that lead to positive cost-benefit balances (Troisi and McGuire, 1991).

Alcohol Dependence

To turn to *DSM-IV* (APA, 1994) again as a starting point, "The essential feature of Substance Dependence is a cluster of cognitive, behavioral, and physiological symptoms indicating that the individual continues use of the substance despite significant substance-related problems" (p. 176). Possible causes of cocaine and heroin dependence were discussed in chapter 5. Here, our focus is on alcohol dependence.

Epidemiological findings suggest that approximately 13% of males and 6% of females abuse alcohol at some time in their life (Kessler et al., 1994). In males, the diagnosis of alcohol dependence is most frequently made between the ages of 21 and

34, and often, there is a prior history of behavioral and school difficulties. In females, dependence develops later in life.

Several not mutually exclusive explanations have been offered for this disorder. Theories postulating genetically influenced predispositions to substance dependence build on pedigree findings that report a higher than chance prevalence of alcohol dependence and other disorders among first-degree relatives of persons who use alcohol excessively (e.g., Bohman et al., 1987; Grove et al., 1990; Kendler, Pedersen, et al., 1993b). Alcohol abuse also overlaps with a host of personality disorders, as well as mood disorders, and interactions between personality and alcohol-seeking behavior have been explained as a form of exploratory appetitive behavior (e.g., Cloninger, 1987b). Learning theorists have argued that the anxiety-reducing effects of alcohol are reinforcing, and it is likely that these effects influence both the degree and the style of alcohol consumption. Effects of alcohol on the serotonin, dopamine, and GABA systems and, in turn, possible interactions with opioid systems have also been put forth: Alcohol is thought to temporarily increase CNS serotonin function and mediate a pleasurable feeling state, possibly due to serotonin itself, possibly due to its effects on other systems such as the dopamine system and, in turn, on CNS reward systems (LeMarquand, Pihl, and Benkelfat, 1994). These findings are consonant with the view that there is a strong chemical contribution to addiction. The serotonin hypothesis is also in agreement with findings of low baseline CSF 5-HIAA levels in a percentage of persons who abuse alcohol (e.g., Virkkunen et al., 1994). Once serious addiction occurs, the pleasurable effects of alcohol are less pronounced and shift more to those of numbing psychological pain and reducing attention to social details.

Developmental, cultural, and social influences are also contributing factors. Clinically, adverse upbringing environments and alcohol abuse positively correlate, a relationship that also holds for adults in adverse environments or who experience repeated losses. Cultural contributions are suggested by the fact that different cultures have different prevalence rates of alcoholism and that short-term changes in the amount of alcohol use interact with cultural upheavals (e.g., wars, economic depressions). Social effects result from desired behavioral changes, such as inhibition release.

Evolutionary Analysis

An evolutionary analysis of alcoholism offers several key points: (1) It decreases anxiety; (2) it occurs more frequently among males than among females; (3) in males it is most frequently diagnosed during the prime reproductive and resource acquisition years; (4) it is frequently associated with a history of competitive loss or an inability to achieve social goals (e.g., poor school performance); (5) it may temporarily lead to desirable psychological-physiological states, such as brief periods of elevated self-esteem and positive self-assessment; (6) it facilitates behavior (e.g., by decreasing inhibition and increasing intimacy) that is otherwise unlikely; and (7) there is a high rate of relapse. Exceptions to these points are found in a subset of individuals in whom excessive alcohol intake is associated with an increase in aggressive behavior. When the preceding points are combined, alcohol abuse can be understood in part as a strategy to offset the undesirable consequences associated with actual or perceived failures.

While resource allocation programs and motivations-goals are assumed to be normal, automatic systems may be either suboptimal or dysfunctional, and these differences influence long-term outcomes of the disorder. The initial effects of alcohol result in both automatic system and algorithm change, and this change is signaled by alterations in one's emotional state. (Heroin and cocaine have similar automatic system effects.) In males, the higher prevalence of the disorder during the key reproductive years points to suboptimality of the reproductive behavior system: Dependence is often observed among persons who fail to acquire a mate. Other behavior systems may sometimes be involved, however, as when survival is threatened (e.g., chronic diseases), when one is ostracized by kin, and when one is lonely and unable to establish nonkin relationships. In females, behavior system dysfunctionality is less often confined to the reproductive behavior system. Women consume alcohol excessively for a number of reasons: because they are lonely (nonkin behavior system), because they have lost contact with or influence over kin (kin-investment behavior system), and because of failures to acquire mates and reproduce.

Scenario development may be minimally compromised, and novel scenarios are frequently voiced by persons who drink excessively. However, turning "healthy" scenarios into sustained action is another matter, a point suggesting that behavioral strategy development is compromised. Capacities to monitor the effects of drinking on oneself and others are often limited and refractive to change. The greater the refractiveness, the less likely the remission.

The high rate of relapse can be attributed to the combined effects of compromised infrastructures and the pleasurable and (later) numbing effects of alcohol. Fluctuations in abuse and the high relapse rate are consistent with the view that dysregulation and reduced biological goal achievement are critical factors in remission rates. The success of Alcoholics Anonymous (AA) is relevant to this formulation. Persons who join AA and discontinue drinking usually participate in AA sessions several times each week. Persons enter an empathic and positive feedback environment, one in which helping behavior is built in. Frequent interactions of this type are essential to optimize physiological and psychological regulation (see chapter 14).

Suicide

Although suicide is not listed as a disorder in *DSM-IV* (APA, 1994), it is of interest because an evolutionary analysis offers insights into its multiple causes.

For this discussion, suicide is defined as an attempt to kill oneself in a way that has a high probability of success. Epidemiological reports suggest that the lifetime rates of suicide differs across *DSM* diagnostic categories: For bipolar disorder, unipolar disorder, and all other *DSM-III*-defined Axis 1 disorders (APA, 1980), lifetime rates are reported to be 29.2%, 15.9%, and 4.2%, respectively (Chen and Dilsaver, 1996). These findings suggest that suicide and depression closely interact and that higher rates of suicide should be present in those disorders that are associated with chronic reduced goal achievement. Further, the previously noted finding that persons placed in solitary confinement attempt suicide far more frequently than controls (age-, sex-, and crime-matched prisoners not in solitary confinement) implicates dysregulation as a risk-increasing variable. Family studies do not suggest that there are genes for suicide, although genetic information may increase the probability of conditions that are associated with increased suicide risk. Developmental studies are not implicated when suicides are looked at as a group, although individual cases provide exceptions. Physiological measures among persons who have committed suicide frequently point to dysfunction of the serotonin system (e.g., Brown et al., 1982; Arranz et al., 1994; Stein and Stanley, 1994). Evidence suggesting reduced dopamine metabolism has also been reported (Roy, Karoum, and Pollack, 1992).

Evolutionary Analysis

To follow on an evolutionary line of thinking, there are good reasons to predict suicide in a variety of circumstances. For example, its incidence should increase when an individual perceives that his or her costs to kin exceed the benefits to kin of his or her continued existence. In such instances, suicide may be understood as a form of kin-related altruism. This is an obvious prediction from kin selection theory, and it can explain a certain percentage of the suicides of persons who are terminally ill or who view their existence as highly costly to relatives. Findings from a number of studies are consistent with this interpretation: 58% of the variance in suicidal ideation is reported to be due to family-social variables or perceived burdensomeness on the family (deCatanzaro, 1980, 1991). These types of suicides do not point to strategy failures or atypical resource allocations; rather, they may reflect an adaptive strategy. Automatic system or algorithm functionality therefore need not be compromised.

Failure to achieve reproductive goals is another reason to expect an increase in suicide: Women who have never married or who are unable to reproduce should have higher rates of suicide than married parous or nonparous women. Findings support this prediction (Hoyer and Lund, 1993). Extending this line of reasoning, the risk of suicide should correlate negatively with the number of living offspring. Findings also support this prediction (Hoyer and Lund, 1993). Sexual differences, particularly greater female identity with offspring production, is implicated in these findings.

Suicides are also associated with specific conditions. For example, in adolescent males who commit suicide, narcissistic and schizoid traits, as well as major depression, are frequently present (Diekstra, 1989; Apter et al., 1993). Compromised infrastructural function, strategy failures, and reduced goal achievement are suggested by these findings; for example, unattainable goals or distorted recognition is likely to increase the chances of misinterpreting one's value to others and to reduce expectations of future satisfaction.

A further prediction is that more living monozygotic than dizygotic cotwins of twin suicide victims will themselves have attempted suicide. Several studies have addressed this question (e.g., Roy et al., 1991; Roy, Segal, and Sarchiapone, 1995). In a recent study of 26 living monozygotic cotwins and 9 living dizygotic cotwins of twins who had committed suicide, 10 of the 26 surviving monozygotic cotwins and none of the surviving dizygotic cotwins had themselves attempted suicide (Roy et al., 1995). Although different developmental histories may partly explain these findings, they are consistent with previous studies of suicide in twins that implicate genetic contributions to disorders associated with an increased probability of suicide but not to suicide itself (Roy et al., 1995). A related factor is cross-person differences in capacities to regulate: Suboptimal capacities render some persons more vulnerable to suicide in certain environments, and fluctuating degrees of dysregulation may explain the episodic nature of suicide attempts.

Spousal Abuse

Spousal abuse is also not a formally classified disorder in *DSM-IV* (APA, 1994). While its reported frequency may seem perplexing, an evolutionary analysis offers reasons for its high prevalence rate.

Evolutionary Analysis

Following evolutionary reasoning, the physical abuse of one's wife, which is far more common than physical abuse of one's husband, is predicted when males with compromised infrastructures are strongly motivated to control female sexuality for one or more of the following reasons: to make intimacy predictable, paternity certain, and child care assured, as well as to improve self-esteem and social status (e.g., Smutz and Smuts, 1993). Suboptimality of the reproductive behavior system is implicated in these instances. Algorithm suboptimality is also likely with causal modeling and scenario development, which are the systems that are most likely to be involved. Among females, an increased frequency of spousal abuse is expected in situations of male infidelity, resource squandering, substance abuse, and rejection. As noted, spousal abuse may be deplored socially. However, it is worth recognizing that both males and females invest considerable time and energy in relationships and in their attempts to control each other's behavior for their respective advantages. In such circumstances, failures to control are not necessarily easily tolerated.

Extreme forms of violence result in spousal homicide, and the data dealing with homicide are consistent with evolutionary predictions: Young wives are at greater risk as homicide victims than older wives; wives who are estranged from their husbands are at greater risk than coresiding wives; and wives in de facto marital unions are at greater risk than wives in legally registered marriages (M. Wilson and Daly, 1992).

Not surprisingly, spousal abuse (and child abuse) extends to family violence and physical intimidation of family members (Wolfner and Gelles, 1993). Gains in inclusive fitness, which individuals receive from associating with one another, depend in part on their degree of relatedness and in part on the degree to which they aid one another. When one invests in nonrelated others (the case among husbands and wives as well as among step-parents and step-children), and one's investments are not offset with benefits, an increased frequency of dysregulation is likely. Evidence is compatible with this prediction. Family violence is characterized by higher than average levels of intrafamilial stress (e.g., value disagreements), cross-member coercion (Burgess and Draper, 1989), disagreements over reproductive strategies, and mental disorders.

Child Abuse

Child abuse, a term that is most often used to refer to the physical rather than the psychological abuse of children, is not a formally classified disorder in *DSM-IV* (APA,

1994). However, it is identified as a behavior that may require medical and psychiatric attention.

Evolutionary Analysis

In evolutionary analysis, an increased incidence of child abuse is predicted in situations in which the abuser is *locked in* to his or her social environment and unable to extricate himself or herself from interactions that will result in negative cost-benefit outcomes (Daly and Wilson, 1985). Males and females both contribute to child abuse (Gelles and Harrop, 1991), and a high percentage of the variance in rates of abuse correlates with structural variables. For example, preschool children living with one natural parent and one stepparent (almost always a male) are far more likely to be victims of child abuse than are same-age children living with two natural parents (Daly and Wilson, 1985, 1987a, 1987b, 1988a, 1988b). Within the one-natural-parent-one-stepparent category, two prototypes are observed. In both, a male lives with a female and her offspring, and the male is not the child's biological father. Requirements of intimacy and caretaking between the adults may reduce investment in the child. The child objects. In turn, the child may be abused. Or a male may be required to invest in a child as a condition of intimacy with the child's mother. From the male's perspective, this requirement amounts to allocating resources so that the costs will increase while benefits will be minimal, that is, investing in another male's genes. From the child's perspective, demands may be made on the mother because of the absence of a biological or a socially desirable father. From the mother's perspective, rejection or abuse of the child may be necessary to preserve her relationship with the male with whom she is living. Suboptimality of the reproductive automatic system is usually implicated in these circumstances, and possibly also the kin investment system. Reproductive system compromises are more likely among males who are living in households where the children are not their own. And suboptimality of the kin-investment automatic system is more likely in females who need adult male companionship.

The preceding points need to be tempered by two findings: (1) In most male-female living settings in which the male is not the father of children in the household, or in single-parent households, child abuse does not occur, and (2) not all disorders are associated with an increased frequency of aggressive or abusive behavior (Troisi and Marchetti, 1994).

Abuse also occurs in single-parent households, and when it does, it is usually associated with low social status and poverty (Gelles, 1989). (Recall that the incidence of referrals of preadolescents for psychiatric care is seven times greater in one-natural-parent-one-stepparent households than in two-biological-parent households; see chapter 3.)

In addition to the preceding points, there are a variety of other incidence-contributing factors. Trait variation may be responsible for reduced capacities to tolerate others' requirements of attention and care. That child abuse represents a distinct form of violence rather than an extension of normal aggressive behavior has been suggested (Gelles, 1991). This possibility implicates genetic contributions. Dysregulation due to conflicts between household adults is a likely contributing factor in episodic abuse. And certain disorders, particularly those of which impulsivity is a significant component, will alter the frequency of abuse.

Concluding Comments

We have discussed a number of disorders in evolutionary context, with particular emphasis on the importance of strategies, the impact of environmental contingencies, and the consequences of different types of compromised infrastructures. Each of the disorders appears to have different causal profiles. While not strongly emphasized, the possibility that different disorder types within single diagnostic categories have different causal profiles is an obvious implication from our analysis.

14

Dysthymic Disorder: A Study of Infrastructural Suboptimality

Chronic conditions are conditions in which compromised infrastructures, functional capacities, and the social environment interact. In this chapter, we shift away from the approach taken in previous chapters (primarily, the introduction of evolutionary concepts into psychiatry) to discuss a study of dysthymic disorder. The study was designed to collect data relevant to an evolutionary approach and to narrow the number of possible disorder-causing hypotheses.

Most clinicians view dysthymic disorder (DD) as a complex web of overlapping conditions, as one of the least specific *DSM* diagnoses, and as one of the less interesting disorders. Despite these views, DD turns out to be a very interesting condition, as well as one that is conveniently used to illustrate how an evolutionary perspective can alter how disorders are characterized and explained: (1) The clinical state of persons with DD is not so disorganized that infrastructures cannot be systematically studied; (2) there are clear indications of person-environment-symptom interactions; (3) there is evidence of common final pathway constraints; (4) the disorder is associated with functional consequences; and (5) causal predictions based on evolutionary reasoning are possible. Some findings from the study have been published previously (Essock-Vitale and McGuire, 1990; McGuire et al., in press). Other findings are reported here for the first time.

The Study

According to *DSM-IV* (APA, 1994), the essential feature of DD is chronically depressed mood for most of the day for at least two years. The diagnostic criteria include:

- Depressed mood for most of the day, more days than not, for at least two years.
- The presence of at least two of the following symptoms: (a) poor appetite or overeating, (b) insomnia or hypersomnia, (c) low energy or fatigue, (d) low self-esteem, (e) poor concentration or difficulty making decisions, and/or (f) feelings of hopelessness.
- During a two-year period never without the symptoms above for more than two months at a time.
- No major depressive episode during the first two years of the disturbance.
- No history of a manic or hypomanic episode and criteria have never been met for Cyclothymic Disorder.
- Signs and symptoms are not superimposed on a chronic psychotic disorder.
- The symptoms are not due to the direct physiological effects of a substance.
- The symptoms cause clinically significant distress or impairment in social, occupational, or other important areas of functioning. (p. 349)

Lifetime prevalence rates are estimated at 4.8% for males and, in some studies, as high as 8% for females (*DSM-IV*; Kessler et al., 1994). The description of DD in *DSM-IV* varies minimally from the description of DD in *DSM-III* (APA, 1980). *DSM-III* criteria were used for the study reported here.

Selection of the Study Population

Potential DD subjects and control subjects were solicited through newspaper advertisements in Los Angeles and two adjacent cities. The advertisements contained a description of the clinical features of the disorder. Potential subjects who met the advertised criteria were interviewed and given psychological and physical examinations. To be accepted into the study, the subjects had to be female, between 22 and 45 years old, meet the *DSM-III* diagnostic criteria for DD, and not be suffering from a second mental disorder or a medical disorder. Particular care was taken to exclude persons with *DSM-III* diagnosable personality disorders, which often are associated with DD (Spalletta et al., 1996). Controls were accepted into the study if they were female, between 22 and 45 years old, and not suffering from either a mental or a medical disorder. English was the required first language for all subjects. The DD subjects had to agree to participate in the study for 18 months, to commit themselves to approximately 150 hours in clinical and experimental evaluations, and not to use psychotropic medications during the study unless prescribed by one of the study's physicians. None of the DD subjects required drug treatment during the study. And at the end of the study, they were paid $150. Controls participated in the study for 6 weeks, spent 40 hours as subjects, and were paid $50.

All subjects who met the initial screening criteria received two structured interviews. Because the available structured interviews did not address many of the topics of interest to the investigators, special interviews were designed for the study. Trial interviews were refined and finalized by use with subjects not in the study. Prestudy testing of the interviews with two investigators revealed a high degree of agreement (>.80) on those questions that were finally used. Once the study had begun, answers to questions dealing with behavior and symptoms (items with which relatives or friends were likely to be familiar) were verified by phone contact. Verification revealed that the DD subjects had described their symptoms, their behavior, and their life conditions accurately, but not others' behavior.

Structured Interview 1

The first of the structured interviews was designed to obtain basic symptom, sign, and functional information. Selected findings from this interview are summarized in Table 14.1 and reveal clear differences between the DD subjects and the control subjects, as well as a clinical profile of the DD subjects that is consistent with both *DSM-III* and *DSM-IV* diagnostic criteria for DD. Not shown in the table is the number of subjects who had received treatment. Prior to entering the study, 37 of the 42 DD subjects had received some type of professional help (e.g., medications, counseling, psychotherapy) for their condition. At best, the treatments had been moderately effective, none completely so. Of the 22 control subjects, 3 had received some type of treatment for periods of distress. None had been diagnosed as suffering from a disorder.

Structured Interview 2

All subjects underwent a second structured interview for the purposes of identifying disorder-related features and their consequences. Findings for the DD subjects and the

Table 14.1 Demographic and clinical features of DD subjects and control subjects.

Subject characteristics	Dysthymic disorder subjects ($n = 42$)	Control subjects ($n = 22$)
Age range	22–45	21–39
Attended college	91	99
Married	12	32
Divorced	32	9
Single	56	59
Insomnia or hypersomnia >6 months	79	18
Low energy or chronic tiredness >6 months	90	9
Feelings of inadequacy, low self-esteem, or self-depreciation >6 months	84	9
Decreased productiveness at home, school, or work >6 months	100	0
Decreased ability to concentrate or to think clearly >6 months	70	0
Loss of interest in or enjoyment of activities >6 months	60	0
Irritability or excessive anger >6 months	13	0
Inability to respond with pleasure to desirable events >6 months	30	0
Less active or talkative >6 months	58	9
Tearfulness or crying >6 months	49	0
Recurrent thoughts of death or suicide >6 months	63	9
Dyspnea, palpitations, chest pain >6 months	46	0
Choking or smothering sensations >6 months	9	0
Dizziness, vertigo, unsteadiness >6 months	36	0
Feelings of unreality >6 months	2	0
Paresthesias >6 months	31	9
Hot and cold flashes >6 months	2	0
Sweating or faintness >6 months	21	9
More than or less than 10% weight change during last 6 months	42	9

Note. The table presents demographic, sign, and symptom profiles for the DD subjects ($n = 42$) and the control subjects ($n = 22$). Except for age range, all figures are in percentages, which are rounded to the nearest whole number.

control subjects are combined in this section. Percentages are rounded to the nearest whole number.

Pedigree influences: 36% of the DD subjects and 9% of the control subjects reported that first-degree relatives had conditions similar to DD or clusters of signs and symptoms associated with disorders, for example, impulsive behavior, excessive irritability, or depression.

Developmental disruptions: 40% of the DD subjects reported that they had been emotionally abused as children, and 9% reported that they had been abused repeatedly. One control subject reported that she had repeatedly been physically beaten as a child.

Social function capacities: 81% of the DD subjects and 14% of the control subjects reported functional difficulties (e.g., adjusting to school and developing satisfactory social support networks) prior to adolescence. Compared to those DD subjects whose symptoms had first appeared during either adolescence or early adulthood, DD subjects who had had an early onset of their symptoms had had fewer childhood friends and more sporadic adult work histories and were more economically disadvantaged at the time of the study.

Responses to minor undesirable events: 74% of the DD subjects reported that irritating events (e.g., a friend arriving 30 minutes late for a social engagement, money from a family member arriving a day late, failure to find a desired item while shopping) led to an increase in the number of their symptoms. Control subjects reported that similar events often caused brief periods of frustration but rarely precipitated symptoms.

Reading others' behavior rules: DD subjects expressed limited views of others' motivations and behavior. Statements such as "She screws everyone she can," "All she does is lookout for herself," and "All he wants is sex" were typical of the ways in which DD subjects characterized others. In contrast, control subjects offered more complex views of others' motivations and behavior, for example, "She wants to help her brother and his wife, but she thinks that her offer will be misinterpreted" or "She loves her kids, even thought her son gives her a lot of trouble." Among the controls, others' motivations were viewed more often as socially positive than socially negative. The opposite was true among the DD subjects.

Causal modeling: The DD subjects developed causal models and attributed causes to others' behavior using minimal information. For example, one DD subject disliked a neighbor because "She washes her car every weekend." Another disliked her sister because "She only listens to classical music." DD subjects seldom changed their views of others despite new or disconfirming information. The controls also developed causal models and attributed causes using minimal information. However, they frequently changed their models in response to new information: "I didn't like her when I first met her, but I changed my mind when I saw the way she treated others."

Novel behavioral strategies: DD subjects frequently repeated the same behavioral strategy, for example, communicating to kin or friends that they were ill or struggling financially. Repetition continued even when these strategies were minimally effective in eliciting empathy or acquiring financial support. In contrast, the control subjects seldom reused a strategy if it had been ineffective. None of the controls reported that they had presented themselves to others as ill.

Self-monitoring: DD subjects were aware that many of their strategies were ineffective, and *the self-monitoring capacities of the majority of them were indistinguishable from those of the control subjects.* However, over 85% of the DD subjects failed to use self-monitored information to develop novel strategies, while only 9% of the control subjects reported similar failures (see Figure 14.2).

Identification of biological goals: Day-to-day management of their lives and developing relationships with others who could provide assistance were the high-priority goals of the DD subjects. Low-priority goals included having offspring and helping kin. Offspring, potential offspring, and kin were often viewed as actual or potential burdens. Further, the majority of DD subjects felt they had been unfairly treated by their kin and that their kin did not deserve special consideration. High-priority goals among the control subjects were the same as those listed in the four basic functional categories in chapter 4: survival, reproduction, kin investment, and trading favors with friends. Offspring were not viewed as actual or potential burdens, although those control subjects who were parents acknowledged that raising children is often frustrating.

Explanation of symptoms: 72% of the DD subjects viewed external events or the behavior of others as the primary causes of their symptoms, their social relationship difficulties, and their unsatisfactory work histories (if applicable). The remaining 28% saw themselves as suffering from a disorder over which they had minimal control. Of the control subjects, 9% blamed others or external events for their frustrations and difficulties. *The remaining 91% saw their own behavior as contributing to their difficulties.*

Others' response to distress: The majority of DD subjects reported that others sometimes responded when they were distressed, although many persons who had formerly done so had stopped. The DD subjects hesitated to respond to others' requests for assistance, often feeling that others were "better off" and "didn't need help." The control subjects reported that others responded when they requested help, and vice versa.

Subjects' social environment: The control subjects interacted with kin and nonkin, with members of both sexes, and with persons of different ages. Their likes and dislikes of others were based primarily on experience. The DD subjects interacted most often with females of the same age, and where possible, they avoided social and work environments that were associated their with symptoms, although they were seldom entirely successful. The majority of the DD subjects also viewed males as insensitive and exploitive, and as persons to be avoided. Compared to the control subjects, the DD subjects reported a threefold greater frequency of being deceived by males with whom they had interacted sexually. The majority of DD subjects also reported that they had often suffered losses (e.g., loss of jobs, friends, and kin support, loss of a potential mate to another female), as well as declines in social status (e.g., decline in income and social influence). The control subjects did not view males as excessively deceptive, and only 9% reported histories of repeated losses or significant declines in social status.

Self-esteem: While the majority of the control subjects respected themselves, they also acknowledged that there were ways in which they could improve their self-esteem (e.g., being more responsive to others). A different picture emerged among the DD subjects: One quarter of them viewed themselves as superior to others, particularly close kin and former friends; 40% disliked themselves; and the remainder did not differ from the control subjects.

Psychological, Psychophysiological, and Neuropsychological Tests

In addition to the interviews, the DD subjects were given extensive psychological, psychophysiological, and neuropsychological tests at Months 1, 6, 12, and 18 of the study. The control subjects were assessed only once.

The DD subjects and the controls differed significantly in their scores on the Minnesota Multiphasic Personality Inventory (MMPI) and the Symptom Check List-90 (SCL-90). The DD subjects had MMPI profiles and SCL-90 test scores typical of depression, while the controls had normal MMPI profiles and SCL-90 scores (A. Rosen, unpublished data, 1991). As the study progressed, the MMPI profiles and the SCL-90 scores for the DD subjects more closely approximated normal MMPI profiles and SCL-90 profiles; however, at the 18-month assessment, the profiles and scores still distinguished the DD subjects and the controls, and the DD subjects still met the criteria for depression. Psychophysiological tests revealed clear differences between the two groups, with DD subjects showing significantly higher levels of anxiety and greater physiological responsiveness to provocative stimuli (D. Shapiro, unpublished data, 1991). Psychophysiological findings about the DD subjects *did not* change over the 18 months of the study. Only 2 of the 51 neuropsychological tests (including IQ tests) differentiated the two groups, the primary finding being impairment in memory consolidation in the DD subjects. On the 49 neuropsychological tests on which DD subjects and the controls did not differ, no trends suggesting group differences were apparent (J. and G. Marsh, unpublished data, 1991). The neuropsychological test scores of the DD subjects did not change from the first to the last (18-month) assessment.

Ethological Studies

Table 14.2 presents findings from an ethological evaluation of the first 20 members of each subject group. The data presented in the table were collected by means of direct observation techniques described elsewhere (Polsky and McGuire, 1980, 1981).

Table 14. 2 Ethological analysis of DD subjects and control subjects

	Control subjects	DD subjects	p values
Social behaviors			
Send verbal	19.67	19.23	
Verbal long	0.64	0.57	
Verbal short	1.72	1.01	<.005
Smile	1.13	1.24	
Laugh	0.90	0.43	<.001
Head nod	1.92	1.18	<.05
Eyebrow flash	0.33	0.27	
Head even	93.42	90.88	
Look at person	88.84	78.52	<.003
Illustrator	0.24	0.28	
Self-adaptor behaviors			
Touch self	0.49	0.59	
Groom	0.37	0.44	
Pathological behaviors			
Pathological	1.40	4.61	<.001

Note. Behavioral frequencies that reached significance ($p < .05$, t tests) are listed in the right-hand column.

All groups were balanced equally for DD subjects ($n = 4$) and control subjects ($n = 4$), and there was no designated leader. Each group session lasted for 64 minutes, and groups with the same membership met for eight sessions over an eight-week period. Observations were made at each session. Observers were located behind a one-way mirror, unable to hear what the subjects said, and blind to which individuals were DD subjects and control subjects. Each subject was observed for one minute eight separate times (order random); a 10 scoring method was used for the behaviors listed in the table. Coded behaviors were defined in common language terms, with the following exceptions: verbal long means continuous verbalization >5 seconds; verbal short means continuous verbalization <5 seconds; pathological means atypical postures or acts, such as leg shake (>5 seconds) and talking to oneself (>5 seconds); and atypical illustrators means excessive facial grimaces and self-observation (>15 seconds). Behavioral frequencies are adjusted to show the mean frequency per behavior per subject per hour.

For a number of behaviors in Table 14.2 (e.g., send verbal), frequency measures do not distinguish DD subjects and controls. However, four of the social behaviors (verbal short, laugh, head nod, and look at person) show significant frequency differences between DD subjects and controls, each behavior occurring less frequently among DD subjects than among controls. The reduced frequency of these behaviors among DD subjects is similar to findings reported by other investigators for persons with similar clinical profiles (e.g., Grant, 1968). The frequency of pathological behaviors also differed between the two groups, the DD subjects engaging in pathological behaviors three times as often as the controls. That pathological behaviors are often recognized by persons outside clinical settings was noted earlier (Table 4.3).

Symptom Change during the Study

Table 14.3 presents findings from symptom assessments of the first 22 DD subjects (the only DD subjects tested with the symptom protocol) at Months 1, 6, 12, and 18 of the study.

In the collection of data for Table 14.3, each of the 40 symptoms listed in the table was printed on a card, along with the symptom's common language definition. The cards were shuffled. The DD subjects were asked to go through the cards and to select those cards that described their symptoms. To qualify for selection, a symptom had to have been present each day for the 14 days prior to the assessment, and it had to have resulted in a change in a subject's living routine (e.g., cancellation of a social engagement). For each reported symptom, investigators reviewed the cards and verified both the 14-day and the alteration-of-living-routine requirements. An independent (blind) assessment (structured interview) of symptoms was conducted by two experienced clinicians on six DD subjects and six controls. This assessment resulted in essentially the same symptom profiles as those shown in the table.

In Table 14.3, the numbers listed for each of the four assessment periods refer to the number of DD subjects reporting the presence of the symptoms listed in the left-hand column of the table at each assessment. For each cell, 22 is the maximum possible number. At the first assessment, 17 of 22 DD subjects reported that they had suffered from fatigue; 19 reported the same symptom at the 6-month assessment; 15 at the 12-month assessment; and 13 at the 18-month assessment. The symptoms are

Table 14.3 Change in symptom frequency in DD subjects over the 18 months of the study

Symptoms	Assessment periods				Row totals
	1	2	3	4	
1. Fatigue	17	19	15	13	64
2. Specific worries	19	15	14	10	58
3. Feelings of inadequacy	18	16	10	10	54
4. Decreased effectiveness	17	13	9	10	49
5. Pessimism	16	13	9	9	47
6. Avoidance of situations	13	12	13	8	46
7. Feeling slowed down	13	14	10	6	43
8. Insomnia/hypersomnia	9	12	14	6	41
9. Edginess, irritability	14	9	13	5	41
10. Generalized worries	15	12	7	6	40
11. Difficulty concentrating	11	9	8	8	36
12. Poor memory	11	9	8	8	36
13. Less talkativeness	12	10	10	4	36
14. Tearfulness, crying	14	11	5	5	35
15. Obsessions	12	9	7	5	33
16. Guilt	12	7	5	7	31
17. Diminished interest	13	9	7	2	31
18. Social withdrawal	10	9	6	6	31
19. Decreased pleasure	12	8	6	4	30
20. Pain in back, joints	9	8	5	5	27
21. Decreased interest in sex	8	6	6	6	26
22. Urinary frequency	9	5	4	5	23
23. Diarrhea, constipation	4	7	5	5	21
24. Paresthesias	7	6	2	4	19
25. Palpitations	9	7	1	1	18
26. Dizziness, vertigo	6	5	2	3	16
27. Feelings of unreality	3	6	4	1	14
28. Chest pain, discomfort	7	3	1	2	13
29. Feeling that can't accomplish	6	3	3	1	13
30. Trembling, shaking	5	4	3	1	13
31. Hyperalertness	4	1	4	3	12
32. Fear of dying	7	2	1	1	11
33. Choking, smothering	3	3	2	1	9
34. Sweating	4	2	2	0	8
35. Shortness of breath	2	2	3	0	7
36. Poor recall	2	2	3	0	7
37. Self-mutilation	2	3	0	1	6
38. Recurrent dreams	2	2	1	1	6
39. Faintness	1	3	2	0	6
40. Hot and cold flashes	2	1	0	0	3
Column totals =	360	287	230	173	

listed from the most to the least frequent when scores for all four assessment periods are totaled (row totals).

A comparison of the first (Month 1) and the fourth (Month 18) entries in each row shows that over the 18-month course of the study, the number of symptoms declined for each of the 40 symptoms. Recall that in this study, (1) DD subjects did not have diagnosable personality disorders; (2) they did not take or receive psychotropic medications; (3) prior to entering the study, 37 of the 42 DD subjects had received some type of treatment for their condition; and (4) previous treatments had been only moderately effective in reducing their symptoms.

The decline in symptoms shown in Table 14.3 is consistent with predictions of the RDT (chapter 8). When the DD subjects entered the study, they identified themselves as afflicted by symptoms and suffering (Table 14.1). Over the course of the study, they spent more than 150 hours participating in tests and interacting with investigators and staff. The DD subjects had a special room with coffee and doughnuts where they could relax and socialize. They were aware of their importance to the study, and they were well treated and respected by research personnel; for example, research schedules were posted several days in advance, and the schedules were altered to meet the DD subjects' extra study needs and responsibilities. Prior to testing, each of the experimental protocols was explained in detail. Soon after the study began, the DD subjects began socializing among themselves both at the research site and elsewhere.

As the study progressed, the DD subjects less frequently identified themselves as suffering from a disorder or as symptomatic. In effect, they entered a social environment that was supportive, nonjudgmental, minimally demanding socially, and structured. It was also one in which they were important and their importance was frequently confirmed. As noted in chapter 8, such environments positively correlate with infrastructural regulation and symptom reduction. Elsewhere, other investigators have noted that nonspecific interventions correlate with clinical improvement (Wolpe, 1988).

Within limits, the findings in Table 14.3 are consistent with the competitive yielding hypothesis for depression (chapter 7): In the supportive environment of the study, yielding behavior (essentially, behaving submissively) had minimal utility. However, this hypothesis does not explain the full range of clinical features among the DD subjects: Their symptoms did not entirely disappear, and as discussed below, *their infrastructures and functional capacities remained compromised.*

Evidence of Suboptimal Algorithms

One of the most striking findings of the study was the compromised ability of the DD subjects to efficiently solve routine social navigation problems. Both causal modeling and behavioral strategy development are implicated in these difficulties. Figure 14.1 presents a flowchart showing how the DD subjects and the control subjects modeled events causally. Figure 14.2 presents findings from a task designed to test self-monitoring capacities and the influence of monitored information on scenario development and behavioral strategy change.

The top half of Figure 14.1 shows the characteristic ways in which the control subjects modeled dissatisfying interpersonal events. The bottom half of the figure does the same for the DD subjects. When dissatisfied, the controls attributed their responses

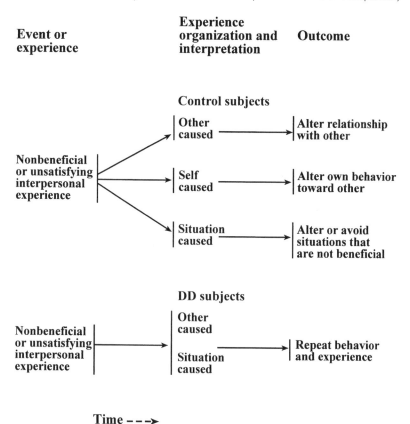

Figure 14.1 Flowchart of causal modeling and causal attributions among DD subjects and control subjects.

to either their own behavior, others' behavior, or nonpersonal situational variables (e.g., the illness of a relative). In contrast, the majority of the DD subjects saw others' behavior or situational variables (which they often personalized) as the primary sources of their dissatisfactions. Among the controls, there was a positive correlation between viewing oneself as a source of one's dissatisfactions and the development of novel strategies. For the control subjects, this relationship is shown in the outcome column of the figure. The DD subjects presented a different picture. Not only did they usually fail to consider themselves a possible source of their difficulties, but they also seldom initiated novel strategies in response to their dissatisfactions; for example, they repeated strategies that failed to accomplish specific short-term goals. Significant consequences accompanied their behavior: 80% of the DD subjects reported frequent social-interaction, work-related, and personal achievement difficulties. Less than 10% of the controls reported similar difficulties or failures. Other studies of like populations have shown that symptomatic improvement may occur even though features such as rigidity, level of activity, and interpersonal dominance do not abate (e.g., Henderson et al., 1980; Hirschfeld et al., 1983; Kocsis, 1993).

As noted, the DD subjects were aware of their inability to successfully alter their strategies. Moreover, prior to entering the study, the majority of the DD subjects had attempted to do so, although such efforts had usually failed, often because they were associated with an increase in symptoms. These points are illustrated in Figure 14.2, which depicts the ways in which the DD subjects and the controls managed an experimental task.

In the task, both the DD subjects and the controls were required to develop strategies for carrying out a number of specific acts (A to E) in a finite period of time in a familiar urban area close to the research site, for example, going to the store to buy shoe polish, changing money at the bank, purchasing a sandwich, and returning to a starting point within a specified time. The task required that the subjects develop a plan that included a cognitive map of a familiar urban area. In developing their maps, the subjects had to take into account the time required to get from one point to another, make estimates of the time involved in accomplishing each task, and develop a strategy that would optimize their chances of completing all of the tasks within the specified time. The subjects were then asked to undertake the task, all at the same time of day on the same day of the week.

In Figure 14.2, capital letters and horizontal lines with arrows (original plan) identify typical strategies developed by the subjects. Boxes around letters indicate that the estimated time required to accomplish a specific task was exceeded. Downward-pointing arrows attached to boxes indicate that subjects changed their original strategy and pursued an alternative strategy. Interrupted lines indicate that subjects did not change their initial strategy, even in situations where these subjects believed that doing so might facilitate completing the task within the specified time period.

The pretask plans for the DD subjects and the control subjects were indistinguish-

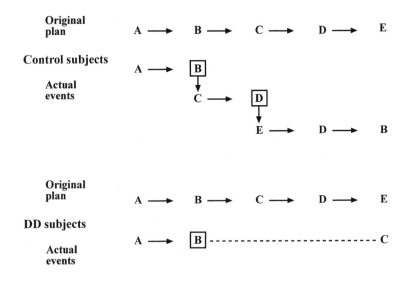

Figure 14.2 Self-monitoring and novel strategy development by DD subjects and control subjects.

able. However, once the task began, things changed. For example, when DD subjects arrived at Point B and encountered a delay, such as a long line of people waiting to cash checks in a bank, the majority continued waiting. The DD subjects often did not alter their strategies even though they recognized that the waiting time was likely to be excessive, and that waiting would diminish the likelihood of their completing the remaining tasks within the specified time.

The majority of the DD subjects did not complete all the tasks during the allotted time, while the majority of the controls did. Posttest interviews established that when the DD subjects considered strategy changes, they became uncertain and anxious, states that several subjects described as "emotionally freezing." This finding points to suboptimal automatic system functionality in the DD subjects, an increase in symptoms being triggered by contingencies requiring a choice when the outcome of the choice was uncertain (e.g., a subject might find an equally long line at the next place the subject visited). The difficulties the DD subjects encountered in developing new behavioral strategies also implicate algorithm suboptimality for behavioral strategy development, although not for self-monitoring. In contrast, the control subjects frequently altered their strategies during the task. Although they were frustrated by such events as long lines at the bank, they nonetheless changed their plans and their behavior. The controls considered the possibility that alternative strategies might be counterproductive, but this outcome was not thought to be more probable than the conditions in which the controls found themselves.

The findings in Figures 14.1 and 14.2 can be viewed in terms of costs and benefits. Persons who experience difficulty in developing and executing novel strategies will experience increasing cost-benefit deficits in their efforts to achieve short-term goals. In turn, dysregulation and the onset of symptoms are likely. Comparable findings have been reported in studies of similar subjects (Roy-Byrne et al., 1986). From this perspective, a state of chronic, moderate depression (costs > benefits) appears to have been the optimal resolution for the majority of the DD subjects. They were motivated to achieve short-term goals in their social environment (the *locked-in* principle), yet they were unable to do so proficiently. Nor were they able to develop a ready resolution to their predicament. For example, they frequently withdrew from environments in which they were not achieving goals. Yet, withdrawal often proved to be counterproductive because it reduced their social options. Blaming others for their troubles further contributed to their problems. Both interview and test findings were consistent with the preceding points. Anxiety increased when the DD subjects found themselves in situations in which they failed to achieve goals. In turn, they withdrew, and, the intensity of their depression increased when they were alone.

Functional Consequences

Table 14.4 lists some of the functional consequences of DD.

Clear differences between the two subject groups on a variety of functional measures are shown in the table. Perhaps the most informative findings relate to the reproductive behavior system (identified by asterisks). For the DD subjects, there was (1) a reduced probability of being married; (2) a greater likelihood of being raped; (3) a reduced likelihood of having had a live birth; (4) a greater likelihood of having had

Table 14.4 Functional consequences of DD subjects

Functional measure	Outcome
Likelihood of ever being married	DDS < CS
Income	DDS < CS
Likelihood of having a religious affiliation	DDS < CS
Number of sexual partners	DDS > CS*
Likelihood of having been raped	DDS > CS*
Likelihood of having had a live birth	DDS < CS*
Likelihood of voluntary abortion	DDS > CS*
Frequency of dating	DDS < CS*
Satisfaction with recreational activities	DDS < CS
Number of friends	DDS < CS
Number of special friends	DDS < CS
Likelihood of having received physical punishment as a child	DDS > CS
Likelihood of having received severe physical punishment as a child	DDS > CS
Perceived importance of self	DDS < CS
Number of major life events	DDS > CS
Mental-health-severity score	DDS > CS

Note. All differences = $p < .05$; DDS = DD subjects; CS = control subjects. The asterisks indicated a relation to the reproductive behavior system. The mental-health severity score is from a paper and pencil test; the higher the score, the worse one's mental health. Adapted from Essock-Vitale and McGuire (1990).

a voluntary abortion; and (5) a reduced frequency of dating. The greater number of sexual partners of DD subjects may have several explanations. One is that males may have viewed the DD subjects as short-term sexual partners, not as long-term mate prospects. Elevated mental health severity scores and the higher frequency of life events among the DD subjects were likely contributing factors to these outcomes. An alternative explanation is that the DD subjects engaged in sex to attract males. A number of the DD subjects reported that they had employed this strategy even though they did not enjoy sex. These explanations are not mutually exclusive. The high frequency of voluntary abortions among the DD subjects is in agreement with a point discussed in chapter 2: A reduced capacity for reproduction is not a consistent finding in females with disorders.

A Causal Analysis of DD

When the preceding findings are put in perspective, the most consistent and striking features among the DD subjects are (1) the repetitive nature of the behavior leading to cost-benefit deficits, (2) the chronicity of their symptoms, (3) their constricted causal modeling capacities, and (4) their limited capacity to develop novel behavioral strategies. These findings point to infrastructural suboptimality, and they can serve as the basis for evaluating both evolutionary and prevailing-model explanations of suboptimality. Possible explanations include:

1. Pedigree (genetic information) influences.
2. Trait variation.

3. Pedigree influences + trait variation.
4. Pedigree influences + adverse developmental environment.
5. Trait variation + adverse developmental environment.
6. Pedigree influences + trait variation + adverse developmental environment.
7. Normal genetic makeup + adverse developmental environment.
8. Normal genetic makeup + adverse current environment.
9. DD is a core adaptation.
10. DD is a condition in which persons attempt to adapt within constraints.

This list is not exhaustive. Further, because the accuracy of some findings (e.g., details of developmental histories) could not be verified, some hypotheses could not be disproved. Nevertheless, some explanations are more plausible than others.

Provided there is an absence of evidence of adverse developmental environments, Explanations 1 (Pedigree influences) and 2 (Trait variation) can explain instances in which suboptimal infrastructures, goal achievement failures, and symptoms appear early in life. Of the DD subjects in this study, 36% had pedigree histories in which first-degree relatives had some disorder-related signs and symptoms. Thus, these subjects can be tentatively assigned to Explanation 1 (pedigree influences). Approximately the same percentage of the DD subjects had early life histories compatible with suboptimal infrastructural function without positive pedigree histories or histories of adverse environments. These subjects can be tentatively assigned to Explanation 2 (trait variation). Explanation 2 assumes that the chance mixing of genetic information contributes to suboptimal traits. Because of the chance factor, more diverse clinical profiles would be expected compared to Explanation 1, unless common final pathway constraints are operative.

Explanation 3 (Pedigree + trait variation) can explain clinical pictures similar to those in Explanations 1 and 2, but greater disorder severity would be expected. This explanation could apply to those few DD subjects whose capacities were most severely compromised, who were angry, and who were uncooperative at different points during the study. Explanation 4 (Pedigree + adverse development environment) requires the onset of symptoms following adverse circumstances. This explanation is consistent with the histories provided by four DD subjects who had experienced repeated emotional abuse during early childhood and whose symptoms had begun in close association with the abuse. Explanation 5 (Trait variation + adverse development environment) invites reasoning similar to that for Explanation 4 but focuses on trait variation rather than pedigree influences. While some DD subjects had developmental histories that were compatible with this explanation, clinical interviews did not permit a clear distinction between possible trait and environmental contributions to their condition. Explanation 6 (Pedigree influences + trait variation + adverse development environment) represents a combination of Explanations 1 and 2, plus adverse environments. As with Explanation 3, greater disorder severity would be expected than was observed among the DD subjects.

Explanation 7 (Normal genetic makeup + adverse developmental environment) was not consistent with any of the histories provided by the DD subjects in this study. Those subjects whose relatives did not have disorder-related signs or symptoms, in which case one might assume normal genetic makeup, also had histories suggesting that their behavior had contributed to their aversive experiences. Moreover, because many persons experience adverse developmental environments similar to those reported by the DD subjects yet do not develop conditions, this type of explanation must be applied conservatively: Symptom onset should be closely tied to the time when one was mistreated. Explanation 8 (Normal genetic makeup + adverse current environment) would be supported if positive changes in the social environment correlated with the disappearance of symptoms and the use of more effective strategies. As noted in Table 14.3, symptoms did not entirely disappear among the DD subjects, and no DD subject was entirely symptom-free at the end of the study. Further, compromised capacities to develop novel strategies persisted among the DD subjects throughout the study. Thus, explanation 8 is unlikely. None of the explanations is incompatible with the postulate that persons should be unresponsive to environmental information, a point underscored by the finding that symptoms declined among all the DD subjects during the study (Table 14.3). In effect, despite the persistence of suboptimal infrastructures, some degree of symptom reduction can occur when individuals are in positive and supportive social environments (e.g., Howland and Thase, 1991).

To summarize, Explanations 1, 2, 3, and 4 appear to be the likely explanations for DD, with Explanations 1 and 2 tentatively accounting for more than 70% of our DD subjects. We give greater weight to Explanation 2 than to 1 because (1) our pedigree data revealed a mixture of disorder-related signs and symptoms among relatives rather than signs and symptoms consistent with DD; (2) our DD subjects were in many ways indistinguishable from our control subjects, a point suggesting that trait variation was confined to some but not all traits; and (3) our DD subjects were capable of utilizing unafflicted or minimally afflicted infrastructures to minimize some of the consequences of their condition and partially achieve a limited number of short-term goals in particular environments.

The characteristic psychophysiological, neuropsychological, and algorithm (causal modeling and strategy development) assessments of the DD subjects changed minimally over the course of the study. Thus, despite a reduction in symptom number, clear indications of both suboptimal automatic systems and algorithms were present throughout the study (Howland and Thase, 1991). The degree of symptom reduction in the DD subjects is particularly informative in that it illustrates the degree to which symptoms can fluctuate across environments. When symptom reduction is combined with findings showing enduring psychological, physiological, and functional features, distinctions between suboptimal and dysfunctional infrastructures are possible (e.g., the findings from the study suggest that dysregulated physiological systems were secondary, not primary, contributing factors to the symptom states of the DD subjects).

The preceding points take us back to behavior systems and functional units. Clearly, there were functional consequences associated with DD (Table 14.4). Moreover, a large percentage of the consequences were associated with reproduction-related behavior. This finding suggests that the reproductive behavior system may be the primary suboptimal system in this condition (or group of conditions), and that the symptoms largely reflect a failure to achieve reproduction-related goals.

Our analysis does not fully take account the 15% principle discussed in chapter 7. For example, studies suggest that two branches of the sympathetic nervous system (neural and adrenal) are hyperactive in persons with DD (Lechin et al., 1995), and these and other DD-related findings have not been discussed. Further, the reasons for symptoms in many of those with DD may differ from those for major depression

(Lechin et al., 1995); that is, the symptoms of depression may not be best character-ized on a continuum. Evidence compatible with this point was found in the response of our DD subjects to drugs. None of the DD subjects who had received drug treat-ment prior to entering the study experienced a full reduction in their symptoms, only a partial reduction. However, drugs often lead to significant symptom reductions in persons with major depression.

Was the behavior of our DD subjects adaptive? On the whole, we would answer no. Can DD be characterized as a *core adaptation* as discussed in chapter 7? Again, we would answer no. Rather, both the data and the interpretations point to multiple causes of conditions, common final pathway constraints, and limits on the degree to which core explanations apply. *Thus, Explanation 9 is unlikely.* However, different features of the behavior of the DD subjects in this study do qualify as attempts to adapt, for example (1) avoiding social and work environments associated with cost-benefit deficits and increased symptoms; (2) attempting to interact with individuals who provide more help than they require in return; (3) increasing their social support network while in the study; and (4) obtaining abortions when they had neither the resources nor the social environment to raise children. On the latter point, these DD subjects were surprisingly realistic about the costs associated with having offspring, a point that again suggests that their self-monitoring capacities did not differ signifi-cantly from those of the control subjects. The DD subjects can thus be characterized as attempting to adapt within the constraints of their condition: When features such as emotional freezing, limited strategy development capacities, and rigid causal mod-eling are combined with accurate self-monitoring, the DD subjects were left with little choice but to utilize strategies that were marginally adaptive for achieving short-term biological goals. That they persisted in their attempts to achieve biological goals sug-gests that their motivations-goals and resource allocations were normal; for example, ultimate cause contributions to behavior (e.g., goal seeking, engaging others) persisted while the consequences of proximate disorder-contributing factors (e.g., pedigree in-fluences, trait variation) compromised the ability to carry out ultimately caused behav-iors. The chronicity of these features, when tied to reduced goal achievement, can be viewed as the primary reason for chronic symptoms among those with DD. *Thus, Explanation 10 is likely.*

Finally, it is worth asking if the reported male-female prevalence differences in DD are elucidated by this study. Given that the social navigation capacities of the DD subjects were suboptimal, and given that reproduction-related goals and the develop-ment and maintenance of social support networks are more central to female than to male identity, a greater prevalence of DD among females would be predicted.

Concluding Comments

This chapter has focused on findings from a study of subjects with dysthymic disorder, which have been interpreted in an evolutionary context. Evolutionary explanations were used to narrow possible causal explanations, and our analysis led to the view that pedigree influences, trait variation, and developmental environments are key ex-planatory factors in this condition. Different causal profiles were emphasized. The overlap in symptoms and signs was viewed as a consequence of common final pathway constraints. The analysis is consonant with the view that an expanded understanding of disorders requires the collection of information often overlooked by the prevailing models, that trait variation is a more likely disorder-contributing factor than is normally appreciated, and that viewing disorders in an evolutionary context offers novel insights into disorder causes and consequences.

PART IV

TREATMENT IN EVOLUTIONARY CONTEXT

15

Intervention Strategies

"Is psychiatry terminally ill?" one psychiatrist pondered (Genova, 1993). Probably not. Mental afflictions are ever present, as they have been for centuries. Those who are mentally ill desire and deserve treatment. And the causes of their afflictions, as well as their optimal treatment, remain subjects of clinical, public, and scientific concern. Because of its long history of caring for and studying persons with conditions, psychiatry is well positioned to address these concerns. Yet, if psychiatry wishes to maintain its position, changes are in order. Its approach to classification creates as many problems as it solves. Much the same may be said about the priorities it assigns to symptoms and signs relative to function. The future of psychiatry hinges largely on its ability to treat individuals effectively. But better treatment presupposes an understanding of the multitude of factors that contribute to and alter conditions. How do we approach treatment?

The most obvious way is to evaluate the intervention outcome literature. What one finds is literally hundreds of reports, many of which chronicle impressive results. Drugs reduce the symptoms of anxiety and depression, they modulate the mood fluctuations of bipolar disorder, and in some instances, they improve function and reduce symptoms of persons with schizophrenia. Specific phobias can be ameliorated by behavioral interventions. Personal and socially unproductive strategies can be altered with psychoanalysis, psychotherapy, and cognitive therapy. Electroconvulsive therapy (ECT) is effective in improving cognitive functions and reducing depressed mood in drug-resistant depressions. And environmental change leads to symptom and sign reductions, as well as to improvements in short-term goal achievement (e.g., Bowden and Rhodes, 1996; Herz, 1996; Stein and Jobson, 1996). When the preceding findings are viewed from afar, they seem to point to one conclusion: Treatment has improved and is likely to continue to do so.

Nonetheless, disconcerting facts remain:

1. Thus far, treatment is effective for only a percentage of the currently classified disorders.
2. For many disorders, it is not clear which intervention is most effective, which is to say that more than one intervention may be effective.
3. Despite advances in both the choice and the use of interventions, there is a growing sense that many disorders are lifelong, not transient. It matters little which study one cites (e.g., Paykel and Weissman, 1973; Dilling and Weyerer, 1980; Klerman, 1980, 1989; Schulsinger, 1980; Tsuang, 1980; Shea et al., 1990; Hirschfeld, 1996), the findings are essentially the same: A percentage of mental afflictions turn out to be enduring even though multiple interventions have been tried. When the outcome findings from such studies are lumped together, one comes up with the following approximations: Treatments are highly effective 20% to 30% of the time, partially effective 40% to 50% of the time, and minimally effective or ineffective 30% of the time.
4. Recidivism rates remain high for many disorders that initially respond to treatment.
5. How therapies work is poorly understood.
6. A conceptual framework that permits predictions about which treatments are likely to work and why they work when they do remains to be developed.

The preceding points simply restate a fact familiar to all clinicians: prevailing treatments are far from ideal. They also lead to the question: Would treatment improve if an evolutionary framework for therapy is put to use? Our answer is yes. Outlining such a framework and illustrating its clinical applications are the two primary objectives of this chapter.

We begin with a discussion of the prevailing-model interventions, so that we can establish benchmarks for assessing the similarities and differences between the prevailing models and the evolutionary approach we have developed. A listing of key points integral to an evolutionary-based framework for therapy follows. Next, we discuss principles of an evolutionary approach to treatment. We conclude the chapter with illustrative case histories.

Several points help set the context for this chapter:

1. It would be surprising if many effective treatments had not been discovered and refined, as the treatment of mental conditions extends back to at least the earliest moments of recorded history. Thus, an evolution-based framework can be expected to incorporate many of the prevailing forms of therapy.
2. It would also be surprising if specific therapies were uniformly effective in the treatment of specific disorders. This point is an obvious inference from our assessment of current classification practices: Definitions are porous and invite imprecision. It is also an obvious inference from the 15% principle and common final pathway postulates: In effect, if an important contributing factor to a condition is the use of a counterproductive strategy, drugs are unlikely to change the strategy, although the symptoms associated with the condition may be ameliorated; conversely, if an important contributing factor to the same condition is an insufficient amount of a neurotransmitter, and if psychotherapy does not change neurotransmitter levels, there is no reason to expect that psychotherapy will fully resolve the insufficiency. What an evolutionary approach to treatment offers is, in part, the possibility that the number of instances in which ineffective interventions are initiated will be reduced.

3. It would be surprising if there were not conditions that failed to respond to treatment. Both trait variation postulates and the available data favor this view. The use of an evolutionary framework may not be able to alter this situation, although it can help explain it, and perhaps, it will lead to greater tolerance and more empathic responses to individuals with such afflictions.

Prevailing-Model Interventions

Each of psychiatry's prevailing models is associated with a set of preferred interventions. For the most part, these preferences are tied to views about the contributing causes.

Psychoanalytic and Psychotherapeutic Interventions

Psychoanalytic and psychotherapeutic interventions, which for this discussion include psychoanalysis, psychotherapy, and cognitive therapy, focus on intrapsychic conflicts and distortions or cognitive malfunctions as primary condition-contributing factors. Signs, symptoms, unproductive behavior, and counterproductive strategies are viewed as epiphenomena. Changing ways of processing information and changing counterproductive strategies are the primary aims of these interventions, which may require addressing conflicts, distortions, cognitive processes, and their emotion-related features. Verbal techniques, such as clarification, confrontation, explanation, and interpretation are used to identify and alter the targets of treatment. Education about effective ways of navigating the social environment (new models) may be provided. A key factor influencing success rates is an individual's capacity to utilize the interventions. For selected conditions, the interventions correlate with behavioral and symptom change, although the reported success rates vary across studies as well as across condition type (e.g., Elkin et al., 1989; Klerman, 1989; Shea et al., 1990).

Behavioral Interventions

In behavioral interventions, techniques such as desensitization, reinforcement, and flooding are the treatments of choice. Signs and symptoms are *not* viewed as epiphenomena. Atypical or inadequate learning, not underlying dysfunctional processes, are postulated as the primary condition-contributing factors. Altering what individuals have learned and facilitating new learning are the primary aims of therapy: With a change in strategies, behavior also changes. Studies confirm that behavioral interventions can be effective in treating blood phobias, tics, some features of obsessive-compulsive disorder, agoraphobia, possibly polydipsia, and suicidal thoughts (e.g., Krone, Himle, and Nesse, 1991; Viewig, 1993).

Sociocultural Interventions

In the sociocultural model, adverse social environments—both direct (e.g., interactions with others) and indirect (e.g., limited employment and social opportunities) environmental factors—are the postulated condition-contributing factors. Signs and symptoms are sometimes viewed as epiphenomena, sometimes not; sometimes as indices of an individual's capacities to manage the environment; and sometimes as the inevitable consequences of environmental perturbations. Frequently used interventions include teaching alternative interaction styles, encouraging persons to change environments, developing social support networks, and utilizing supportive public services. These techniques are often effective; for example, symptom intensity frequently declines when individuals find themselves in supportive social environments, as occurred among the dysthymic disorder subjects described in (chapter 14). Further, social environments can positively or negatively affect the clinical course of a large number of conditions, including schizophrenia, the dementias, social phobia, and posttraumatic stress disorder (Rea et al., 1991).

Biomedical Interventions

In biomedical interventions, signs and symptoms are viewed as epiphenomena of the underlying pathogenic processes. Pharmacological agents and genetic counseling are the primary intervention tools. Although there are exceptions (e.g., lithium carbonate treatment of bipolar disorder), among the prevailing models biomedical interventions are tied most directly to putative disorder causes; for example, a disorder that is thought to be due to CNS dopamine dysfunction is treated with drugs to normalize dopamine function or to offset dysfunctional effects. Currently, interventions using drugs probably exceed the number of all other interventions combined.

Because of their frequent use, it is worth taking a closer look at drug treatments. Many of their positive outcomes have already been noted in earlier chapters. When these outcomes are given a closer look, however, they apply primarily to signs and symptoms, and less often to functional improvement. Like all interventions, drugs have their limits (e.g., Bronish et al., 1985; Aguglia et al., 1993). For example, reports indicate that only about half of elderly depressed persons respond to interventions, the majority of which are pharmacological (Burvill et al., 1991; J. M. Murphy et al., 1991). A number of conditions are thought to have a significant physiological component (e.g., drug-resistant unipolar depression) for which drug interventions are not effective (Elkin et al., 1989; Guscott and Grof, 1991; Roth, 1991; Vallejo et al., 1991; Igbal and van Praag, 1993; Borison, 1995; Sharma and Janicak, 1996; Stein and Jobson, 1996). Drugs may also have undesirable side effects, such as tardive dyskinesia (e.g., from neuroleptics), blood dyscrasia, and social interaction reduction (e.g., from chlorpromazine), all of which influence compliance and may negatively influence function. Multicause hypotheses predict this array of outcomes: Physiological contributions to conditions are only part of the causal package, and contributions vary from person to person and within and across conditions.

Further, in many instances, medications are not thought to alter the primary contributing factors (vitamin supplements are possible exceptions) but may alter secondary systems. The possibility that drugs are influencing secondary systems is one of several interpretations when the time between commencing treatment and detectable clinical effects is excessive. If secondary systems are targeted, certain conditions are likely to require long-term drug treatment; otherwise, signs and symptoms will reappear. There is also the issue of multiple dysfunctional physiological systems. An axiom of clinical psychiatry is that it is counterproductive to try to treat all dysfunctional physiological systems simultaneously. The axiom makes sense. Physiological systems are exceedingly complex, and there are well-known interactions between different systems, such as neurotransmitters and second messengers (Fuller, 1992; S. Rose, 1995). Thus, no matter how specific their initial effects, all drugs influence nontargeted systems, and such influences may or may not be clinically beneficial. Further, some persons with some conditions (personality and reading disorders that are asymptomatic) are likely to have statistically normal physiological profiles, in which case drugs would be expected to have minimal effects. However, when persons with personality disorders are symptomatic, comorbidity often enters the picture: A consistent theme in the treatment literature since the mid-1980s is that persons with these disorders have a poorer outcome prognosis for *DSM-III* (APA, 1980) and *DSM-III-R* (APA, 1987) Axis I disorders than persons without personality disorders who also have Axis I disorders (Shea et al., 1990; Reich and Green, 1991). In short, there are factors that constrain the potential effectiveness of drugs.

This discussion *does not* imply a criticism of either the biomedical model or pharmacological agents. *An analysis of the other prevailing model interventions would lead to similar conclusions.* Moreover, to argue that therapy is far from perfect and that the prevailing models lack a framework for therapy *is not* to suggest that many condition-contributing factors postulated by the prevailing models are invalid: (1) Genetic mistakes do occur; (2) atypical physiological states exist; (3) persons are predisposed to conditions; (4) adverse developmental environments contribute to condition vulnerability and risk; (5) intrapsychic conflicts and atypical or dysfunctional learning take place and affect behavior; and (6) social environments are often stressful and the basis of dysregulating and condition-triggering events. But do these events satisfactorily explain conditions? Only in a limited number of instances. Far more is involved in causal explanations, as well as in treatment. Said another way, the prevailing models are too focused both in their explanations and in their interventions. As we have emphasized, two of the reasons for this focus are the method of classifying disorders and the strong tendency of the prevailing models to limit their explanations and interventions to putative proximate causes. These practices have a number of treatment-related consequences:

- Conditions, and their features, are typically viewed as epiphenomena.
- Similar phenotypes that result from different causal packages are given the same diagnosis and often receive the same initial treatment.
- Single-type rather than multitype intervention strategies are favored.
- Ultimate cause contributions to conditions are largely ignored.
- The possibility that some conditions, and their features, are adaptations or reflect attempts to adapt is seldom considered.
- Findings showing that conditions, and their features, change across social environments and that treatments that are effective in one environment (e.g., a hospital) may not be effective in other environments (e.g., at home) are inconsistently addressed.
- The full implications of trait variation are infrequently considered in either causal explanations or intervention designs.
- Detailed functional assessments (as distinguished from global assessments) are seldom used to narrow the possible causal factors or to assess the intervention outcome.

Basic Points in an Evolutionary Framework for Therapy

In our view, the essential points to incorporate into an evolutionary framework for therapy are the following:

- Like all other species, *Homo sapiens* is imperfectly developed.
- Normal phenotypes as well as conditions have multiple contributing factors (the 15% principle).
- Individuals are best characterized as mosaics of independent and semi-independent traits.
- Trait variation is applicable to all traits, and cross-person trait variation is far more pervasive than is usually recognized.
- Both ultimate and proximate causes contribute to conditions.
- The majority of conditions correlate positively with constricted behavioral plasticity and function.
- Emotions provide information about one's past, ongoing, or expected goal achievement states.
- Motivations-goals and resource allocation programs are normal in all but a few conditions.
- Individuals with and without conditions act to optimize the achievement of short-term goals, and their behavior reflects interactions between their strategies, their functional capacities, and environmental contingencies.
- The social environment can be a significant factor contributing to condition onset and to fluctuations in the clinical course.
- Self-correcting systems offset many of the effects of compromised infrastructural systems.

Intervention Principles in an Evolutionary Context

In the framework developed here, the primary goal of therapy designed in an evolutionary context is to improve individuals' capacities to achieve their short-term biological goals. This is not a new idea. Regardless of their theoretical persuasion, most clinicians have the same objective, although often, it is not explicit. It is explicit among those therapists who have discussed interventions in an evolutionary context (e.g., McGuire, 1979a; Sloman et al., 1979; Gut, 1982, 1989; Wenegrat, 1984, 1990; Glantz, 1987; Marks, 1987; Bailey, 1988; Glantz and Pearce, 1989; Williams and Nesse, 1991; Bailey, Wood, and Nava, 1992; Gilbert, 1992, 1995; Nesse and Williams, 1994; Price et al., 1994).

Behavior systems serve as the organizing system within which intervention principles are applied. To review key points from chapter 4, behavior systems are functionally and causally related behavior patterns that are mediated by infrastructural systems. To the degree that conditions are associated with compromised behavior systems, interventions that take into account the workings of these systems have a greater chance of being effective. For example, if interventions designed to ameliorate the symptoms of agoraphobia are limited to drugs and do not attempt to address an individual's inability to achieve goals because of compromised algorithms, limited therapeutic effects are to be expected, and the probability of an extended period of drug treatment is increased. In an evolutionary context, the signs and symptoms associated with agoraphobia represent various combinations of factors: (1) compromised automatic systems that contribute to misperceptions of the social environment, and particularly of environmental dangers; (2) limited capacities to develop novel behavior strategies; (3) the use of avoidance strategies to minimize the chances of being in social situations that are perceived as dangerous; (4) signals to others that the signaler is unable to manage certain social contingencies; and (5) associated psychological and physiological sequelae.

A therapeutic relationship is a high-priority goal among most clinicians, irrespective of their views about the causes of conditions. The need for clinicians to be warm, considerate, and responsive is rarely debated. A therapeutic relationship is critical for a number of reasons:

1. Warmth and understanding confirm patients' hopes in those who will treat them. Patients enter treatment with expectations about the social role of clinicians, and to the degree that these expectations are fulfilled, the chances of successful treatment are improved.
2. Considerate and caring behavior reduces the uncertainty that patients feel when receiving treatment.
3. Warmth and understanding have psychological and physiological regulating effects; for example, *empathy is not a weak-kneed, shallow concept, but a therapeutic tool.* The discussions of RDT (chapter 8) and the environment-symptom interaction among subjects suffering from dysthymic disorder (chapter 14) illustrate this point.

Treatment Principles

An evolutionary framework for therapy can be viewed as a set of principles that can be conveniently divided into three groups: information collection, causal analysis, and intervention options and strategies.

Group 1: Information Collection

PRINCIPLE 1: FUNCTIONAL EVALUATION   Functional evaluations focus on the development of functional impairment and the sign, symptom, and behavior profiles that accompany impairment. Historical data are essential to determining the time of onset, the duration, the intensity, and the consequences of impaired functionality in order to (1) identify functions that are and *are not* impaired; (2) assess strategies that are used to compensate for functional limitations; and (3) identify contributing infrastructural and environmental factors. For most conditions, physiological measures are minimally informative. Exceptions include disorders that strongly correlate with specific physiological states, such as hyperthyroidism, Korsakoff's syndrome, and substance abuse. In a similar vein, many investigative techniques, such as magnetic resonance imaging (MRI), may be of value for a small percentage of conditions (e.g., MRI, when changes in CNS anatomical structures are suspected), but not for the vast majority.

PRINCIPLE 2: FUNCTIONAL CAPACITY EVALUATION   Functional capacity evaluation involves the assessment of the items listed in Table 4.2. Historical data provide critical information about which capacities were compromised prior to the onset of a condition. For example, in persons with personality disorders, compromised capacities are often present prior to diagnosis, while in persons with posttraumatic stress disorder, compromised capacities often develop following a traumatic event. Such comparisons are necessary to decisions about whether interventions should attempt to restore prior capacities or attempt to improve longstanding limitations.

PRINCIPLE 3: EVALUATION OF BIOLOGICAL GOALS AND THEIR CURRENT PRIORITIES   Assessing current goal priorities is an essential step both in identifying the functional effects of compromised automatic systems and algorithms and in determining intervention choice. For example, adjustment disorder occurs at different ages and in association with different environmental perturbations. A child may develop this condition in association with punitive and restrictive parents, while a recently married female in her mid-20s may develop the condition in association with her husband's infidelity. Goal priorities will differ across these situations, as will environmental contingencies and intervention options.

PRINCIPLE 4: RESOURCE ALLOCATIONS   Like motivations-goals, and with few exceptions, resource allocation programs are assumed to be similar among persons with and without conditions. Further, the upper limits of allocation are seldom reached. Nevertheless, how persons allocate resources is important to specify. Allocations reveal goal priorities, influence behavior strategies, and frequently contribute to conditions. For example, persons with narcissistic personality disorder frequently persist in self-interested attempts to control others in ways that others resist; that is, goal priorities and associated resource allocations endure even though the goals are not achieved. Symptoms often follow.

PRINCIPLE 5: THE COSTS AND BENEFITS OF ACHIEVING BIOLOGICAL GOALS   Pursuing goals has costs, and it may have benefits. One can estimate cost-benefit outcomes by evaluating the time and effort (costs) required to achieve goals relative to the benefits received. The use of time budgets is helpful in such assessments. For example, if a person spends an excessive number of hours courting a potential mate yet fails in his or her efforts, the costs accumulate and the benefits are not forthcoming. Or if a person spends several hours deliberating over a simple decision, such as whether to have lunch with a friend, and the customary amount of time for such decisions is one to two minutes, excessive costs have been incurred irrespective of the decision. Similar points apply to attempts to adapt when one is symptomatic, such as efforts to reduce anxiety through increased social bonding: One's strategies may be efficient or inefficient. Assessments in this category serve to narrow the types of interventions that may be effective; for example, automatic system dysfunctionality, which leads to misperceptions invites a different type of intervention strategy from algorithm dysfunctionality that leads to inaccurate self-monitoring.

PRINCIPLE 6: PERSON-ENVIRONMENT INTERACTIONS   This assessment has several parts:

1. One part deals with social contingencies. Social structure, social options, and the demand features of the social environment interact with the degree of infrastructural functionality, as well as the probability and the course of conditions.
2. A second part addresses how well persons understand their environmental options. Conditions in which recognition systems are compromised strongly correlate with misreadings of options.
3. A third part assesses in what kinds of environments individuals are capable of achieving biological goals and whether individuals are locating themselves in such environments. The outcome of these assessments not only determines if environmental change is advisable but also further specifies which infrastructural system(s) requires treatment.

PRINCIPLE 7: CONDITION-CONTRIBUTING TRAITS AND STATES   An evaluation of condition-contributing traits and states provides information about possible causes and treatment options. Evaluations of recognition and signaling similar to the one described in Figures 6.1 and 6.2 can serve as a prototype for this assessment; for example, missing connections point to infrastructural deficits and set limitations on therapeutic options, while inhibited and hyperactive connections suggest dysfunctional infrastructures and invite specific yet different types of interventions.

PRINCIPLE 8: STRATEGIES   Viewing conditions as, in part, manifestations of strategies influences therapeutic choices. Strategies may be efficient or inefficient. They may or may not be relevant to achieving goals in specific environments. And in many instances, they may not be apparent; for example, the symptom-sign-behavior complex of schizophrenia is often so striking that the efforts of schizophrenic individuals to avoid specific types of social or physical environments are overlooked. When strategies are not apparent, detailed observations of behavior may be required. When strategies are ineffective, they are often the optimal targets of treatment.

Group 2: Causal Analysis

PRINCIPLE 9: ULTIMATE CAUSE CONTRIBUTIONS   This assessment involves the identification of interactions between conditions and ultimately caused behavior. *Ultimate cause contributions to conditions are most apparent when motivations-goals persist despite the failure to achieve goals.* This type of interpretation was offered for personality disorders (chapter 9), somatoform disorders (chapter 13), and anorexia nervosa (chapter 10). Ultimately caused behavior, such as parent-offspring conflict, sibling-sibling rivalry, and kin investment conflicts, is sometimes the source of conditions, but equally often, conditions are the source of conflicts. Sorting out these possibilities is essential, and treatment will vary depending on the outcome of the sort.

PRINCIPLE 10: INFRASTRUCTURAL FUNCTIONALITY   Automatic system assessments are made by evaluations of information biases and distortions (how information is selected, organized, and prioritized), as well as of the intensity and longevity of moods, particularly those for which there are no obvious precipitating factors.

Algorithm assessments are made by evaluations of signature features of individual algorithms, for example, an individual's capacity to develop causal models.

PRINCIPLE 11: PROXIMATE STATES AND EVENTS   Evaluations of dysfunctional psychological states, dysfunctional physiological systems, the consequences of atypical genetic information, and interactions between the environment and physiological and psychological states are examples of proximate state and event assessments. Putative proximate causes are typically the preferred targets in interventions, and depending on which proximate states and events are thought to be contributory, interventions will differ. Still, as we have stressed, identifying ultimate contributions is as essential as identifying proximate contributions, and to the degree that ultimate causes are incorporated into intervention formulations, the chances of successful treatment increase.

PRINCIPLE 12: THE 15% PRINCIPLE   Application of the 15% principle requires a consideration of each of the likely condition-contributing factors (e.g., predispositions, within-trait variation, infrastructural suboptimality and dysfunctionality, environmental contingencies) so as to develop relevant causal explanations that will both narrow and specify treatment targets.

Group 3: Intervention Options and Strategies

PRINCIPLE 13: IDENTIFYING INTERVENTION GOALS   Given the preceding principles, as well as our view that the most important goal of psychiatric interventions is to improve short-term goal achievement, intervention goals can be conceptualized in terms of options and constraints:

Restore prior trait or state functionality. Interventions designed to restore prior trait or state functionality assume that (1) precondition states or traits can be identified; (2) restoration is a desirable therapeutic objective; and (3) the factors contributing to current dysfunctional states or traits can be reversed. If these three conditions are met, proximate factors are often at work. Therapies favored by any of the prevailing models may be effective, for example, removal of a person from a stressful environment, behavioral modification, or psychotropic medications. Further, sequential treatments may be necessary; for example, an initial aim of therapy may be to establish a relatively stable biological substrate as a prerequisite for psychosocial or behavioral interventions. In such instances, drugs may be the initial intervention choice (e.g., Abroms, 1983). To the degree that behavior is a form of knowledge, and that behavioral change leads to increased regulation, behavior can be an intervention target.

Undo constrained maturational programs. The goal of this type of intervention is to liberate constrained or unrefined infrastructural systems. In practice, this approach usually amounts to altering a patient's characteristic ways of feeling and socially relating. Key steps in this process are (1) refining algorithm capacities so that persons more accurately model both their environments and themselves; (2) improving functional capacities; and (3) enlarging a person's range of experiences and social options. Initially, psychotherapy may be most effective, but drugs may be helpful in reducing symptoms at critical points during treatment. Environmental change may be essential at later stages of treatment.

Improve chronically compromised infrastructural functionality. Persons differ in their infrastructural functionality, and these differences contribute directly to conditions; for example, an inability to develop and execute novel behavior strategies is frequently the basis of dysregulation. However, altering minimally adaptive capacities is usually time-consuming. New experiences, which often require environmental change, combined with psychotherapy, cognitive therapy, or skill training, are frequently essential if significant improvement is to be achieved and lasting. Drugs may be helpful at specific points during treatment.

Ameliorate enduring symptoms and signs due to suboptimal infrastructures. For signs and symptoms that are enduring and result from suboptimal infrastructures that are largely refractory to refinement, drugs and environmental placement are most effective.

More specifically:

- For signs and symptoms that individuals find undesirable and that are minimally adaptive, treatment should attempt to reduce the undesirable features, provided that infrastructural systems and goal achievement are minimally compromised. Examples include the reduction of excessive anxiety or depression, hallucinations, concentration difficulties, excessive fears that lack a basis in reality, and undesirable persistent thoughts.
- For behavior that is socially undesirable and that compromises goal achievement, therapy should attempt to reduce the socially undesirable behavior, provided that existing infrastructural capacities and goal achievement are not compromised. Examples include reductions of threatening behavior, impulsivity, manipulativeness and deception, stubbornness, fetishes, perversions, addictions, and self-preoccupation. When infrastructures are suboptimal, this type of intervention often requires the development of alternative ways of achieving the same goals, and thus, treatment often is extended.
- For suboptimal social functional capacities that are due to trait variation, therapy should attempt either to refine traits or to foster the use of alternative capacities that improve the likelihood of achieving high-priority short-term goals.
- For behavior that is adaptive in some environments but minimally adaptive in others (e.g., assertiveness that is valuable in the marketplace but is often counterproductive in personal interactions), therapy should facilitate the development of revised models of the social environment and should attempt to improve patients' capacities to selectively exercise behaviors associated with goal achievement.
- For behavior that is adaptive, no change is required unless a behavior negatively affects others (e.g., antisocial personality disorder).

The latter point deserves some clarification, particularly regarding the treatment of conditions that can be adaptive by evolutionary criteria (e.g., instances of antisocial and histrionic personality disorders, malingering, ADHD, adjustment disorder) but that are often socially undesirable. Attempts to treat these conditions are socially justified and appropriate. However, justification does not take away from two important facts: (1) By evolutionary criteria these phenotypes may represent evolved high-risk strategies, and (2) successful treatment may require that they be understood and treated differently from conditions that have their origins in compromised anatomical or physiological systems.

PRINCIPLE 14: TREATING CONDITION-CONTRIBUTING FACTORS, NOT SECONDARY EFFECTS   Treating primary contributing factors makes more sense than treating secondary systems. Moreover, grouping conditions by behavior systems (Table 7.1) is likely to facilitate the identification of contributing factors. The clinical cases that follow illustrate this point. However, treating primary factors is not always possible, as in instances in which causes are unknown or traits are suboptimal and highly resistant to change, for example, in schizoid and schizotypal personality disorders. In such instances, secondary effects become primary treatment targets.

The evolutionary approach avoids designing interventions on the basis of DSM-*type disorder classifications. Until otherwise established, it does not assume that conditions, or their features, are epiphenomena, and it does assume for all but a few conditions that persons are* locked in *to their social environment. Improving the chances of achieving short-term biological goals by altering the functionality of infrastructural systems and optimizing a person's environment, while not compromising adaptive capacities, constitutes the primary therapeutic aims. Systems, not mechanisms, are the targets of treatment.*

Case Histories in an Evolutionary Context

Three types of cases illustrate the clinical use of the framework: (1) those in which evolutionary analyses expand prevailing model approaches; (2) those in which evolutionary analyses do not enhance the prevailing-model approaches; and (3) those in which an evolutionary approach is essential. As in the earlier clinical examples, the full complexity of individual case histories is not addressed here. Rather, evolutionary interpretations are emphasized to illustrate how an evolutionary approach enriches both our understanding of treatment and our treatment options.

Cases in Which an Evolutionary Analysis Expanded the Prevailing-Model Treatment Approaches

Case 1: Changing the Social Environment and Decreasing Medications by Using Intact Algorithms to Foster Regulation-Enhancing Behavior

The patient was a 25-year-old unemployed female who, two years prior to treatment, had begun suffering from chronic anxiety and one or two severe anxiety attacks per week. The onset of the anxiety correlated with her graduation from college and the emergence of fears that she would fail to obtain a job or to marry. One year prior to treatment she had begun taking excessive amounts of antianxiety medications. The medications reduced the severity of her anxiety attacks, but not her chronic anxiety. Several self-initiated attempts at discontinuing medications failed. For the eight months prior to treatment her social life had deteriorated, and when therapy began, it consisted primarily of "running around from bar to bar and boy to boy," activities in which she experienced minimal pleasure. Clinical evaluation suggested that the patient was bright and aware of both her deteriorating condition and her unproductive behavior (i.e., intact self-monitoring capacities); that selected algorithms (e.g., novel strategy development) were dysfunctional; that she was motivated to reduce her dependence on drugs; and that her capacities to assess the costs and benefits of using alternative short-term goal achievement strategies were intact.

Evolutionary Analysis

The patient's anxiety was interpreted as an indication that she anticipated that she would not achieve her reproductive and resource acquisition goals. Treatment was formulated in terms of utilizing her capacities to self-monitor and to accurately evaluate her condition in order to reduce the dysfunctionality of her algorithms for selecting optimal environments, improving her capacities to develop novel behavior strategies, and progressively reducing her drug dependence.

Intervention

Therapy, which included psychotherapy, drug tapering, and changing the social environment, was initiated. Psychotherapy began with an assessment of her social environment and its negative cost-benefit outcomes. After three months of therapy, the patient began making new friends and spending evenings in recreational activities in preference to visiting bars. After four months, both the severity and the frequency of her anxiety attacks had begun to decline, and tapering of medication was initiated. Her anxiety attacks became more frequent for three weeks following the initial reduction in medications but then declined in frequency. Medication tapering continued, and the medications were discontinued at Month 6. From then until the end of therapy at Month 14, the patient experienced three anxiety attacks, each one milder than the previous one, and her chronic anxiety disappeared. During this period, psychotherapy focused on improving her capacity to develop novel strategies. At Month 8 of treatment, she began dating the same person, and at Month 10, she obtained a job and applied to graduate school. At Month 13, she was accepted into graduate school. A two-year follow-up revealed that she had continued to develop new friends, had established an enduring relationship with a male, and was continuing graduate school, and that her anxiety attacks had ceased except for occasional periods of anticipatory anxiety associated with school examinations.

Comments

The intervention was designed to take advantage of the patient's capacity to accurately assess her situation and her motivation to change her behavior so that she could achieve her high-priority goals (a relationship with a man and a good job). The initiation of infrastructural regulation, associated with environmental change, offset the unpleasant effects of discontinuing medication and eventually made the medication unnecessary.

As noted, the case covers only selected features of the patient's past history, current life circumstances, and treatment. During treatment, other factors had to be addressed, for example, the possible causes of her fears, her compromised ability to develop novel strategies, her willingness to take excessive amounts of drugs, her failure to discontinue drug use, and her repeated use of strategies that she found minimally satisfying.

Case 2: Treating Proximate Mechanisms That Contribute to Algorithm Dysfunctionality by Using Psychotherapy and Environmental Alteration

The patient was a 55-year-old married male with five children. He was a successful businessman who had a lifelong commitment to kin investment through resource acquisition—holding down a good job. Eighteen months prior to entering treatment, he had significantly expanded his business. His work week had shifted from approximately 45 hours per week to 75 hours because of additional management responsibilities and increasing financial losses in his business.

Approximately six months prior to seeking therapy, he had begun suffering from depression, severe enough so that he would sometimes stay home from work. He became extremely critical of his wife and employees. Medications provided by a friend and several physicians were unsuccessful in reducing his symptoms.

Evolutionary Analysis

The patient's depression was interpreted as an indication that he was experiencing a severe cost-benefit deficit with respect to his kin investment goals. The treatment was formulated to address the patient's strong need to achieve resource acquisition goals to benefit kin, and to address his compromised capacities to develop strategies for decoupling himself from a situation in which his goals were not being achieved. The specific goals of the therapy were to develop more realistic causal models of 'his capacities to invest in kin, the behavior of those in his social environment, and his own behavior.

Intervention

Drugs were discontinued. A combination of psychotherapy and environmental alteration was initiated. For the first four months, psychotherapy focused on the consequences of his depression, his inability to extricate himself from his work, and the deteriorating quality of his life. In the fifth month, the therapist recommended that the patient consult an accountant to review the financial state and operation of his business. The patient accepted the recommendation reluctantly. The accountant advised the patient to downsize his business. After much deliberation, the patient agreed to do so, and for several weeks 'his depression increased. Over the subsequent three months, the patient began to carry out the accountant's recommendations. Six months later, the patient's business again became profitable. 'His depression began to decline, and his relationship with his family and his employees improved. Therapy was discontinued at Month 13, at which time the patient was suffering from occasional brief periods of depression. A two-year follow-up revealed that his signs and symptoms were no longer present and his business continued to be profitable.

Comments

The intervention illustrates the use of techniques to reduce the environmental and personal constraints that interfere with achieving high-priority goals in a highly motivated individual with limited capacities to alter his behavior in response to negative cost-benefit outcomes. Psychotherapy brought into focus his inability to develop novel ways of handling his business difficulties and, eventually, an increased willingness to seek and utilize professional advice. The return of his business to a profitable state (increased goal achievement), along with the reduced work hours, contributed to his physiological regulation, a reduction in his symptoms, and improved algorithm functionality.

Again, the presentation does not do justice to the complexity of the case. For example, there were indications that the patient's self-worth was closely tied to his ability to meet the financial requests of his wife, children, and kin, and that their requests were excessive. Further, his wife and children were minimally helpful once he became depressed. The fact that the patient responded to his circumstances with severe depression rather than other symptoms suggests a vulnerability to depression. And there is a strong possibility that similar circumstances in the future could lead to depression.

Case 3: Combined Interventions to Alter Algorithm Functionality

The patient was a 31-year-old married female, a mother of three children, and an employed schoolteacher. Her past was characterized by above-average commitments to the care of her children, the management of her home, and performance as a good wife and successful teacher. The patient was loved by her family and appreciated by her students and peers. However, her failure to meet her own expectations resulted in periods of self-depreciation, fears of rejection, and nonspecific anger. Six months prior to entering therapy, the patient's mother, with whom she had had a close relationship, had died unexpectedly. The patient had responded with a severe depression, coupled with the feeling that she had somehow failed her mother. Apprehension about teaching followed. Her relationships with her husband, children, and friends deteriorated. Three months prior to therapy, she had become phobic about school, and a medical leave of absence was arranged. Medications provided by a family doctor had had minimal effects on her symptoms. At her husband's insistence, she agreed to consider psychiatric treatment.

Evolutionary Analysis

The patient's symptoms were interpreted in terms of existing cost-benefit deficits (depression) for kin investment and reciprocation goals, and of a fear of increasing the deficit in social encounters (phobia). The treatment was formulated to refine the algorithms associated with her cost-benefit assessments of kin investment and reciprocation. The specific goals of therapy were to improve the patient's capacity to accurately assess others' feelings, particularly those associated with her fear of shame and fears of rejection if she failed to meet her own perfectionist expectations, and to address her misperceptions of having failed her mother.

Intervention

Two therapists participated; one provided psychotherapy, and the other provided behavior therapy. The patient's medications were discontinued. The psychotherapy focused on others' responses to the patient (others' needs and feelings). The behavior therapy focused on her phobia about teaching. After three months, her phobic symptoms began to decline. Signs of an increasing capacity to accurately perceive others' feelings were first apparent at five months, when the patient began utilizing novel social interaction and self-assessment models that were being discussed in psychotherapy. At Month 6, she began interpreting her response to her mother's death. Because her family remained supportive, environmental change was not required. Following seven months of treatment, her symptoms began to decline, and her relationship with her family started to improve. Soon after, she began participating in household activities. At nine months, she applied to return to her teaching job. Three months later, she was reinstated, and treatment was discontinued. Improvement continued through a two-year follow-up.

Comments

This case illustrates an intervention strategy designed to reduce automatic system dysfunctionality and thus improve algorithm function. Behavior therapy led to a reduction in the patient's fears about social encounters and an improvement of her dysregulated state. Psychotherapy led to her refining her capacities to more accurately assess others' thoughts and feelings and her self-worth. In turn, the patient began to perceive her social environment differently, particularly her importance to others.

From a more detailed perspective, the case illustrates how a person who is vulnerable to developing depression and anxiety (phobia) can become relatively free of symptoms in an "ideal" environment—loved by her family and pupils—which she had worked hard to create (e.g., her above-average commitment to her family and teaching). The capacity to carry out such commitments implies that many infrastructural systems were intact and functioning.

Cases in Which an Evolutionary Approach Did Not Improve Intervention Outcomes

Case 4: Failure of Combined Therapy to Change a Trait

The patient was a 23-year-old single male with a long history of antisocial behavior and several minor convictions for criminal behavior. When he was 19, his family had legally disowned him. Up to that time, the family had provided a supportive environment. The patient had frequently engaged in petty criminal behavior, and he had traveled from city to city in order to avoid the police. He had suffered from intermittent periods of anxiety, usually associated with periods of being alone, and he had a history of multiple unsuccessful social relationships, coupled with an active, diverse, and reasonably satisfying sexual life. Except for police interference, he *was not* displeased with his behavior or his life. Following a criminal conviction, he entered treatment at the insistence of the court.

Evolutionary Analysis

The patient's anxiety was interpreted as an indication that he anticipated failing to achieve his reproductive and resource acquisition goals by using his present strategy. The intervention was formulated in terms of the patient's strong need to achieve short-term goals associated with sexual conquests and resource acquisition; his suboptimal capacities for reading others' behavior rules; and his poor self-monitoring (e.g., assessing the effects of his behavior on others). The specific goals of treatment were to increase his capacities to seek long-term goals and to refine algorithm capacities associated with his attachment and affiliative behaviors. The prognosis was guarded.

Intervention

Psychotherapy and pharmacological therapy were initiated. Antianxiety drugs were prescribed for ad hoc use during periods of anxiety. The psychotherapy focused on the effects of the patient's behavior on others. Initially, the treatment appeared to succeed, in that the patient temporarily discontinued his criminal behavior and engaged in efforts to reunite with his family. However, after three months, the frequency and intensity of his anxiety attacks returned to their pretreatment level, excessive drug use was apparent, and the patient engaged in a series of minor criminal behaviors. After five months, the patient discontinued treatment. A two-year follow-up revealed that the patient had been in jail for 18 months following a series of thefts.

Comments

From an evolutionary perspective, the intervention failed largely because the costs associated with normalizing behavior exceeded the perceived or experienced benefits of doing so and because the patient was not distressed by his behavior. The case illustrates the therapeutic difficulties associated with attempts to treat individuals who engage in high-risk strategies despite strategy failures.

This case closely mirrors the description of antisocial personality disorder (chapter 9), except that this patient was less successful in his strategies than many others who engage in similar high-risk strategies. In addition, the treatment of persons who utilize such strategies is often complicated by an intrinsic distrust of others, which reduces the potential influence of therapists.

Case 5: Combined Drug Treatment, Behavior Therapy, and Environment Change, and a Failure to Alter Algorithm Function

The patient was 33-year-old male with a history of fleeing intimate relationships with females. He had been engaged several times but never married. Evaluation revealed there was a strong correlation between the degree of intimacy, the intensity of his anxiety, and the fear that women would hurt him. The fear of being hurt mirrored an experience with his mother when the patient was a year and a half old. She had briefly left him in a train station only to die moments later in a freak accident. Evaluation also revealed that the patient had devoted considerable effort to overcoming his fears; that he had tried a number of different therapeutic approaches, none of which had been successful; and that he nearly always interacted with females "on their territory."

Evolutionary Analysis

The patient's anxiety was interpreted as information that he would fail to achieve reproduction-related goals. The treatment was formulated to address the developmental disruption (his mother's death) that had constrained the patient's ability to refine the capacities associated with mate acquisition and, in particular, to tolerate and enjoy interpersonal and sexual intimacy. The specific goal of the treatment was to undo developmental constraints without compromising the patient's capacities to develop novel scenarios and behavior strategies.

Intervention

Antianxiety drugs were prescribed for use in situations in which the patient anticipated anxiety. Behavior therapy focused on reducing his fears of intimacy and of abandonment. The patient was encouraged to meet with his girlfriends either at neutral locations or in his own apartment. He followed these recommendations. However, seven months of therapy led to no appreciable change in the patient's behavior or fears. Therapy was discontinued. A two-year follow-up revealed that he remained essentially unchanged from his pretreatment state.

Comments

While pharmacological intervention partially reduced the patient's anxiety, the combination of behavior therapy and environmental change did not significantly offset the patient's fears of intimacy. Thus, the patient was left with the option of not engaging in intimate relationships if he wished to avoid undesirable symptoms. The case illustrates the limitations of the available interventions for dealing with severely disruptive events and their maturational consequences.

The effects of some developmental disruptions are difficult to alter despite subsequent positive experiences and therapy. This case illustrates the consequences of a negative model of relationships (closeness leads to loss) developed at a critical period during maturation. Such models are closely akin to lifelong phobias that are manifested only in particular circumstances and that override intellectual capacities suggesting that one's fears should be disregarded.

Cases in Which an Evolutionary Approach Was Essential

Case 6: Environmental Alteration and Symptom Decline

The patient was a 37-year-old unmarried female who was unable to bear children because of a constricted uterus. She had begun suffering from depression at age 28. Her depression had remained mild through her early 30s. While she had been able to maintain a job and participate in social relationships, she had avoided close relationships with men because of her fear of rejection should they become aware that she could not have children. At age 36, her depression had worsened and had begun interfering with her capacity to work efficiently. After several warnings about her job performance, she sought therapy.

Evolutionary Analysis

The patient's symptoms were interpreted as resulting from her having failed to have children. The treatment was formulated to address the dysregulating effects of her inability to reproduce (depression) and the compromised algorithms that led to her inability to develop novel ways of dealing with her reproductive incapacity. The specific goal of therapy was to facilitate kin investment.

Intervention

Combined psychotherapy and pharmacological therapy were used. The psychotherapy focused on the causes of her dissatisfactions. Concurrent antidepressant medication was moderately effective in reducing her symptoms. However, medications did not result in improvement in her job performance. During the fifth month of therapy, the patient was encouraged to visit her five siblings, who lived in another part of the country. She took a three-month leave of absence from work and made the visit. While visiting, she experienced a significant decline in her depression and a strong desire to help her siblings raise their children. She returned to therapy and discussed her desire to relocate near her brothers and sisters so that she could assist them with their children. The move was encouraged. Two months later, the patient moved. Within six months, she had discontinued her medications and was essentially symptom-free. A two year follow-up revealed that she remained symptom-free, felt that she was a useful family member, and had found a new job.

Comments

This case illustrates how important reproduction is to female identity, the often dys-regulating effects of reproductive failure, and the importance of environmental change to facilitate goal achievement (e.g., investment in collateral kin). It is doubtful that this patient's compromised capacities to develop novel strategies improved, but her new environment did not require such a change.

Other factors are relevant to this case. Many women who are unmarried or who are unable to have offspring manage their situation without developing a mental condition. In this case, however, the patient was vulnerable to depression, and the depression might well have developed even in other circumstances (e.g., divorce, rejection by siblings). Moreover, it is doubtful that this vulnerability was altered though her symptoms declined. Thus, further periods of depression are likely due to changing circumstances (e.g., her nieces and nephews mature and no longer require her assistance).

Case 7: Hierarchy Manipulation, Changing Social Status, and Improved Algorithm Functionality in Chronic Schizophrenic Patients

This case is a brief report of an experiment conducted by one of the authors (MM) to assess the effects of social status manipulation on the behavior and drug requirements of chronic schizophrenic patients. Prior to the intervention, male outpatients (subjects) with the diagnosis of residual schizophrenia were evaluated for their social capacities and their medication requirements. The intervention consisted of bringing together six groups of eight subjects to work on a Christmas toy project for economically disadvantaged children. Parts for toys were provided, and the toys could be either assembled by one person or constructed in "assembly-line" fashion, each person in the assembly line performing a special function. Deadlines were set for productivity, and rewards were promised to members of the most productive groups. The project was designed to continue for 20 weeks. The initial leader of each group was an experienced psychiatric nurse who was instructed only to be present, not to manage the project. Medical management of the patients was provided by physicians not familiar with the details of the study. The physicians were instructed to reduce the medications if reductions were clinically indicated. Once members of the group became familiar with the details of the project, all six groups chose to assemble the toys in assembly-line fashion. Following the third group session, the nurse-leader for each group was assigned to another job, leaving the groups without a staff member. In five of the six groups, a leader from among the subjects emerged to direct the project. The medication requirements of all five subjects who became leaders declined, and their social capacities improved. There was essentially no change in these measures for those subjects who did not become group leaders.

Evolutionary Analysis

The intervention was formulated on the basis of data showing that social status is associated with specific physiological states, and that high-status individuals are more physiologically regulated than low-status individuals (chapters 5 and 8). The goal of therapy was to maximally reregulate physiological states within the constraints of the patients' disorder.

Intervention

The intervention is described in the introduction to this case and did not involve a traditional type of treatment.

Comments

One important point of this case is that persons with severe, debilitating, chronic conditions often remain responsive to certain types of interventions—social manipulation in this example—even though such interventions only partially influence their conditions. This point is consistent with an implication of the 15% principle: Different systems, but not all systems, may be responsive to specific interventions.

Concluding Comments

This chapter has focused on evolutionarily designed interventions, their rationale, and their use. Particular emphasis has been placed on the importance of identifying and designing therapeutic goals that are consistent with a theory of behavior, and on optimizing interventions by using multiple intervention techniques in which the components complement one another. In considering the cases described here, a number of points deserve reemphasis. (1) Both the formulations and goals of evolutionarily designed treatments and the preferred targets of therapy often differ from the targets of the prevailing models; (2) combined therapies usually have greater therapeutic efficacy than single interventions; (3) interventions designed to facilitate goal achievement introduce a specific outcome measure for therapy; and (4) evolutionarily designed interventions overcome much of the opposition of biological to psychosocial causation that often plagues current psychiatric thinking.

PART V

CONCLUSION

16

Key Points

We have developed arguments, presented hypotheses, and cited findings that are consistent with the idea that evolutionary biology should serve as a basic science for psychiatry (McGuire et al., 1992). Nonetheless, skeptics remain: "Evolutionary explanations alone are rarely, if ever, satisfactory explanations for disease or ill health. Unless evolutionary explanations are tied directly to genetic and physiological knowledge of why some people get sick in certain ways while others do not, they are too vague and general to be useful in medicine" (Guze, 1992, p. 92).

Views like this one are often voiced by advocates of the biomedical model. These views equate evolutionary explanations with phylogenetic reconstruction, and they reflect the belief that behaviors labeled as disordered are epiphenomena that result from the actions of dysfunctional physiological systems, disorder-predisposing genes, or genetic accidents. Such views are limited and outdated. They not only fail to take account of the unique features of *Homo sapiens* but also fail to acknowledge that, at best, genes and physiology are only part of the story of behavior. A theory that can accommodate genes and physiology and that makes distinctions between signs and symptoms as indices of strategies or dysfunctionality, as well as the other condition-contributing factors (e.g., ultimate causes), is required. Evolutionary theory provides a framework for collecting, interpreting, and utilizing its own novel explanations of behavior (e.g., ultimate causation, sexual selection), as well as explanations developed by the prevailing models. Where relevant, evolutionary explanations *are* tied directly to genetic and physiological knowledge. Further, the intervention options exceed those currently associated with the prevailing models. *However vague they may be, evolutionary explanations are less vague, more comprehensive, and more promising than the prevailing-model explanations, which, thus far, have satisfactorily explained only a limited number of conditions and their features.*

A Review of Key Points

Several key points from the preceding chapters are worth recapping:

1. *Evolutionary theory is the best available theory for explaining both ordered and disordered behavior.* Evolutionary theory builds from a set of assumptions and concepts, as well as a growing body of evidence, that lead to novel and testable explanations of both normal and disordered behavior. *The theory is grounded in the evolutionary history of the species.* It addresses such causal variables as ultimate causation (e.g., preferential investment in kin, constraints on learning), sexual selection, trait variation, factors influencing trait refinement and expression (e.g., the developmental environment), condition predispositions, past and current environmental effects, proximate events, and interactions between these variables. The theory embraces the inherent complexity of behavior and the systems and contingencies responsible for behavior. It is at home with explanations that posit multiple contributing factors (the 15% principle).

2. *Adopting a theory of behavior is essential.* We estimate that over the past century, about 4,000 books and 80,000 refereed journal articles have been published on condition-related topics. If, on average, a book requires a year to write (preparation and writing) and an article takes a month—these are conservative estimates—these publications represent approximately 10,666 human years of work. Still, an understanding of behavior that is considered disordered is little more than a distant hope. Further, the causal findings developed by advocates of one prevailing model are often overlooked by the advocates of other models; for example, physiological-behavioral correlations are largely ignored by advocates of the psychoanalytic, behavioral, and sociocultural models, just as findings dealing with psychic conflicts, dysfunctional learning, and environmentally induced stress effects are largely ignored by biomedical advocates. Such findings won't be adequately culled, explained, and integrated unless there is a theoretical framework that both facilitates obtaining species-relevant data and fosters hypothesis testing. Such a framework requires a *theory of behavior.*

3. *Behavior systems are the basic data-organizing systems.* When conditions are interpreted in the context of behavior systems, infrastructures, environmental contingencies, the ways in which findings are organized, prioritized, and interpreted differ significantly from those in interpretations developed by the prevailing models. Moreover, search strategies for identifying causal variables change dramatically; for example, assessments of function, resource allocations, and environmental structure are fundamental to all explanations of conditions, while genes, physiology, and psychic conflicts may or may not be important.

4. *Evolutionary concepts are a source of novel insights into why individuals behave as they do.* Ultimately caused behavior (e.g., kin investment and reciprocal altruism), unavoidable evolutionary outcomes (e.g., trait variation and sex differences), and evolved high-risk strategies are often major contributing factors to conditions. Not only do evolutionary concepts realign thinking so that it takes into account species-characteristic behavior, but they can also explain much of the variance of behavior in persons with and without conditions; for example, the clinical manifestations of schizophrenia in a female who is 18 years old are different from those in a female who is 36 years old. While evolutionary hypotheses introduce a new level of interpretive complexity (the 15% principle), they also offer the possibility of clarifying our understanding of a variety of conditions and their features that have eluded explanation so far (e.g., postmenopausal depression, ADHD, malingering, personality disorders). Similar points apply to therapy, where evolutionary concepts serve to narrow the kinds of interventions that are likely to be effective and add precision to explanations of intervention outcomes (e.g., an individual resisting change because she or he is engaging in an evolved high-risk strategy).

5. *Evolutionary theory seriously addresses strategies.* Irrespective of how it is classified, behavior partly reflects strategies to achieve biological goals. In discussing this point, we have sometimes described familiar, normal species-typical strategies that have their origins in the species genome, such as increased efforts to attract members of the opposite sex during adolescence, moral indignation at being cheated, and preferential investment in kin. There are, however, numerous other strategies, and many are apparent in conditions, for example, social withdrawal to reduce painful social input, behaviors that reflect attempts to adapt by persons with personality disorders, weight reduction to avoid competition for males and delay ovulation, deception, and the use of symptoms as signals. Internal strategies might include hallucinations to create a private and safer world, delusions to compartmentalize dysfunctional CNS activity, self-deception, and infrastructural self-correction. To view such behaviors only as epiphenomena is to ignore their function.

When clinical experience and research data are combined, we are unable to find any evidence suggesting that persons with conditions are any less involved in enacting strategies than persons without conditions. What often differentiates the two groups is the goal-related effectiveness of the strategies, their social appropriateness, and the presence of capacities to make midstream strategy adjustments in response to changing contingencies. It follows that strategies themselves, the reasons for choosing one strategy rather than another, and the capacities and constraints that influence midstream strategy changes need to be brought into sharper focus.

6. *The social environment has critical condition-contributing attributes, as well as ameliorative properties.* Each of psychiatry's prevailing models gives some credence to the social environment. The biomedical model gives the least, the sociocultural model gives the most, and the behavioral and psychoanalytic models fall somewhere between. However, none address the complexity of the environment and its interactions with behavior, regulation-dysregulation, or condition-triggering effects and amelioration found in evolutionary biology. *Homo sapiens is a highly social species that significantly influences and is significantly influenced by the social environment.* The environment is a source of stimuli (e.g., social demands, hostility, praise) and goal-related options (e.g., opportunities for mates and friendships). It is also competitive and often depriving. Environments respond differently to different strategies; for example, moderate changes in the social environment are advantageous for some strategies but disadvantageous for others. From another perspective, the social environment may be an inadvertent contributor to condition prevalence in that persons with genetically influenced suboptimal infrastructures may partition themselves socially and mate within those boundaries.

There is nothing magical about the social environment. It doesn't cure dementia, schizophrenia, autism, or mental retardation, just as it doesn't cause them. However, it is the major contributor to some conditions, such as posttraumatic stress disorder, adjustment disorders, amnesia, anorexia nervosa, seasonal disorders, phobias, dysregulation, and the symptoms of anxiety and depression. It can reduce or intensify signs and symptoms (e.g., as in dysthymic disorder), influence goal achievement, and alter condition prevalence and recidivism rates. The contributions of the social environment to conditions are both partial and differential, and because they are, their importance is comparable to dysfunctional physiological and psychological states, inadequate learning, and genetic mistakes, which are also partial and differential. While such points are not foreign to clinicians, it is a fact of clinical practice that the social environment is seldom studied in detail. Psychiatric evaluation and treatment take place in offices and hospitals, rarely in patients' natural environments, and the constraints that such practices introduce into diagnostic and treatment efforts are significant. There is a not-so-subtle arrogance in the view that one can accurately understand patients' social environments simply by discussing these environments within the confines of one's office.

7. *Function is critical in identifying, classifying, and narrowing possible condition causes.* As we have repeatedly stressed, signs and symptoms only partly characterize conditions, and with few exceptions, conditions are associated with compromised function. Optimal functioning for individuals requires not only maximally regulated infrastructures, but also that persons interact socially in ways that facilitate goal achievement. Explanations that focus primarily on signs and symptoms and the putative reasons for their production (e.g., the norepinephrine hypothesis of depression), and only secondarily on function, limit themselves to explaining only the features of disordered behavior. Further, interventions designed to reduce signs and symptoms often do so at the expense of function. For example, persons receive drugs to bring about symptom relief, but function may be compromised in the process.

At a more technical level, the precise measurement of function is essential to an accurate assessment of functional capacities, to the development of precise outcome measures and valid diagnostic categories (if these are possible), to contributions to precise inferences about infrastructural function, and to the minimization of ineffective interventions. There is also the issue of working backward from functional assessments to possible causes. The more detailed the behavioral and functional characterizations, the more effective this process. Working backward was illustrated in chapters 8 to 14, and in each example, the process led to a narrowing of the possible causes.

8. *An appreciation of traits and trait variation is essential to an understanding of conditions.* Two of the most important ideas in the preceding chapters are that persons are best conceived of as mosaics of independent and semi-independent traits, and that there is significant within- and cross-trait variation. Traits are differentially influenced by genetic information and experience. And conditions consist of clusters of suboptimal, dysfunctional, and functional traits. Yet, determining whether an individual's condition is primarily a consequence of trait variation, whether trait variation is a secondary consequence of nontrait causal factors, or whether trait variation is unrelated to a condition remains a task that psychiatry has yet to address seriously. *Trait variation in persons with and without conditions is a finding that won't go away.* Not to recognize this finding invites misunderstanding.

It is important to emphasize once again that evolution has thriven on variation; that trait variation is one of the outcomes of sexual selection and the mixing of the genes at each conception; that trait variation is both increased and decreased by events influencing maturation; and that trait variation applies as much to automatic systems, algorithms, and functional capacities as to observable phenotypes. A clear implication of these points is that, at best, "normal behavior" is an arbitrary concept or set of statistical measures and, at worst, no more than a vague concept with idiosyncratic interpretations and applications.

9. *In evolutionary biology, genetic explanations do not lead to reductionistic inter-pretations of behavior.* Genetic information is only part of the story of conditions. Equally clearly, genetic information is not always expressed, as may be the case among nonafflicted monozygotic twins when one twin suffers from schizophrenia or bipolar disorder. Multiple intervening variables and social contingencies influence phenotypic expression, and these influences may be more or less important than genetic information. An in-depth understanding of conditions presupposes an understanding of each of these potential contributing factors.

10. *Some conditions are adaptive.* Viewed in evolutionary context, some conditions are adaptive. Undetected or difficult-to-detect forms of antisocial and histrionic personality disorders are examples, but the point can also apply to a certain percentage of persons with ADHD and other types of personality disorders, as well as to time-limited anorexia nervosa, adjustment disorders, childhood phobias, and the symptoms of depression and anxiety. There is no evidence that persons with these conditions or symptoms have fewer offspring than persons without conditions. (The more striking finding is that persons with severe and debilitating mental conditions, such as schizophrenia, reproduce at nearly the same rate as age- and sex-matched individuals without conditions.)

11. *Some features of conditions represent attempts to adapt.* The social withdrawal of persons who are depressed is perhaps the most obvious and least contested example of an attempt to adapt. Yet, evidence is compatible with this interpretation for amnesia, features of personality disorders and schizophrenia, adjustment disorders, phobias, delusions, and obsessive-compulsive behavior.

12. *An evolutionarily based theory of behavior incorporates and provides an integrative context for the features of the prevailing models.* Much of the confusion in psychiatry and related disciplines (Freeman, 1992; Gilbert, 1995) stems from the absence of a theoretical framework that would facilitate the culling and integration of focused theories and their findings. Without such a framework, psychiatry will continue its production of reams of data and will continue to miss opportunities for the optimal interpretation and use of its strong findings. For example, a finding such as that serotonin reuptake blockers increase the number of serotonin molecules in the synaptic patch is usually described and interpreted without recourse to evolutionary insights. However, when thousands of such findings exist, the need for a theoretical framework that facilitates the integration and analysis of these findings becomes clear. The studies of serotonin changes in male vervet monkeys (chapters 5 and 8) illustrate this point. The initial finding of a broad range in peripheral serotonin measures suggested little about the factors that might contribute to the range. Further studies tied high social status to high serotonin measures and low status to low measures, but how status contributes to these physiological differences remained to be explained.

Subsequent studies which showed that serotonin levels reflect the frequency of received submissive displays narrowed the explanatory possibilities, as did studies indicating that serotonin differences between high- and low-status males disappeared when females were removed from a multimale social group. Had these studies not been guided by a theory of behavior that could simultaneously accommodate variables such as social status, physiological change, social information, and ultimate cause motivations (the reproductive behavior of males), the original physiological findings might simply be a set of numbers in another report buried somewhere in the literature. In short, bringing an evolutionary perspective to psychiatry brings the promise of a far more comprehensive understanding of conditions and their contributing factors than now exists.

13. *There are therapeutic options.* A clear implication of the 15% principle is that there are multiple therapeutic avenues. Said another way, if A causes the X part of a condition, B the Y part, and C the Q part, treating A, B, and C, rather than the whole condition, makes sense. The downside of this view is that there will not be any miracle cures for the vast majority of conditions. Rather, different combinations of interventions will be required, and the requirements will change over the course of an illness.

The Immediate Implications of the Evolutionary Approach for Psychiatry

1. *New types of data must be collected.* Data dealing with strategies, behavioral capacities, sex differences, and function are just a few examples of the kinds of data psychiatric research will need to address. This is not to say that data that are of interest to the prevailing models should not be collected. As we have noted throughout the book, many prevailing-model hypotheses and their associated data can be integrated into an evolutionary framework. Thus, what is called for is both an expansion and a reprioritization of the types of data that are most likely to have high utility.

2. *Psychiatry's preferred method of classifying conditions must undergo revision.* Time and again, we have argued in favor of a functional approach to classification. This approach is consistent with parts of *DSM-IV* (APA, 1994), but there is a major difference: The approach we favor places primary, not secondary, emphasis on function. Psychiatry has contributed to its own confusion by trying to develop a taxonomic system that attempts to identify the core features of conditions by using quasi-factor-analytic techniques, while simultaneously attempting to overlook individual differences. We have also argued that there is no use in trying to develop a theory-free, objective classification system for human behavior. Attempts to do so have not worked, and one cannot be optimistic about their future. Further, our recommendation—that classification systems reflect theory—need not be feared. While theories undergo revision, in which case the classification of conditions would also undergo revision, theory-driven taxonomies are as likely to lead to valid taxonomic categories as currently prevailing practices.

3. *Psychiatry's explanations of conditions must be developed within a theory of behavior.* A theory of behavior is a theoretical system that organizes, prioritizes, interrelates, and gives meaning to key findings and explanations and that optimally utilizes information from hypothesis-testing research. In the models developed in the preceding chapters, conditions are primarily manifestations of infrastructural suboptimality or dysfunctionality, different social contexts, and attempts to adapt. A theory of behavior is also essential to research. Although it is undoubtedly true that many research studies are carried out because of "hunches," and not because of theory-based predictions, the capacity to develop testable hypotheses from theory is invaluable, both in directing research and in deciding which research questions are likely to be most (and least) worthwhile trying to answer. The potential utility of the preceding point becomes clear if one takes into account the time, effort, and resources psychiatry has devoted to attempting to identify condition-contributing factors.

4. *Psychiatry's preferred methods of treating conditions must undergo revision.* This recommendation includes several key points: (1) Therapy should aim to facilitate the goals of patients; (2) it should improve how persons regulate themselves physiologically and psychologically; (3) it should use functional assessments as the primary measures of intervention outcomes; (4) it would recognize that different modes of therapy can have similar functional outcomes; and (5) it should recognize that multiple therapies are more likely to be effective than single therapies.

That many therapists already engage in multiple therapies is a clinical fact. Clinicians who prescribe drugs also talk with patients, give advice about changing lifestyles, and so forth. But providing ancillary advice is not the same as initiating multiple therapies, each with a specific objective. To the degree that different therapies supplement one another, interventions should be designed to optimize supplementation. Further, different types of interventions are more likely to be effective at different points during the course of a condition. Thus, for some individuals, drug interventions may precede discussions about changing social environments, other persons may benefit from the opposite sequence.

FINALLY, A PERSONAL OPINION. In the models developed in the preceding chapters, an evolutionary approach to mental illness turns out to reflect a more humanistic and tolerant view of *Homo sapiens* than any of the prevailing models. Its emphasis on individual variation (e.g., trait variation) and multicause hypotheses, its sensitivity to the effects of the environment, its postulates about adaptive behavior and attempts to act adaptively, its emphasis on strategies, and its focus on the species and its history— all are factors underlying this opinion.

Appendix

The appendix outlines three behavior systems discussed in Chapter 4 but not included in Table 4.1. The legend is the same for each system: *Biological motivations-goals, their primary functions, and their associated features.* Only a limited number of physiological events are included. Because the details of very few physiological-behavior relationships are fully understood, an asterisk (*) is placed in front of three examples of physiological systems (the norepinephrine, dopamine, and opioid systems) that have been reported or postulated to interact with behavior, and that require further study. The up arrows (↑) signify an increase, and the down arrows (↓) indicate a decrease in the associated factors or outcomes. An expanded list of possible interacting physiological systems can be found in McGuire (1988).

The Survival System

Motivation-goal: SEEK PROXIMITY
  Functional-psychological events: ↑ psychological regulation, joy, satisfaction, sense of closeness; ↓ anxiety, fear, uncertainty
  Physiological events: ↑ physiological regulation, serotonin activity; ↓ stress-induced hormone activity; (*) norepinephrine, dopamine, and opioid activity
  Responses associated with goal-related failure: ↑ dysregulation, anxiety, fear, agitation, vigilance; ↓ serotonin activity; enact search strategies to establish proximity and minimize dysregulation; (*) norepinephrine, dopamine, and opioid activity

Motivation-goal: MAINTAIN INTERPERSONAL CONSTANCY

Functional-psychological events: ↑ psychological regulation, predictable intimacy, knowledge of self and others; ↓ anxiety, fear, uncertainty

Physiological events: ↑ physiological regulation, serotonin activity; ↓ stress-induced hormone activity; (*) norepinephrine, dopamine, and opioid activity

Responses associated with goal-related failure: ↑ dysregulation, irritation; ↓ serotonin activity; search for others who will provide constancy and minimize dysregulation; (*) norepinephrine, dopamine, and opioid activity

Motivation-goal: AVOID STRANGERS

Functional-psychological events: ↑ sense of safety; ↓ fear, vigilance, uncertainty, possibility of psychological dysregulation

Physiological events: ↓ possibility of physiological dysregulation, stress-induced hormone activity; (*) norepinephrine, dopamine, and opioid activity

Responses associated with goal-related failure: ↑ dysregulation, anxiety, fear, vigilance, possibility of flight; ↓ serotonin activity; act to avoid strangers and minimize dysregulation; (*) norepinephrine, dopamine, and opioid activity

Motivation-goal: ESTABLISH SEPARATENESS

Functional-psychological events: ↑ knowledge and mastery of self and environment when alone

Physiological events: ↑ knowledge of tolerable physiological dysregulation associated with separateness

Responses associated with goal-related failure: ↑ dysregulation, frustration; ↓ serotonin activity; enact alternative strategies to achieve tolerable separateness and minimize dysregulation; (*) norepinephrine, dopamine, and opioid activity

Motivation-goal: IDENTIFY AND PROTECT SAFE ENVIRONMENTS

Functional-psychological events: ↑ psychological regulation, certainty; ↓ anxiety, vigilance

Physiological events: ↑ physiological regulation, serotonin activity; ↓ stress-induced hormone activity; (*) norepinephrine, dopamine, and opioid activity

Responses associated with goal-related failure: ↑ dysregulation, anxiety, vigilance; ↓ health, competitive advantage, serotonin activity; enact alternative strategies to identify safe environments and minimize dysregulation; (*) norepinephrine, dopamine, and opioid activity

Motivation-goal: MAINTAIN PHYSICAL AND MENTAL HEALTH

Functional-psychological events: ↑ psychological regulation, satisfaction, self-esteem

Physiological events: ↑ physiological regulation; ↓ stress-induced hormone activity; (*) norepinephrine, dopamine, and opioid activity

Responses associated with goal-related failure: ↑ dysregulation, depression, anxiety; ↓ serotonin activity; enact alternative strategies to enhance health and minimize dysregulation; (*) norepinephrine, dopamine, and opioid activity

Motivation-goal: IDENTIFY RESOURCES
  Functional-psychological events: ↑ psychological regulation, satisfaction; ↓ vigilance, anxiety, deception
  Physiological events: ↑ physiological regulation, serotonin activity; ↓ stress-induced hormone activity; (*) norepinephrine, dopamine, and opioid activity
  Responses associated with goal-related failure: ↑ dysregulation, anxiety, vigilance; ↓ social status, competitive advantage; enact alternative strategies to identify resources and minimize dysregulation; (*) norepinephrine, dopamine, and opioid activity

Motivation-goal: ACQUIRE, RETAIN, AND USE RESOURCES
  Functional-psychological events: ↑ psychological regulation, satisfaction; ↓ vigilance, deception, anxiety
  Physiological events: ↑ physiological regulation, serotonin activity; ↓ stress-induced hormone activity; (*) norepinephrine, dopamine, and opioid activity
  Responses associated with goal-related failure: ↑ dysregulation, frustration, anger, depression, probability of aggression; ↓ competitive advantage, serotonin activity; enact alternative strategies to acquire and retain resources and minimize dysregulation; (*) norepinephrine, dopamine, and opioid activity

The Help Kin System

Motivation-goal: SEEK PROXIMITY
  Functional-psychological events: ↑ closeness, intimacy, protection, investment options, psychological regulation
  Physiological events: ↑ physiological regulation, serotonin activity; ↓ stress-related hormone activity; (*) norepinephrine, dopamine, and opioid activity
  Responses associated with goal-related failure: ↑ dysregulation, frustration, anger, depression; ↓ serotonin activity; enact alternative strategies to attain proximity and minimize dysregulation; (*) norepinephrine, dopamine, and opioid activity

Motivation-goal: MAINTAIN KIN INTERPERSONAL CONSTANCY
  Functional-psychological events: maintain kin support network
  Physiological events: ↑ physiological regulation, serotonin activity; ↓ stress-related hormone activity; (*) norepinephrine, dopamine, and opioid activity
  Responses associated with goal-related failure: ↑ dysregulation, frustration, depression; ↓ serotonin activity; enact alternative strategies to maintain interpersonal constancy and minimize dysregulation; (*) norepinephrine, dopamine, and opioid activity

Motivation-goal: IDENTIFY AND PROTECT SAFE ENVIRONMENTS FOR KIN
  Functional-psychological events: ↑ kin safety, health, social status, psychological regulation; ↓ uncertainty
  Physiological events: ↑ physiological regulation, serotonin activity; ↓ stress-related hormone activity; (*) norepinephrine, dopamine, and opioid activity

Responses associated with goal-related failure: ↑ dysregulation, frustration, anger, depression, probability of aggression; ↓ competitive advantage, serotonin activity; enact alternative strategies to identify safe environments and minimize dysregulation; (*) norepinephrine, dopamine, and opioid activity;

Motivation-goal: IDENTIFY KIN-RELATED RESOURCES
   Functional-psychological events: identify kin-relevant resources; ↑ psychological regulation
   Physiological events: ↑ physiological regulation, serotonin activity; ↓ stress-related hormone activity; (*) norepinephrine, dopamine, and opioid activity
   Responses associated with goal-related failure: ↑ dysregulation, frustration, anger, depression; ↓ competitive advantage, serotonin activity; enact alternative strategies to identify kin-related resources and minimize dysregulation; (*) norepinephrine, dopamine, and opioid activity

Motivation-goal: ACQUIRE, RETAIN, AND OPTIMALLY USE KIN-RELATED RESOURCES (KIN INVESTMENT)
   Functional-psychological events: acquire kin-relevant resources, invest resources optimally; ↑ kin health, social status, competitive advantage, psychological regulation
   Physiological events: ↑ physiological regulation, serotonin activity; ↓ stress-related hormone activity; (*) norepinephrine, dopamine, and opioid activity
   Responses associated with goal-related failure: ↑ dysregulation, frustration, anger, depression; ↓ competitive advantage, serotonin activity; enact alternative strategies to acquire, retain, and invest resources and minimize dysregulation; (*) norepinephrine, dopamine, and opioid activity

Motivation-goal: PROTECT KIN FROM ATTACK
   Functional-psychological events: ↑ safety of kin, competitive advantage, psychological regulation
   Physiological events: ↑ physiological regulation, serotonin activity; ↓ stress-related hormone activity; (*) norepinephrine, dopamine, and opioid activity
   Responses associated with goal-related failure: ↑ dysregulation, anger, probability of aggression; ↓ competitive advantage, serotonin activity; develop and possibly enact retaliation strategies; (*) norepinephrine, dopamine, and opioid activity

The Nonkin Reciprocation System

Motivation-goal: IDENTIFY GOOD AND BAD RECIPROCATORS
   Functional-psychological events: identify good and bad reciprocators; ↑ psychological regulation
   Physiological events: ↑ physiological regulation, serotonin activity; ↓ stress-related hormones; (*) norepinephrine, dopamine, and opioid activity
   Responses associated with goal-related failure: ↑ dysregulation, frustration, anger, depression, probability of retaliation; ↓ competitive advantage, seroto-

nin activity; enact alternative strategies to identify good and bad reciprocators and minimize dysregulation; (*) norepinephrine, dopamine, and opioid activity

Motivation-goal: TRADE FAVORS WITH GOOD RECIPROCATORS
 Functional-psychological events: invest in and receive favors from good reciprocators; ↑ health, social status, competitive advantage, psychological regulation
 Physiological events: ↑ physiological regulation, serotonin activity; ↓ stress-related hormones; (*) norepinephrine, dopamine, and opioid activity
 Responses associated with goal-related failure: ↑ dysregulation, frustration, anger (moral indignation), depression, probability of aggression; ↓ competitive advantage, serotonin activity; enact alternative strategies to identify good and bad reciprocators and minimize dysregulation; (*) norepinephrine, dopamine, and opioid activity

Motivation-goal: MAINTAIN NONKIN INTERPERSONAL CONSTANCY (SOCIAL SUPPORT NETWORKS)
 Functional-psychological events: maintain social support network; ↓ uncertainty, anxiety; ↑ psychological regulation
 Physiological events: ↑ physiological regulation, serotonin activity; ↓ stress-related hormone activity; (*) norepinephrine, dopamine, and opioid activity
 Responses associated with goal-related failure: ↑ dysregulation, frustration, anger, depression; ↓ competitive advantage, serotonin activity; enact alternative strategies to maintain constancy and minimize dysregulation; (*) norepinephrine, dopamine, and opioid activity

Motivation-goal: PROTECT CLOSE FRIENDS FROM ATTACK
 Functional-psychological events: ↑ safety of friends, competitive advantage, psychological regulation; social status
 Physiological events: ↑ physiological regulation, serotonin activity; ↓ stress-related hormone activity; (*) norepinephrine, dopamine, and opioid activity
 Responses associated with goal-related failure: ↑ dysregulation, frustration, anger, depression, probability of aggression; ↓ competitive advantage, serotonin activity; enact alternative strategies to protect friends and minimize dysregulation; (*) norepinephrine, dopamine, and opioid activity

References

Abroms, E. M. Beyond eclecticism. *American Journal of Psychiatry. 140:*740–745, 1983.

Adler, G., and W. F. Gattaz. Pain perception threshold in major depression. *Biological Psychiatry. 34:*687–689, 1993.

af Klinteberg, B., D. Schalling, G. Edman, L. Oreland, and M. Asberg. Personality correlates of platelet monoamine oxidase (MAO) activity in female and male subjects. *Neuropsychobiology. 18:*89–96, 1987.

Agras, W. S. *Behavior Modification: Principles and Clinical Applications,* 2nd ed. Boston: Little, Brown, 1978.

Agras, W. S. Learning theory. In H. I. Kaplan and B. J. Sadock (eds.), *Comprehensive Textbook of Psychiatry,* Vol. 5. Baltimore: Williams and Wilkins, 1989, pp. 262–271.

Aguglia, E., M. Casacchia, G. B. Cassano, C. Faravelli, G. Ferrari, P. Giordano, P. Pancheri, L. Ravizza, M. Trabucchi, F. Bolino, A. Scarpato, D. Berardi, G. Provenzano, R. Brugnoli, and R. Rozzini. Double-blind study of the efficacy and safety of sertraline versus fluoxetine in major depression. *International Clinical Psychopharmacology. 8:*197–202, 1993.

Ainsworth, M. D. S., M. C. Blehar, E. Waters, and S. Wall. *Patterns of Attachment.* Hillsdale, NJ: Erlbaum, 1978.

Akiskal, H. S. Interaction of biologic and psychologic factors in the origin of depressive disorders. *Acta Psychiatricia Scandinavica. 71:*131–139, 1985.

Akiskal, H. S. The classification of mental disorders. In H. I. Kaplan and B. J. Sadock (eds.), *Comprehensive Textbook of Psychiatry,* Vol. 5. Baltimore: Williams and Wilkins, 1989, pp. 583–598.

Albin, R. L. The pleiotropic gene theory of senescence: Supportive evidence from human genetic disease. *Ethology and Sociobiology. 9:*371–382, 1988.

Alexander, R. D. *Darwinism and Human Affairs.* Seattle: University of Washington Press, 1979.

Alexander, R. D. *The Biology of Moral Systems.* New York: Aldine de Gruyter, 1987.

Alexander, R. D. Epigenetic rules and Darwinian algorithms: The adaptive study of learning and development. *Ethology and Sociobiology. 11:*241–303, 1990a.

Alexander, R. D. *How Did Humans Evolve? Reflections on the Uniquely Unique Species.* University of Michigan Museum of Zoology, Special Publication #1. 1990b.

Alexander, R. D., and K. M. Noonan. Concealment of ovulation, parental care, and human social evolution. In N. A. Chagnon and W. Irons (eds.), *Evolutionary Biology and Human Social Behavior: An Anthropological Perspective.* North Scituate, MA: Duxbury Press, 1979, pp. 436–453.

Allen, J. S., and V. M. Sarich. Schizophrenia in an evolutionary perspective. *Perspectives Biology Medicine. 32:*132–153, 1988.

Alnaes, R., and S. Torgersen. DSM-III personality disorders among patients with major depression, anxiety disorders, and mixed conditions. *Journal of Nervous Mental Disease. 178:* 693–698, 1990.

Altmann, S. A. (ed.). *Social Communication among Primates.* Chicago: University of Chicago Press, 1967.

American Psychiatric Association. *Diagnostic and Statistical Manual of Mental Disorders,* 3rd ed. (*DSM-III*). Washington, DC: Author, 1980.

American Psychiatric Association. *Diagnostic and Statistical Manual of Mental Disorders,* 3rd ed., rev. (*DSM-III-R*). Washington, DC: Author, 1987.

American Psychiatric Association. *Diagnostic and Statistical Manual of Mental Disorders,* 4th ed. (*DSM-IV*). Washington, DC: Author, 1994.

Anderson, J. L., and C. B. Crawford. Modeling costs and benefits of adolescent weight control as a mechanism for reproductive suppression. *Human Nature. 3:*299–334, 1992.

Anderson, J. R. Is human cognition adaptive? *Behavioral Brain Sciences. 14:*471–517, 1991.

Andreasen, N. C. Thought, language, and communication disorders: 1. Clinical assessment, definition of terms, and evaluation of their reliability. *Archives of General Psychiatry. 36:* 1315–1321, 1979a.

Andreasen, N. C. Thought, language, and communication disorders: 2. Diagnostic significance. *Archives of General Psychiatry. 36:*1325–1330, 1979b.

Andreoli, A., S. E. Keller, M. Rabaeus, L. Zaugg, G. Garrone, and C. Taban. Immunity, major depression, and panic disorder comorbidity. *Biological Psychiatry. 31:*896–908, 1992.

Andrews, G., G. Stewart, R. Allen, and A. S. Henderson. The genetics of six neurotic disorders: A twin study. *Journal of Affective Disorders. 19:*23–29, 1990.

Andrews, G., G. Stewart, A. Morris-Yates, P. Holt, and S. Henderson. Evidence for a general neurotic syndrome. *British Journal of Psychiatry. 157:*6–12, 1990.

Apter, A., A. Bleich, R. A. King, S. Kron, A. Fluch, M. Kotler, and D. J. Cohen. Death without warning? A clinical postmortem study of suicide in 43 Israeli adolescent males. *Archives of General Psychiatry. 50:*138–142, 1993.

Archer, J. Sex differences in social behavior: Are the social role and evolutionary explanations compatible? *American Psychologist. 51:*909–917, 1996.

Argyle, M. Nonverbal communication in human social interaction. In R. A. Hinde (ed.), *Nonverbal Communication.* London: Royal Society and Cambridge University Press, 1972a, pp. 243–269.

Argyle, M. *Social Behavior and Mental Disorder: The Psychology of Interpersonal Behavior,* 2nd ed. Harmondsworth, UK: Penguin, 1972b.

Argyle, M. *The Psychology of Happiness.* London: Methuen, 1987.

Argyle, M. *Cooperation.* London: Routledge, 1991.

Argyle, M., V. Salter, H. Nicholson, M. Williams, and P. Burgess. The communication of inferior and superior attitudes by verbal and non-verbal signals. *British Journal of Social and Clinical Psychology. 9:*222–231, 1970.

Arndt, S., R. J. Alliger, and N. C. Andreasen. The distinction of positive and negative symptoms: The failure of a two-dimensional model. *British Journal of Psychiatry. 158:*317–322, 1991.

Arranz, B., A. Eriksson, E. Mellerup, P. Plenge, and J. Marcusson. Brain 5-HT1A, 5-HT1D, and 5-HT2 receptors in suicide victims. *Biological Psychiatry. 35:*457–463, 1994.

Asher, S. R. Recent advances in the study of peer rejection. In S. R. Asher and J. D. Coie (eds.), *Peer Rejection in Childhood.* New York: Cambridge University Press, 1990, pp. 3–14.

Asnis, G. M., L. K. McGinn, and W. C. Sanderson. Atypical depression: Clinical aspects and noradrenergic function. *American Journal of Psychiatry. 152:*31–36, 1995.

Ast, D., and M. R. Gross. Status dependent sexual deception: Which men lie? Paper given at the Human Behavior and Evolution Society meeting, Santa Barbara, CA, June 28–July 2, 1995.

Avital, E., and E. Jablonka. Social learning and the evolution of behaviour. *Animal Behavior. 48:*1195–1199, 1994.

Badcock, C. *Essential Freud,* 2nd ed. Oxford: Blackwell, 1992.

Badcock, C. *PsychoDarwinism.* London: HarperCollins, 1994.

Baenninger, M., R. Baenninger, and D. Houle. Attractiveness, attentiveness, and perceived male shortage: Their influence on perception of other females. *Ethology and Sociobiology. 14:* 293–304, 1993.

Bailey, K. Mismatch theory: 1. Basic principles. *Across Species Comparison and Psychopathology. 9:*7–9. 1996.

Bailey, K. G. *Human Paleopsychology.* Hillsdale, NJ: Erlbaum, 1987.

Bailey, K. G. Psychological kinship: Implications for the helping professions. *Psychotherapy. 25:*132–141, 1988.

Bailey, K. G., H. E. Wood, and G. R. Nava. What do clients want? The role of psychological kinship in professional helping. *Journal of Psychotherapy Integration. 2:*125–147, 1992.

Barash, D. P. *Sociobiology and Behavior,* 2nd ed. New York: Elsevier, 1982.

Bargh, J. A., M. Chen, and L. Burrows. Automaticity of social behavior: Direct effects of trait construct and stereotype activation on action. *Journal of Personality and Social Psychology. 71:*230–244, 1996.

Barkow, J. The distance between genes and culture. *Journal of Anthropological Research. 40:* 367–379, 1984.

Barkow, J., L. Cosmides, and J. Tooby (eds.). *The Adapted Mind.* New York: Oxford University Press, 1992.

Barkow, J. H. *Darwin, Sex, and Status.* Toronto: University of Toronto Press, 1989.

Baron, M. Genes, environment and psychopathology. *Biological Psychiatry. 29:*1055–1057, 1991.

Baron, M. Genetic linkage and mental disorders: An update on analytic methodologies. *Biological Psychiatry. 36:*1–4, 1994a.

Baron, M. Novel strategies in molecular genetics of mental illness. *Biological Psychiatry. 35:* 757–760, 1994b.

Barsky, A. J., J. D. Goodson, R. S. Lane, and P. D. Cleary. The amplification of somatic symptoms. *Psychosomatic Medicine. 50:*510–519, 1988.

Barsky, A. J., and G. Wyshak. Hypochondriasis and somatosensory amplification. *British Journal of Psychiatry. 157:*404–409, 1990.

Barton, R. A., A. Whiten, S. C. Strum, R. W. Byrne, and A. J. Simpson. Habitat use and resource availability in baboons. *Animal Behaviour. 43:*831–844, 1992.

Baucom, D. H., P. K. Besch, and S. Callahan. Relation between testosterone concentration, sex role identity, and personality among females. *Journal of Personality and Social Psychology. 48:*1218–1226, 1985.

Baxter, L. R., M. E. Phelps, J. C. Mazziotta, J. M. Schwartz, R. H. Gerner, C. E. Selin, and R. M. Sumida. Cerebral metabolic rates for glucose in mood disorders. *Archives of General Psychiatry. 42:*441–447, 1985.

Baxter, L. R., Jr., J. M. Schwartz, K. S. Bergman, M. P. Szuba, B. H. Guze, J. C. Mazziotta, A. Alazraki, C. E. Selin, H. Ferng, P. Munford, and M. E. Phelps. Caudate glucose metabolic rate changes with both drug and behavior therapy for obsessive-compulsive disorder. *Archives of General Psychiatry. 49:*681–689, 1992.

Beahrs, J. O. *The Limits of Scientific Psychiatry.* New York: Brunner/Mazel, 1986.

Beahrs, J. O. The evolution of post-traumatic behavior: Three hypotheses. *Dissociation. 3:* 15–21, 1990.

Beatty, J. Fitness: Theoretical contexts. In E. F. Keller and E. A. Lloyd (eds.), *Keywords in Evolutionary Biology.* Cambridge, MA: Harvard University Press, 1992, pp. 115–119.

Beauchamp, A. J., J. P. Gluck, H. E. Fouty, and M. H. Lewis. Associative processes in differentially reared rhesus monkeys (*Macaca mulatta*): Blocking. *Developmental Psychobiology. 24:*175–189, 1991.

Beck, A. T. *Cognitive Therapy and Emotional Disorders.* New York: Meridian, 1976.

Beck, A. T. Cognitive therapy of depression: New Perspectives. In P. J. Clayton and J. E. Barratt (eds.), *Treatment of Depression: Old Controversies and New Approaches.* New York: Raven Press, 1983, pp. 37–48.

Beck, A. T., A. Freeman, and Associates (eds.). *Cognitive Therapy of Personality Disorders.* New York: Guilford Press, 1990.

Beecher, M. D. Signature systems and kin recognition. *American Zoologist. 22:*477–490, 1982.

Beecher, M. D. Signalling systems for individual recognition: An information theory approach. *Animal Behaviour. 38:*248–261, 1989.

Belcher, A. M., A. B. Smith III, P. C. Jurs, B. Lavine, and G. Epple. Analysis of chemical signals in a primate species (*Saguinus fuscicollis*): Use of behavioral, chemical, and pattern recognition methods. *Journal of Chemical Ecology. 12:*513–531, 1986.

Bellack, A. S., R. L. Morrison, J. T. Wixted, and K. T. Mueser. An analysis of social competence in schizophrenia. *British Journal of Psychiatry. 156:*809–818, 1990.

Belmaker, R. H., and J. Biederman. Genetic markers, temperament, and psychopathology. *Biological Psychiatry. 36:*71–72, 1994.

Bench, C. J., K. J. Friston, R. G. Brown, L. C. Scott, R. S. J. Frackowiak, and R. J. Dolan. The anatomy of melancholia—Focal abnormalities of cerebral blood flow in major depression. *Psychological Medicine. 22:*607–615, 1992.

Benjamin, J., and E. S. Gershon. Genetic discoveries in human behavior: Small effect genes loom large. *Biological Psychiatry. 40:*313–316, 1996.

Benson, P. L., J. Dehority, L. Garman, E. Hanson, M. Hochschwender, C. Lebold, R. Rohr, and J. Sullivan. Interpersonal correlates of nonspontaneous helping behavior. *Journal of Social Psychology. 110:*87–95, 1980.

Berger, M., S. Bossert, J. Krieg, G. Dirlich, W. Ettmeier, W. Schreiber, and D. von Zerssen. Interindividual differences in the susceptibility of the cortisol system: An important factor for the degree of hypercortisolism in stress situations. *Biological Psychiatry. 22:*1327–1339, 1987.

Berkson, G. Social responses to abnormal infant monkeys. *American Journal of Physical Anthropology. 38:*383–386, 1973.

Bertalanffy, L. General systems theory and psychiatry. In S. Arieti (ed.), *American Handbook of Psychiatry,* 2nd ed. New York: Basic Books, 1974, pp. 1095–1117.

Bick, P. A., and M. Kinsbourne. Auditory hallucinations and subvocal speech in schizophrenic patients. *American Journal of Psychiatry. 144:*222–225, 1987.

Bickhard, M. H. *Cognition, Convention, and Communication.* New York: Praeger, 1980.

Billings, A. G., and R. H. Moos. Chronic and nonchronic unipolar depression. *Journal of Nervous and Mental Diseases. 172:*65–75, 1984.

Birtchnell, J. *How Humans Relate.* Westport, CT: Praeger, 1993.

Birtchnell, J. Detachment. In C. G. Costello (ed.), *Personality Characteristics of the Personality Disordered.* Baltimore: Wiley, in press-a.

Birtchnell, J. Personality set within an octagonal model of relating. In R. Plutchik and H. R. Conte (eds.), *Circumplex Models of Personality and Emotions.* Washington, DC: American Psychological Association Press, in press-b.

Biver, F., S. Goldman, V. Delvenne, A. Luxen, V. DeMaertelear, P. Hubain, J. Mendlewicz, and F. Lotstra. Frontal and parietal metabolic disturbances in unipolar depression. *Biological Psychiatry. 36:*381–388, 1994.

Blackwell, B. Chronic pain. In H. I. Kaplan and B. J. Sadock (eds.), *Comprehensive Textbook of Psychiatry,* Vol. 5. Baltimore: Williams and Wilkins, 1989, pp. 1264–1271.

Blomberg, S. Influence of maternal distress during pregnancy on postnatal development. *Acta Psychiatrica Scandinavica. 62:*405–417, 1980.

Blum, K., J. G. Cull, E. R. Braverman, and D. E. Comings. Reward deficiency syndrome. *American Scientist. 84:*132–145, 1996.

Blurton Jones, N. G. (ed.). *Ethological Studies of Child Behavior.* Cambridge, England: Cambridge University Press, 1972.

Blurton Jones, N. G. A selfish origin for human food sharing: Tolerated theft. *Ethology and Sociobiology. 5:*1–4, 1984.

Bock, W. J. The definition and recognition of biological adaptation. *American Zoologist. 20:* 217–227, 1980.

Bohman, M., R. Cloninger, S. Sigvardsson, and A.-L. von Knorring. Predisposition to petty criminality in Swedish adoptees: 1. Genetic and environmental heterogeneity. *Archives of General Psychiatry. 39:*1233–1241, 1982.

Bohman, M., R. Cloninger, S. Sigvardsson, and A.-L. von Knorring. The genetics of alcoholism and related disorders. *Journal of Psychiatric Research. 21:*447–452, 1987.

Bond, M. R. Psychological and psychiatric aspects of pain. *Anaesthesia. 33:*355–361, 1978.

Boon, S., and N. Draijer. Multiple personality in The Netherlands: A clinical investigation of 71 patients. *American Journal of Psychiatry. 150:*489–494, 1993.

Borgerhoff Mulder, M. Adaptation and evolutionary approaches to anthropology. *Man. 22:* 25–41, 1987a.

Borgerhoff Mulder, M. On cultural and reproductive success: Kipsigis evidence. *American Anthropologist. 81:*617–634, 1987b.

Borison, R. L. (ed.). New antipsychotics for schizophrenia. *Psychiatric Annals. 25:*283–313, 1995.

Bouchard, T. J., Jr. The genetic architecture of human intelligence. In P. A. Vernon (ed.), *Biological Approaches to the Study of Human Intelligence.* Norwood, NJ: Ablex, 1993, pp. 33–93.

Bouchard, T. J., Jr. Genes, environment, and personality. *Science. 264:*1700–1701, 1994.

Bouchard, T. J., Jr., D. T. Lykken, M. McGue, N. L. Segal, and A. Tellegen. Sources of human psychological differences: The Minnesota study of twins reared apart. *Science. 250:*223–228, 1990.

Bouhuys, A. L., D. G. M. Beersma, R. H. van den Hoofdakker, and A. Roossien. The prediction of short- and long-term improvement in depressive patients: Ethological methods of observing behavior versus clinical ratings. *Ethology and Sociobiology. 8:*117S–130S, 1987.

Bouhuys, A. L., C. J. Jansen, and R. H. van den Hoofdakker. Analysis of observed behaviors displayed by depressed patients during a clinical interview: Relationships between behavioral factors and clinical concepts of activation. *Journal of Affective Disorders. 21:*79–88, 1991.

Bouhuys, A. L., and R. H. van den Hoofdakker. The interrelatedness of observed behavior of depressed patients and of a psychiatrist: An ethological study on mutual influence. *Journal of Affective Disorders. 23:*63–74, 1991.

Bouhuys, A. L., and R. H. van den Hoofdakker. A longitudinal study of interaction patterns of a psychiatrist and severely depressed patients based on observed behaviour: An ethological approach of interpersonal theories of depression. *Journal of Affective Disorders. 27:*87–99, 1993.

Bowden, C. L., and L. J. Rhodes. Mania in children and adolescents: Recognition and treatment. *Psychiatric Annals. 26*(Suppl.):S430–S434, 1996.

Bowlby, J. The nature of the child's tie to his mother. *International Journal of Psychoanalysis. 39:*350–373, 1958.

Bowlby, J. *Attachment and Loss: Vol. 1. Attachment.* London: Hogarth, 1969.

Bowlby, J. *Attachment and Loss: Vol. 2. Separation: Anxiety and Anger.* London: Hogarth, 1973.

Bowlby, J. The making and breaking of affectional bonds. *British Journal of Psychiatry. 130:* 201–210, 1977.

Boyce, P., G. Parker, I. Hickie, K. Wilhelm, H. Brodaty, and P. Mitchell. Personality differences between patients with remitted melancholic and nonmelancholic depression. *American Journal of Psychiatry. 147:*1476–1483, 1990.

Boyd, R., and P. J. Richerson. Why does culture increase human adaptability? *Ethology and Sociobiology. 16:*125–144, 1995.

Braungart, J. M., R. Plomin, J. C. DeFries, and D. W. Fulker. Genetic influence on tester-rated infant temperament as assessed by Bayley's infant behavior record: Nonadoptive and adoptive siblings and twins. *Developmental Biology. 28:*40–47, 1992.

Breier, A., D. S. Charney, and G. R. Heninger. Major depression in patients with agoraphobia and panic disorder. *Archives of General Psychiatry. 41:*1129–1135, 1984.

Breier, A., D. S. Charney, and G. R. Heninger. Agoraphobia with panic attacks. *Archives of General Psychiatry. 43:*1029–1036, 1986.

Brewer, W. J., J. Edwards, V. Anderson, T. Robinson, and C. Pantelis. Neuropsychological, olfactory, and hygiene deficits in men with negative symptom schizophrenia. *Biological Psychiatry. 40:*1021–1031, 1996.

Bronisch, T., and H. Hecht. Major depression with and without a coexisting anxiety disorder: Social dysfunction, social integration, and personality features. *Journal of Affective Disorders. 20:*151–157, 1990.

Bronish, T., H.-U. Wittchen, C. Krieg, H.-U. Rupp, and D. von Zerssen. Depressive neurosis. *Acta Psychiatrica Scandinavica. 71:*237–248, 1985.

Brothers, L. A biological perspective on empathy. *American Journal of Psychiatry. 146:*10–19, 1989.

Brothers, L. The neural basis of primate social communication. *Motivation and Emotion. 14:* 81–91, 1990a.

Brothers, L. The social brain: A project for integrating primate behavior and neurophysiology in a new domain. *Concepts in Neuroscience. 1:*27–51, 1990b.

Brothers, L. Neurophysiology of the perception of intentions by primates. In M. Gazzaniga (ed.), *The Cognitive Neurosciences.* Cambridge, MA: MIT Press, 1995, pp. 1107–1115.

Brothers, L., and B. Ring. A neuroethological framework for the representation of minds. *Journal of Cognitive Neuroscience. 4:*107–118, 1992.

Brothers, L., and B. Ring. Mesial temporal neurons in the macaque monkey with responses selective for aspects of social stimuli. *Behavioural Brain Research. 57:*53–61, 1993.

Brown, G. L., M. H. Ebert, P. F. Goyer, W. J. Jimerson, W. E. Klein, W. E. Bunney, and F. K. Goodwin. Aggression, suicide, and serotonin: Relationships to CSF amine metabolites. *American Journal of Psychiatry. 139:*741–746, 1982.

Brown, G. W., M. N. Bhrolcha'in, and T. Harris. Social class and psychiatric disturbance among women in an urban population. *Sociology. 9:*225–254, 1975.

Brown, G. W., and T. Harris. *Social Origins of Depression.* London: Tavistock, 1978.

Brown, S. L., and D. T. Kenrick. Paternal certainty and female dominance: Should males prefer submissive females? Paper given at the Human Behavior and Evolution Society meeting, Santa Barbara, CA, June 28–July 2, 1995.

Brown, W. M., and B. Palameta. Altruism facilitates the formation of social support networks. Paper given at the Human Behavior and Evolution Society meeting, Santa Barbara, CA, June 28–July 2, 1995.

Brunner, H. G., M. Nelen, X. O. Breakefield, H. H. Ropers, and B. A. van Oost. Abnormal behavior associated with a point mutation in the structural gene for monoamine oxidase A. *Science. 262:*578–580, 1993.

Buck, R. *Human Motivation and Emotion,* 2nd ed. New York: Wiley, 1988.

Buck, R., and B. Ginsburg. Spontaneous communication and altruism. In M. S. Clark (ed.), *Prosocial Behavior: Review of Personality and Social Psychology,* Vol. 12. Newbury Park, CA: Sage, 1991, pp. 149–175.

Buckley, P. F., C. Moore, H. Long, C. Larkin, P. Thompson, F. Mulvany, O. Redmond, J. P. Stack, J. T. Ennis, and J. L. Waddington. [1]H-magnetic resonance spectroscopy of the left temporal and frontal lobes in schizophrenia: Clinical, neurodevelopmental, and cognitive correlates. *Biological Psychiatry. 36:*792–800, 1994.

Budinger, T. F. Critical review of PET, SPECT, and neuroreceptor studies in schizophrenia. *Journal of Neural Transmission. 36*(Suppl.):3–12, 1992.

Burgess, R. L., and P. Draper. The explanation of family violence: The role of biological, behavioral, and cultural selection. In L. Ohlin and M. Torry (eds.), *Sociobiology of Family Violence.* Chicago: University of Chicago Press, 1989, pp. 59–116.

Burley, N. The meaning of assortative mating. *Ethology and Sociobiology. 4:*191–203, 1983.

Burnstein, E., C. Crandall, and S. Kitayama. Some neo-Darwinian decision rules for altruism: Weighing cues for inclusive fitness as a function of the biological importance of the decision. *Journal of Personality and Social Psychology. 67:*773–789, 1994.

Burvill, P. W., W. D. Hall, H. G. Stampfer, and J. P. Emmerson. The prognosis of depression in old age. *British Journal of Psychiatry. 158:*64–71, 1991.

Buss, D. M. Human mate selection. *American Scientist. 73:*47–51, 1985.

Buss, D. M. Sex differences in human mate selection criteria: An evolutionary perspective. In C. Crawford, M. Smith, and D. Krebs (eds.), *Sociobiology and Psychology: Ideas, Issues, and Applications.* Hillsdale, NJ: Erlbaum, 1987, pp. 335–351.

Buss, D. M. The evolution of human intrasexual competition: Tactics of mate attraction. *Journal of Personality and Social Psychology. 54:*616–628, 1988a.

Buss, D. M. From vigilance to violence: Tactics of mate retention in American undergraduates. *Ethology and Sociobiology. 9:*291–317, 1988b.

Buss, D. M. Sex differences in human mate preferences: Evolutionary hypotheses tested in 37 cultures. *Behavioral Brain Sciences. 12:*1–49, 1989.

Buss, D. M. Evolutionary personality psychology. *Annual Review of Psychology, 42:*459–491, 1991.

Buss, D. M. *The Evolution of Desire.* New York: Basic Books, 1994.

Buss, D. M. Evolutionary psychology: A new paradigm for psychological science. *Psychological Inquiry. 6:*1–30, 1995a.

Buss, D. M. The future of evolutionary psychology. *Psychological Inquiry. 6:*81–87, 1995b.

Buss, D. M., R. J. Larsen, D. Westen, and J. Semmelroth. Sex differences in jealousy: Evolution, physiology, and psychology. *Psychological Science. 3:*251–255, 1992.

Buss, D. M., and D. P. Schmitt. Sexual strategies theory: An evolutionary perspective on human mating. *Psychological Review. 100:*1–29, 1993.

Byrne, R. W. The evolution of intelligence. In P. J. B. Slater and T. R. Halliday (eds.), *Behaviour and Evolution.* Cambridge, England: Cambridge University Press, 1994, pp. 223–265.

Cairns, R. B., B. D. Cairns, H. J. Neckerman, S. D. Gest, and J.-L. Gariepy. Social networks and aggressive behavior: Peer support or peer rejection. *Developmental Psychology. 24:* 815–823, 1988.

Cameron, O. G., and R. M. Nesse. Systemic hormonal and physiological abnormalities in anxiety disorders. *Psychoneuroendocrinology. 13:*287–307, 1988.

Cantwell, D. P., and L. Baker. Attention deficit disorder with and without hyperactivity: A review and comparison of matched groups. *Journal of the American Academy of Child and Adolescent Psychiatry. 31:*403–412, 1992.

Cantwell, D. P., and G. L. Hanna. Attention-deficit hyperactivity disorder. *Eighth Annual Review of Psychiatry, 1989.* Washington, DC: American Psychiatric Press, 1990.

Caporael, L. R., R. M. Dawes, J. M. Orbell, and A. J. C. van de Kragt. Selfishness examined: Cooperation in the absence of egoistic incentives. *Behavioral Brain Science. 12:*683–739, 1989.

Caro, T. M., and M. Borgerhoff Mulder. The problem of adaptation in the study of human behavior. *Ethology and Sociobiology. 8:*61–72, 1987.

Carver, C. S., and M. Scheier. *Attention and Self-Regulation.* New York: Springer-Verlag, 1981.

Cashdan, E. Attracting mates: Effects of paternal investment on mate attraction strategies. *Ethology and Sociobiology. 14:*1–24, 1983.

Caspi, A., T. E. Moffitt, D. L. Newman, and P. A. Silva. Behavioral observations at age 3 years predict adult psychiatric disorders. *Archives of General Psychiatry. 53:*1033–1039, 1996.

Cavalli-Sforza, L. L. Genes, peoples and languages. *Scientific American. 265:*104–110, 1991.

Chadwick, P., and M. Birchwood. The omnipotence of voices: A cognitive approach to auditory hallucinations. *British Journal of Psychiatry. 164:*190–201, 1994.

Chagnon, N. A., and W. Irons. *Evolutionary Biology and Human Social Behavior.* North Scituate, MA: Duxbury, 1979.

Chance, M. R. A. (ed.). *Social Fabrics of the Mind.* Hove, UK: Erlbaum, 1988.

Charlesworth, W. R. Darwin and developmental psychology: 100 years later. *Human Development. 29:*1–35, 1986.

Charlesworth, W. R. Darwin and developmental psychology: Past and present. *Developmental Psychology. 28:*5–16, 1992.

Charlesworth, W. R., and P. LaFreniere. Dominance, friendship, and resource utilization in preschool children's groups. *Ethology and Sociobiology. 4:*175–186, 1983.

Charney, D. S., S. W. Woods, L. M. Nagy, S. M. Southwick, J. H. Krystal, and G. R. Heninger. Noradrenergic function in panic disorder. *Journal of Clinical Psychiatry. 51*(Suppl.):5–11, 1990.

Charnov, E. L. Evolution of life history variation among female mammals. *Proceedings of the National Academy Science. 88:*1134–1137, 1991.

Chen, Y.-W, and S. C. Dilsaver. Lifetime rates of suicide attempts among subjects with bipolar and unipolar disorders relative to subjects with other Axis I disorders. *Biological Psychiatry. 39:*896–899, 1996.

Cheney, D. L., and R. M. Seyfarth. *How Monkeys See the World.* Chicago: University of Chicago Press, 1990.

Chomsky, N. *Syntactic Structures.* The Hague: Mouton, 1957.

Chomsky, N. Rules and representations. *Behavioural Brain Sciences. 3:*1–61, 1980.

Clark, A. B. Individual variation in responsiveness to environmental change. In C. Crawford, M. Smith, and D. Krebs (eds.), *Sociobiology and Psychology: Ideas, Issues, and Applications.* Hillsdale, NJ: Erlbaum, 1987, pp. 91–110.

Clark, C. R., G. M. Geffen, and L. B. Geffen. Catecholamines and attention: 1. Animal and clinical studies. *Neuroscience Biobehavioral Review. 11:*341–352, 1987a.

Clark, C. R., G. M. Geffen, and L. B. Geffen. Catecholamines and attention: 2. Pharmacological studies in normal humans. *Neuroscience Biobehavioral Review. 11:*353–364, 1987b.

Clark, L. A., D. Watson, and S. Mineka. Temperament, personality, and the mood and anxiety disorders. *Journal of Abnormal Psychology. 103:*103–116, 1994.

Clarke, A. S., C. M. Kammerer, K. P. George, D. J. Kupfer, W. T. McKinney, M. A. Spence, and G. W. Kraemer. Evidence for heritability of biogenic amine levels in the cerebrospinal fluid of rhesus monkeys. *Biological Psychiatry. 38:*572–577, 1995.

Cleghorn, J. M., and M. L. Albert. Modular disjunction in schizophrenia: A framework for a pathological psychophysiology. In A. Kales, C. N. Stefanis, and J. A. Talbott (eds.), *Recent Advances in Schizophrenia.* New York: Springer-Verlag, 1990, pp. 59–80.

Cloninger, C. R. A unified biosocial theory of personality and its role in the development of anxiety states. *Psychiatric Developments. 3:*167–226, 1986.

Cloninger, C. R. A systematic method for clinical description and classification of personality variants. *Archives of General Psychiatry. 44:*573–588, 1987a.

Cloninger, C. R. Neurogenetic adaptive mechanisms in alcoholism. *Science. 236:*410–416, 1987b.

Cloninger, C. R., S. Sigvardsson, M. Bohman, and A.-L. von Knorring. Predisposition to petty criminality in Swedish adoptees: 2. Cross-fostering analysis of gene-environment interaction. *Archives of General Psychiatry. 39:*1242–1247, 1982.

Cloninger, C. R., D. M. Svrakic, and T. R. Przybeck. A psychobiological model of temperament and character. *Archives of General Psychiatry. 50:*975–990, 1993.

Coccaro, E. F. Central serotonin and impulsive aggression. *British Journal of Psychiatry. 155*(Suppl.):52–62, 1989.

Coccaro, E. F., L. J. Siever, H. M. Klar, G. Maurer, K. Cochrane, T. B. Cooper, R. C. Mohs, and K. L. Davis. Serotonergic studies in patients with affective and personality disorders. *Archives of General Psychiatry. 46:*587–599, 1989.

Cochrane, R., and M. Stopes-Roe. Women, marriage, employment and mental health. *British Journal of Psychiatry. 139:*373–381, 1981.

Cody, M. J., and H. D. O'Hair. Nonverbal communication and deception: Differences in deception cues due to gender and communicator dominance. *Communication Monographs. 50:* 175–192, 1983.

Cofer, D. H., and J. R. Wittenborn. Personality characteristics of formerly depressed women. *Journal of Abnormal Psychology. 89:*309–314, 1980.

Cohen, M. R., R. M. Cohen, D. Pickar, H. Weingartner, and D. L. Murphy. High-dose naloxone infusions in normals. *Archives of General Psychiatry. 40:*613–619, 1983.

Cohen, S., and G. M. Williamson. Stress and infectious disease in humans. *Psychological Bulletin. 109:*5–24, 1991.

Coie, J. D. Toward a theory of peer rejection. In S. R. Asher and J. D. Coie (eds.), *Peer Rejection in Childhood.* New York: Cambridge University Press, 1990, pp. 365–401.

Coie, J. D., K. A. Dodge, and H. Coppotelli. Dimensions and types of social status: A cross-age perspective. *Developmental Psychology. 18:*557–570, 1982.

Coie, J. D., K. A. Dodge, and J. B. Kupersmidt. Peer group behavior and social status. In S. R. Asher and J. D. Coie (eds.), *Peer Rejection in Childhood.* New York: Cambridge University Press, 1990, pp. 17–59.

Colby, K. M., and M. T. McGuire. Signs and symptoms: Zeroing in on a better classification of neuroses. *Sciences. 21:*21–23, 1981.

Coll, C. G., J. Kagan, and J. S. Reznick. Behavioral inhibition in young children. *Child Development. 55:*1005–1019, 1984.

Coon, D. Introduction to Psychology. *Exploration and Application,* 6th ed. New York: West, 1992.

Cords, M. Resolution of aggressive conflicts by immature long-tailed macaques (*Macaca fascicularis*). *Animal Behaviour. 36:*1124–1135, 1988.

Corrigan, M. H. N., G. M. Gillette, D. Quade, and J. C. Garbutt. Panic, suicide, and agitation: Independent correlates of the TSH response to TRH in depression. *Biological Psychiatry. 31:*984–992, 1992.

Coryell, W. A. A blind family history study of Briquet's syndrome. *Archives of General Psychiatry. 37:*1266–1269, 1980.

Cosmides, L. The logic of social exchange: Has natural selection shaped how humans reason? Studies with the Wason selection task. *Cognition. 31:*187–276, 1989.

Cosmides, L., and J. Tooby. From evolution to behavior: Evolutionary psychology as the missing link. In J. Dupre (ed.), *The Latest and the Best: Essays on Evolution and Optimality.* Cambridge, MA: MIT Press, 1987, pp. 277–306.

Cosmides, L., and J. Tooby. Evolutionary psychology and the generation of culture: 2. A computational theory of social exchange. *Ethology and Sociobiology. 10:*51–98, 1989.

Cosmides, L., and J. Tooby. Statistical inference in a multimodular mind: 2. Experiments. Paper given at the Human Behavior and Evolution Society meeting, Albuquerque, NM, July 22–26, 1992.

Cosmides, L., and J. Tooby. Episodic memory, theory of mind, and their breakdown. Paper given at the Human Behavior and Evolution Society meeting, Santa Barbara, CA, June 28–July 2, 1995.

Costello, C. G. Research on symptoms versus research on syndromes. *British Journal of Psychiatry. 160:*304–308, 1992.

Costello, C. G. The advantages of the symptom approach to depression. In C. G. Costello (ed.), *Symptoms of Depression.* New York: Wiley, 1993a, pp. 1–26.

Costello, C. G. *Symptoms of Schizophrenia.* New York: Wiley, 1993b.

Coulter, W. A., and H. W. Morrow. *Adaptive Behavior: Concepts and Measurements.* New York: Grune and Stratton, 1978.

Crawford, C. Sociobiology: Of what value to psychology? In C. Crawford, M. Smith, and D. Krebs (eds.), *Sociobiology and Psychology: Ideas, Issues, and Applications.* Hillsdale, NJ: Erlbaum, 1987, pp. 3–29.

Crawford, C. B. The theory of evolution: Of what value to psychology? *Journal of Comparative Psychology. 103:*4–22, 1989.

Crawford, C. B. The evolutionary significance of true pathologies, pseudopathologies, and pseudonormal conditions. Paper presented at the Human Behavior and Evolution Society meeting, Santa Barbara, CA, June 28–July 2, 1995.

Cronin, H. Sexual selection: Historical perspectives. In E. F. Keller and E. A. Lloyd (eds.), *Keywords in Evolutionary Biology.* Cambridge, MA: Harvard University Press, 1992, pp. 286–293.

Crook, J. H. *The Evolution of Human Consciousness.* Oxford, England: Oxford University Press, 1980.

Crow, T. J. A Darwinian approach to the origins of psychosis. *British Journal of Psychiatry. 167:*12–25, 1995.

Cunnien, A. J. Psychiatric and medical syndromes associated with deception. In R. Rogers (ed.), *Clinical Assessment of Malingering and Deception.* New York: Guilford Press, 1988, pp. 13–33.

Curio, E. Causal and functional questions: How are they linked? *Animal Behaviour. 47:*999–1021, 1994.

Daly, M., and M. Wilson. *Sex, Evolution and Behavior.* North Scituate, MA: Duxbury, 1978.

Daly, M., and M. Wilson. Discriminative parental solicitude: A biological perspective. *Journal of Marriage and the Family. 42:*277–288, 1980.

Daly, M., and M. Wilson. Child abuse and other risks of not living with both parents. *Ethology and Sociobiology. 6:*197–210, 1985.

Daly, M., and M. Wilson. Children as homicide victims. In R. G. Gelles and J. B. Lancaster (eds.), *Child Abuse and Neglect.* New York: Aldine, 1987a, pp. 201–214.

Daly, M., and M. Wilson. Risk of maltreatment of children living with stepparents. In R. G. Gelles and J. B. Lancaster (eds.), *Child Abuse and Neglect.* New York: Aldine, 1987b, pp. 215–232.

Daly, M., and M. Wilson. Evolutionary social psychology and family homicide. *Science. 242:* 519–524, 1988a.

Daly, M., and M. Wilson. *Homicide.* New York: Aldine de Gruyter, 1988b.

Daly, M., M. Wilson, and S. J. Weghorst. Male sexual jealousy. *Ethology and Sociobiology. 3:* 11–27, 1982.

Damasio, A. R., and H. Damasio. Brain and language. *Scientific American. 267:*89–95, 1992.

Darwin, C. *The Origin of Species* (1859). London: Penguin, 1947.

Darwin, C. *The Expression of the Emotions in Man and Animals* (1872). Chicago: University of Chicago Press, 1965.

Davis, D. R. Depression as adaptation to crisis. *British Journal of Medical Psychology. 43:* 109–116, 1970.

Davis-Walton, J. Born too late? Parental investment and birth order in modern Canada. Paper given at the Human Behavior and Evolution Society meeting, Santa Barbara, CA, June 28–July 2, 1995.

Dawkins, R. *The Selfish Gene.* Oxford, England: Oxford University Press, 1976.

Dawkins, R. *The Extended Phenotype.* Oxford, England: Freeman, 1982.

Dawkins, R. *The Blind Watchmaker.* New York: Norton, 1987.

Dawkins, R., and J. R. Krebs. Animal signals: Information or manipulation? In J. R. Krebs and N. B. Davies (eds.), *Behavioural Ecology.* Sunderland, MA: Sinauer, 1978, pp. 282–312.

deCatanzaro, D. Human suicide: A biological perspective. *Behavioral Brain Sciences. 3:*265–290, 1980.

deCatanzaro, D. Evolutionary limits to self-preservation. *Ethology and Sociobiology. 12:*13–28, 1991.

DeLisi, L. E., A. L. Hoff, J. E. Schwartz, G. W. Shields, S. N. Halthore, S. M. Gupta, F. A. Henn, and A. K. Anand. Brain morphology in first-episode schizophrenic-like psychotic patients: A quantitative magnetic resonance imaging study. *Biological Psychiatry. 29:*159–175, 1991.

DeLisi, L. E., P. Stritzke, H. Riordan, V. Holan, A. Boccio, M. Kushner, J. McClelland, O. Van Eyl, and A. Anand. The timing of brain morphological changes in schizophrenia and their relationship to clinical outcome. *Biological Psychiatry. 31:*241–254, 1992.

DeLisi, L. E., W. Twe, S. Xie, A. L. Hoff, M. Sakuma, M. Kushner, G. Lee, K. Shedlack, A. M. Smith, and R. Grimson. A prospective follow-up study of brain morphology and cognition in first-episode schizophrenic patients: Preliminary findings. *Biological Psychiatry. 38:*349–360, 1995.

Delvenne, V., F. Lotstra, S. Goldman, F. Biver, V. DeMaertelaer, J. Appelboom-Fondu, A. Schoutens, L. M. Bidaut, A. Luxen, and J. Mendelwicz. Brain hypometabolism of glucose in anorexia nervosa: A PET scan study. *Biological Psychiatry. 37:*161–169, 1995.

Dennett, D. C. *Darwin's Dangerous Idea.* New York: Simon and Schuster, 1995.

Der, G., S. Gupta, and R. M. Murray. Is schizophrenia disappearing? *Lancet. 335:*513–516, 1990.

Desimone, R. Face-selective cells in the temporal cortex of monkeys. *Journal of Cognitive Neuroscience. 3:*1–8, 1991.

Detterman, D. K., L. A. Thompson, and R. Plomin. Differences in heritability across groups differing in ability. *Behavioral Genetics. 20:*369–384, 1990.

de Waal, F. B. M. Food sharing and reciprocal obligations among chimpanzees. *Journal of Human Evolution. 18:*433–459, 1989.

de Waal, F. B. M., and L. M. Luttrell. Mechanisms of social reciprocity in three primate species: Symmetrical relationship characteristics or cognition? *Ethology and Sociobiology. 9:* 101–118, 1988.

Diekstra, R. F. W. Suicidal behavior and depressive disorders in adolescents and young adults. *Neuropsychobiology. 22:*194–207, 1989.

Dienske, H., J. A. R. Sanders-Woudstra, and G. de Jonge. A biologically meaningful classification in child psychiatry that is based upon ethological methods. *Ethology and Sociobiology. 8:*27S–45S, 1987.

Dilling, H., and S. Weyerer. Psychiatric illness and work capacity. In L. N. Robins, P. J. Clayton, and J. K. Wing (eds.), *The Social Consequences of Psychiatric Illness.* New York: Brunner/Mazel, 1980, pp. 229–247.

Dillon, J. E., M. J. Raleigh, M. T. McGuire, D. Bergin-Pollack, and A. Yuwiler. Plasma catecholamines and behavior in male vervet monkeys. *Physiology Behavior. 51:*973–977, 1992.

Dittmann, J., and R. Schuttler. Disease consciousness and coping strategies of patients with schizophrenic psychosis. *Acta Psychiatrica Scandinavica. 82:*318–322, 1990.

Dixon, A. K., H. U. Fish, C. Huber, and A. Walser. Ethological studies in animals and man, their use in psychiatry. *Pharmacopsychiatry. 22*(Suppl.):44–50, 1989.

Dohrenwend, B. S., and B. P. Dohrenwend. Some issues in research on stressful life events. *Journal of Nervous and Mental Disease. 166:*7–15, 1978.

Donnelly, E. F., I. N. Waldman, D. L. Murphy, R. J. Wyatt, and F. K. Goodwin. Primary affective disorder: Thought disorder in depression. *Journal of Abnormal Psychology. 89:* 315–319, 1980.

Dubrovsky, B. Fundamental neuroscience and the classification of psychiatric disorders. *Neuroscience Biobehavioral Reviews. 19:*511–518, 1995.

Dugatkin, L. A. The evolution of the "Con Artist." *Ethology and Sociobiology. 13:*3–18, 1992.

Duggan, C. F., A. S. Lee, and R. M. Murray. Does personality predict long-term outcome in depression? *British Journal of Psychiatry. 157:*19–24, 1990.

Dunbar, R. I. M. Coevolution of neocortical size, group size and language in humans. *Behavioral Brain Sciences. 16:*681–735, 1993.

Dunbar, R. I. M., and M. Spoors. Social networks, support cliques, and kinship. *Human Nature. 6:*273–290, 1995.

Duval, S., and R. Wicklund. *A Theory of Objective Self-Awareness.* New York: Academic Press, 1972.

Dworkin, R. H., M. F. Lenzenweger, S. O. Moldin, G. F. Skillings, and S. E. Levick. A multidimensional approach to the genetics of schizophrenia. *American Journal of Psychiatry. 145:* 1077–1083, 1988.

Eaton, W. W. Epidemiology of schizophrenia. *Epidemiologic Reviews. 7:*105–526, 1985.

Ebbesson, S. O. E. Evolution and ontogeny of neural circuits. *Behavioral Brain Sciences. 7:* 321–366, 1984.

Ehlers, C. L. Chaos and complexity. *Archives of General Psychiatry. 52:*960–964, 1995.

Eibl-Eibesfeldt, I. Patterns of parent-child interaction in a cross-cultural perspective. In A. Oliverio and M. Zappelia (eds.), *The Behavior of Human Infants.* New York: Plenum Press, 1983, pp. 177–217.

Eibl-Eibesfeldt, I. *Human Ethology.* New York: Aldine de Gruyter, 1989.

Eisemann, M. The relationship of personality to social network aspects and loneliness in depressed patients. *Acta Psychiatrica Scandinavica. 70:*337–341, 1984.

Ekman, P. Universals and cultural differences in facial expressions of emotion. In J. K. Cole (ed.), *Nebraska Symposium on Motivation.* Lincoln: University of Nebraska Press, 1971, pp. 207–283.

Ekman, P. Movements with precise meaning. *Journal of Communication. 14:*14–26, 1976.

Ekman, P. Self-deception and detection of misinformation. In J. S. Lockard and D. L. Paulhus (eds.), *Self-Deception: An Adaptive Mechanism?* Englewood Cliffs, NJ: Prentice-Hall, 1988, pp. 229–250.

Ekman, P. Facial expression and emotion. *American Psychologist. 48:*384–392, 1993.

Ekman, P., and W. V. Friesen. The repertoire of nonverbal behavior: Categories, origins, usage, and coding. *Semiotica. 1:*49–98, 1969.

Ekman, P., W. V. Friesen, and K. S. Scherer. Body movement and voice pitch in deceptive interaction. *Semiotica. 16:*23–27, 1976.

Ekman, P., E. R. Sorenson, and W. V. Friesen. Pan-cultural elements in facial displays of emotion. *Science. 164:*86–88, 1969.

Elkin, I., M. T. Shea, J. T. Watkins, S. D. Imber, S. M. Sotsky, J. F. Collins, D. R. Glass, P. A. Pilkonis, W. R. Leber, J. P. Docherty, S. J. Fiester, and M. B. Parloff. National Institute of Mental Health Treatment of Depression Collaborative Research Program. *Archives of General Psychiatry. 46:*971–982, 1989.

Ellgring, H. *Nonverbal Communication in Depression.* Cambridge, England: Cambridge University Press, 1989.

Ellis, P. M., and C. Salmond. Is platelet imipramine binding reduced in depression? A meta-analysis. *Biological Psychiatry. 36:*292–299, 1994.

Ellis, B. J., and D. Symons. Sex differences in sexual fantasy: An evolutionary psychological approach. Paper given at the Human Evolution and Behavior meeting, Los Angeles, August 18–20, 1991.

Ellyson, S. L., and J. F. Dovidio (eds.). *Power, Dominance, and Nonverbal Behavior.* New York: Springer-Verlag, 1985.

Elowson, A. M., and C. T. Snowdon. Pygmy marmosets, *Cebuella pygmaea,* modify vocal structure in response to changed social environment. *Animal Behaviour. 47:*1267–1277, 1994.

Endler, J. A. Natural selection: Current usages. In E. F. Keller and E. A. Lloyd (eds.), *Keywords in Evolutionary Biology.* Cambridge, MA: Harvard University Press, 1992, pp. 220–224.

Engle, G. L. The clinical application of the biopsychosocial model. *American Journal of Psychiatry. 137:*535–544, 1980.

Erickson, M. T. Rethinking Oedipus: An evolutionary perspective of incest avoidance. *American Journal of Psychiatry. 150:*411–416, 1993.

Erlenmeyer-Kimling, L., and W. Paradowski. Selection and schizophrenia. In C. J. Bajema (ed.), *Natural Selection in Human Populations.* Huntington, NY: Krieger, 1977, pp. 259–275.

Erlenmeyer-Kimling, L., R. A. Wunsch-Hitzig, and S. Deutsch. Family formation by schizophrenics. In L. N. Robins, P. J. Clayton, and J. S. Wing (eds.), *The Social Consequences of Psychiatric Illness.* New York: Brunner/Mazel, 1980, pp. 114–134.

Esser, A. H. Dominance hierarchy and clinical course of psychiatrically hospitalized boys. *Child Development. 39:*147–157, 1968.

Essock-Vitale, S. M. The reproductive success of wealthy Americans. *Ethology and Sociobiology. 5:*45–50, 1984.

Essock-Vitale, S. M., and L. A. Fairbanks. Sociobiological theories of kin selection and reciprocal altruism and their relevance for psychiatry. *Journal of Nervous and Mental Disease. 167:*23–28, 1979.

Essock-Vitale, S. M., and M. T. McGuire. Sociobiology and its potential usefulness to psychiatry. *McLean Hospital Journal. 4:*69–81, 1979.

Essock-Vitale, S. M., and M. T. McGuire. Predictions derived from the theories of kin selection and reciprocation assessed by anthropological data. *Ethology and Sociobiology. 1:*233–243, 1980.

Essock-Vitale, S. M., and M. T. McGuire. Women's lives viewed from an evolutionary perspective: 1. Sexual histories, reproductive success, and demographic characteristics of a random subsample of American women. *Ethology and Sociobiology. 6:*137–154, 1985a.

Essock-Vitale, S. M., and M. T. McGuire. Women's lives from an evolutionary perspective: 2. Patterns of helping. *Ethology and Sociobiology. 6:*155–173, 1985b.

Essock-Vitale, S. M., and M. T. McGuire. Social and reproductive histories of depressed and anxious women. In R. W. Bell and N. J. Bell (eds.), *Sociobiology and the Social Sciences.* Lubbock: Texas Technical University Press, 1990, pp. 105–118.

Essock-Vitale, S. M., M. T. McGuire, and B. Hooper. Self-deception in social-support networks. In J. S. Lockhard and D. L. Paulhus (eds.), *Self-Deception: An Adaptive Mechanism?* Englewood Cliffs, NJ: Prentice-Hall, 1988, pp. 200–211.

Ewald, P. W. Cultural vectors, virulence, and the emergence of evolutionary epidemiology. *Oxford Surveys in Evolutionary Biology. 5:*215–245, 1988.

Ewald, P. W. Transmission modes and the evolution of virulence. *Human Nature. 2:*1–30, 1991a.

Ewald, P. W. Waterborne transmission and the evolution of virulence among gastrointestinal bacteria. *Epidemiology Infection. 106:*83–119, 1991b.

Eysenck, H. J. *Dimensions of Personality.* London: Kegan Paul, 1947.

Eysenck, H. J. *Crime and Personality.* London: Routledge and Kegan Paul, 1964.

Fabre-Nys, C., R. E. Meller, and E. B. Keverne. Opiate antagonists stimulate affiliative behaviour in monkeys. *Pharmacology Biochemistry Behavior. 16:*653–659, 1982.

Fairbanks, L. A. Mother-infant behavior in vervet monkeys. *Behavior Ecology Sociobiology. 23:*157–165, 1988a.

Fairbanks, L. A. Vervet monkey grandmothers: Interactions with infant grandoffspring. *International Journal of Primatology. 9:*425–441, 1988b.

Fairbanks, L. A. Early experience and cross-generational continuity of mother-infant contact in vervet monkeys. *Developmental Psychobiology. 22:*669–681, 1989.

Fairbanks, L. A., and M. T. McGuire. Long-term effects of early mothering behavior on responsiveness to the environment in vervet monkeys. *Developmental Psychobiology. 21:*711–724, 1988.

Fairbanks, L. A., M. T. McGuire, S. R. Cole, R. Sbordone, F. M. Silvers, M. Richards, and J. Akers. The ethological study of four psychiatric wards: Patient, staff, and system behaviors. *Journal of Psychiatric Research. 13:*193–209, 1977.

Fairbanks, L. A., M. T. McGuire, and C. J. Harris. Nonverbal interaction of patients and therapists during psychiatric interviews. *Journal of Abnormal Psychology. 91:*109–119, 1982.

Faraone, S. V., J. Biederman, and S. Milberger. An exploratory study of ADHD among second-degree relatives of ADHD children. *Biological Psychiatry. 35:*398–402, 1994.

Feierman, J. R., and L. A. Feierman. Mapping the human ethogram: Feminine mannerisms and the concept of gender. Paper presented at the Human Behavior and Evolution Society meeting, Albuquerque, NM, July 23–26, 1992.

Fessler, D. M. T. The phylogenetic development of shame and pride. Paper given at the Human Behavior and Evolution Society meeting, Santa Barbara, CA, June 28–July 2, 1995.

Fiddick, L., L. Cosmides, and J. Tooby. Are there really separate reasoning mechanisms for social contracts and precautions? Paper given at the Human Behavior and Evolution Society meeting, Santa Barbara, CA, June 28–July 2, 1995.

Fish, B. Infant predictors of the longitudinal course of schizophrenic development. *Schizophrenia Bulletin. 13:*395–409, 1987.

Fish, B., J. Marcus, S. L. Hans, J. G. Auerbach, and S. Perdue. Infants at risk for schizophrenia: Sequelae of a genetic neurointegrative defect. *Archives of General Psychiatry. 49:*221–235, 1992.

Fisher, R. A. *The Genetical Theory of Natural Selection.* Oxford, England: Oxford University Press, 1930.

Floderus-Myrhed, B., N. Pedersen, and I. Rasmuson. Assessment of heritability for personality, based on a short-form of the Eysenck Personality Inventory: A study of 12,898 twin pairs. *Behavioral Genetics. 10:*153–162, 1980.

Forsyth, D. R. The functions of attributions. *Social Psychology Quarterly. 43:*184–189, 1980.

Forsyth, D. R., R. E. Berger, and T. Mitchell. The effects of self-serving vs. other-serving claims of responsibility on attraction and attribution in groups. *Social Psychology Quarterly. 44:*59–64, 1981.

Fowles, D. C. Schizophrenia: Diathesis-stress revisited. *Annual Review of Psychology. 43:*303–336, 1992.

Fox, R. *The Search for Society.* New Brunswick, NJ: Rutgers University Press, 1989.

Frank, R. H. *Passions within Reasons.* New York: Norton, 1988.

Franzek, E., and H. Beckmann. Season-of-birth effect reveals the existence of etiologically different groups of schizophrenia. *Biological Psychiatry. 32:*375–378, 1992.

Freedman, D. G. *Human Infancy: An Evolutionary Perspective.* Hillsdale, NJ: Erlbaum, 1974.

Freedman, D. G. *Human Sociobiology.* New York: Free Press, 1979.

Freeman, W. Chaos in psychiatry. *Biological Psychiatry. 31:*1079–1081, 1992.

Freud, S. Totem and Taboo (1912–1913). In J. Strachey (ed.), *The Standard Edition of the Complete Psychological Works of Sigmund Freud,* Vol. 13. London: Hogarth, 1968, pp. 11–61.

Freud, S. *Psych-Analysis.* London: Hogarth, 1922.

Friedman, B. H., J. F. Thayer, T. D. Borkovec, R. A. Tyrrell, B. Johnson, and R. Columbo. Autonomic characteristics of nonclinical panic and blood phobia. *Biological Psychiatry. 34:*298–310, 1993.

Frijda, N. H. The place of appraisal in emotion. *Cognition and Emotion. 7:*357–387, 1993.

Friston, K. J. The dorsolateral prefrontal cortex, schizophrenia and PET. *Journal of Neural Transmission. 37*(Suppl.):79–93, 1992.

Frith, C. D. Consciousness, information processing and schizophrenia. *British Journal of Psychiatry. 134:*225–235, 1979.

Fuchs, E., and M. Schumacher. Psychosocial stress affects pineal function in the tree shrew (*Tupaia belangeri*). *Physiology Behavior. 47:*713–717, 1990.

Fulker, D. W., S. B. G. Eysenck, and M. Zuckerman. A genetic and environmental analysis of sensation seeking. *Journal of Personality Research. 14:*261–281, 1980.

Fuller, R. W. The involvement of serotonin in regulation of pituitary-adrenocortical function. Frontiers in *Neuroendocrinology. 13:*250–270, 1992.

Futuyma, D. J. *Evolutionary Biology,* 2nd ed. Sunderland, MA: Sinauer, 1986.

Gaebel, W., and W. Wölwer. Facial expression and emotional face recognition in schizophrenia and depression. *European Archives Psychiatry Clinical Neuroscience. 242:*46–52, 1992.

Gangestad, S. W., and D. M. Buss. Pathogen prevalence and human mate preferences. *Ethology and Sociobiology. 14:*89–96, 1993.

Gangestad, S. W., and R. Thornhill. Human sexual selection, developmental stability, and indicator mechanisms. Paper given at the Human Behavior and Evolution Society meeting, Santa Barbara, CA, June 28–July 2, 1995.

Gangestad, S. W., R. Thornhill, and R. A. Yeo. Facial attractiveness, developmental stability, and fluctuating asymmetry. *Ethology and Sociobiology. 15:*73–86, 1994.

Garcia y Robertson, R., and J. Garcia. Darwin was a learning theorist. In R. C. Bolles and M. D. Beecher (eds.), *Evolution and Learning.* Hillsdale, NJ: Erlbaum, 1987, pp. 17–37.

Gardner, R. Mechanisms of manic-depressive disorder. *Archives of General Psychiatry. 39:* 1436–1441, 1982.

Gazzaniga, M. S. Organization of the human brain. *Science. 245:*947–952, 1989.

Gazzaniga, M. S. *Nature's Mind.* New York: Basic Books, 1992.

Geen, R. G. Social motivation. *Annual Review of Psychology. 42:*377–399, 1991.

Gelles, R. J. Child abuse and violence in single-parent families: Parent absence and economic deprivation. *American Journal of Orthopsychiatry. 59:*492–501, 1989.

Gelles, R. J. Physical violence, child abuse, and child homicide: A continuum of violence, or distinct behaviors? *Human Nature.* 2:59–72, 1991.

Gelles, R. J., and J. W. Harrop. The risk of abusive violence among children with nongenetic caretakers. *Family Relations. 40:*78–83, 1991.

Genova, P. Is American psychiatry terminally ill? *Psychiatric Times.* June 1993, pp. 19–20.

George, M. S., T. A. Ketter, D. S. Gill, J. V. Haxby, L. G. Ungerleider, P. Herscovitch, and R. M. Post. Brain regions involved in recognizing facial emotion or identity: An oxygen-15 PET study. *Journal of Neuropsychiatry Clinical Neurosciences. 15:*384–394, 1993.

George, M. S., T. A. Ketter, P. I. Parekh, B. Horwitz, P. Herscovitch, and R. M. Post. Brain activity during transient sadness and happiness in healthy women. *American Journal of Psychiatry. 152:*341–351, 1995.

Gershon, E. S., M. Martinez, L. R. Goldin, and P. V. Gejman. Genetic mapping of common diseases: The challenges of manic-depressive illness and schizophrenia. *Trends in Genetics. 6:*282–287, 1990.

Gershon, E. S., C. R. Merril, L. R. Goldin, L. E. DeLisi, W. H. Berrettini, and J. I. Nurnberger, Jr. The role of molecular genetics in psychiatry. *Biological Psychiatry. 22:*1388–1405, 1987.

Gert, B. A sex caused inconsistency in DSM-III-R: The definition of mental disorder and the definition of paraphilias. *Journal of Medicine Philosophy. 17:*155–171, 1992.

Gift, T. E., J. S. Strauss, B. A. Ritzler, R. F. Kokes, and D. W. Harder. Social class and psychiatric disorder. *Journal of Nervous and Mental Disease. 176:*593–597, 1988.

Gilbert, P. *Depression: From Psychology to Brain State.* London: Erlbaum, 1984.

Gilbert, P. *Human Nature and Suffering.* Hove, UK: Erlbaum, 1989.

Gilbert, P. *Depression: The Evolution of Powerlessness.* New York: Guilford Press, 1992.

Gilbert, P. Defence and safety: Their function in social behaviour and psychopathology. *British Journal of Clinical Psychology. 32:*131–153, 1993.

Gilbert, P. Biopsychosocial approaches and evolutionary theory as aids to integration in clinical psychology and psychotherapy. *Clinical Psychology Psychotherapy. 2:*135–156, 1995.

Gilbert, P. The evolution of social attractiveness and its role in shame, humiliation, conformity and sex. *British Journal of Medical Psychology,* in press.

Gilbert, P., and S. Allan. Assertiveness, submissive behaviour and social comparison. *British Journal of Clinical Psychology. 33:*295–306, 1994.

Gilbert, P., and S. Allen. Varieties of submissive and subordinate behavior as forms of social defense: Biosocial integration and psychopathology, submitted.

Gilbert, P., S. Allan, L. Ball, and Z. Bradshaw. Overconfidence and personal evaluations of social rank. *British Journal of Medical Psychology. 69:*59–68, 1996.

Gilbert, P., S. Allan, and K. Goss. Parental representations, shame, interpersonal problems, and vulnerability to psychopathology. *Clinical Psychology Psychotherapy. 3:*23–34, 1996.

Gilbert, P., S. Allan, and D. R. Trent. Involuntary subordination or dependency as key dimensions of depressive vulnerability? *Journal of Clinical Psychology. 51:*740–752, 1995.

Gilbert, P., J. Pehl, and S. Allan. The phenomenology of shame and guilt: An empirical investigation. *British Journal of Medical Psychology. 67:*23–36, 1994.

Ginsburg, B. E. Origins and dynamics of social organization in primates and in wolves: Cooperation, aggression and hierarchy. In A. Somit and R. Wildenmann (eds.), *Hierarchy and Democracy.* Baden-Baden: Nomos Verlagsgesellschaft, 1991, pp. 45–62.

Glantz, K. Reciprocity: A possible new focus for psychotherapy. *Psychotherapy. 24:*20–24, 1987.

Glantz, K., and J. Pearce. *Exiles from Eden.* New York: Norton, 1989.

Goedde, H. W., D. P. Agarwal, R. Eckey, and S. Harada. Population genetic and family studies on aldehyde dehydrogenase deficiency and alcohol sensitivity. *Alcohol. 2:*383–390, 1985.

Goedde, H. W., H. G. Benkmann, L. Kriese, P. Bogdanski, D. P. Agarwal, D. Ruofu, C. Liang-zhong, C. Meiying, Y. Yida, X. Jiujin, L. Shizhe, and W. Yongfa. Aldehyde dehydrogenase isozyme deficiency and alcohol sensitivity in four different Chinese populations. *Human Heredity. 34:*183–186, 1984.

Goldberg, L., and S. Breznitz. *Handbook of Stress.* New York: Free Press, 1982.

Goldstein, J. M., S. V. Faraone, W. J. Chen, and M. T. Tsuang. Genetic heterogeneity may in part explain sex differences in the familial risk for schizophrenia. *Biological Psychiatry. 38:*808–813, 1995.

Goodwin, F. K., and K. R. Jamison. *Manic-Depressive Disease.* New York: Oxford University Press, 1990.

Gottesman, I. I., and A. Bertelsen. Confirming unexpressed genotypes for schizophrenia. *Archives of General Psychiatry. 46:*867–872, 1989.

Gottschalk, A., M. S. Bauer, and P. C. Whybrow. Evidence of chaotic mood variation in biopolar disorder. *Archives of General Psychiatry. 52:*947–959, 1995.

Gould, S. J. The confusion over evolution. *New York Review of Books.* November 10, 1992, pp. 84–91.

Gould, S. J., and R. C. Lewontin. The spandrels of San Marco and the Panglossian paradigm: A critique of the adaptionist programme. *Proceedings Royal Society London. 205:*581–598, 1979.

Grammer, K. Age and facial features influencing mate choice. Paper given at the Gruter Institute meeting, Munich, Germany, April 24–26, 1995.

Grammer, K., W. Schiefenhövel, M. Schleidt, B. Lorenz, and I. Eibl-Eibesfeldt. Patterns on the face: The eyebrow flash in crosscultural comparison. *Ethology. 77:*279–299, 1988.

Grant, E. C. An ethological description of non-verbal behaviour during interviews. *British Journal of Medical Psychology. 41:*177–183, 1968.

Grant, E. C. Human facial expression. *Man. 4:*525–536, 1969.

Grant, I., H. L. Sweetwood, J. Yager, and M. Gerst. Quality of life events in relation to psychiatric symptoms. *Archives of General Psychiatry. 38:*335–339, 1981.

Gray, J. A. *The Neuropsychology of Anxiety: An Inquiry into the Functions of the Septo-Hippocampal System.* Oxford, England: Oxford University Press, 1982.

Grinker, R. R., Sr. The relevance of general systems theory to psychiatry. In S. Arieti (ed.), *American Handbook of Psychiatry,* 2nd ed. New York: Basic Books, 1975, pp. 251–272.

Grove, W. M., E. D. Eckert, L. Heston, T. J. Bouchard, Jr., N. Segal, and D. T. Lykken. Heritability of substance abuse and antisocial behavior: A study of monozygotic twins reared apart. *Biological Psychiatry. 27:*1293–1304, 1990.

Gruter, M. *Law and the Mind.* Newbury Park, CA: Sage, 1991.

Guilford, T., and M. S. Dawkins. Receiver psychology and the evolution of animal signals. *Animal Behaviour. 42:*1–14, 1991.

Guscott, R., and P. Grof. The clinical meaning of refractory depression: A review for the clinician. *American Journal of Psychiatry. 148:*695–704, 1991.

Gut, E. Cause and function of the depressed response: A hypothesis. *International Review of Psychoanalysis. 9:*179–189, 1982.

Gut, E. *Productive and Unproductive Depression.* New York: Basic Books, 1989.

Guze, S. B. Biological psychiatry: Is there any other kind? *Psychological Medicine. 19:*315–323, 1989.

Guze, S. B. *Why Psychiatry Is a Branch of Medicine.* New York: Oxford University Press, 1992.

Guze, S. B., D. W. Goodwin, and J. B. Crane. A psychiatric study of the wives of convicted felons: An example of assortative mating. *American Journal of Psychiatry. 126:*1773–1776, 1970.

Hafner, H. The concept of disease in psychiatry. *Psychological Medicine. 17:*11–14, 1987.

Haier, R. J., M. S. Buchsbaum, E. DeMet, and J. Wu. Biological vulnerability to depression:

Replication of MAO and evoked potentials as risk factors. *Neuropsychobiology. 20:*62–66, 1988.

Haier, R. J., B. Siegel, C. Tang, L. Abel, and M. S. Buchsbaum. Intelligence and changes in regional cerebral glucose metabolic rate following learning. *Intelligence. 16:*415–426, 1992.

Haig, D. Genetic conflicts in human pregnancy. *Quarterly Review Biology. 68:*495–532, 1993.

Haig, D. Genetic conflicts in human pregnancy. Paper presented at the Human Behavior and Evolution Society meeting, Santa Barbara, CA, June 28–July 2, 1995.

Haldane, J. B. S. *The Causes of Evolution.* Ithaca, NY: Cornell University Press, 1932.

Hamida, S. B. Mate preferences: Implications for the gender difference in unipolar depression. *ASCAP. 9:*4–29, 1996.

Hamilton, N. G. A critical review of object relations theory. *American Journal of Psychiatry. 146:*1552–1560, 1989.

Hamilton, W. D. The genetical evolution of social behaviour, Parts 1 and 2. *Journal of Theoretical Biology. 7:*1–52, 1964.

Hamilton, W. D., and M. Zuk. Heritable true fitness and bright birds: A role for parasites? *Science. 218:*384–387, 1982.

Hare, E. H., J. S. Price, and E. T. O. Slater. Fertility in obsessional neurosis. *British Journal of Psychiatry. 121:*197–205, 1972.

Hare, R. D. Diagnosis of antisocial personality disorder in two prison populations. *American Journal of Psychiatry. 140:*887–890, 1983.

Harlow, H. F., and M. K. Harlow. Social deprivation in monkeys. *Scientific American. 207:* 136–146, 1962.

Harpending, H. C., and J. Sobus. Sociopathy as an adaptation. *Ethology and Sociobiology. 8:* 63S–72S, 1987.

Harries, M. H., and D. I. Perrett. Visual processing of faces in temporal cortex: Physiological evidence for a modular organization and possible anatomical correlates. *Journal of Cognitive Neuroscience. 3:*9–24, 1991.

Harris, H. W., and K. F. Schaffner. Molecular genetics, reductionism, and disease concepts in psychiatry. *Journal of Medicine Philosophy. 17:*127–153, 1992.

Harris, J., J. L. T. Birley, and K. W. M. Fulford. A proposal to classify happiness as a psychiatric disorder. *British Journal of Psychiatry. 162:*539–542, 1993.

Harris, J. C., and J. D. Newman. Combined opiate/adrenergic receptor blockade enhances squirrel monkey vocalization. *Pharmacology Biochemistry Behavior. 31:*223–226, 1988.

Harris, T., P. Surtees, and J. Bancroft. Is sex necessarily a risk factor to depression? *British Journal of Psychiatry. 158:*708–712, 1991.

Harvey, P. H., and T. H. Clutton-Brock. Life history variation in primates. *Evolution. 39:* 559–581, 1985.

Hegarty, J. D., R. J. Baldessarini, M. Tohen, C. Waternaux, and G. Oepen. One hundred years of schizophrenia: A meta-analysis of the outcome literature. *American Journal of Psychiatry. 151:*1409–1416, 1994.

Henderson, S. Care-eliciting behavior in man. *Journal of Nervous and Mental Disease. 159:* 172–181, 1974.

Henderson, S., D. G. Byrne, P. Duncan-Jones, S. Adcock, R. Scott, and G. P. Steele. Social bonds in the epidemiology of neurosis: A preliminary communication. *British Journal of Psychiatry. 132:*463–466, 1978.

Henderson, S., D. G. Byrne, P. Duncan-Jones, R. Scott, and S. Adcock. Social relationships, adversity and neurosis: A study of associations in a general population sample. *British Journal of Psychiatry. 136:*574–583, 1980.

Henderson, S., P. Duncan-Jones, H. McAuley, and K. Ritchie. The patient's primary group. *British Journal of Psychiatry. 132:*74–86, 1978.

Hepper, P. G. (ed.). *Kin Recognition.* Cambridge: Cambridge University Press, 1991.

Hermans, H. J. M. Voicing the self: From information processing to dialogical interchange. *Psychological Bulletin. 119:*31–50, 1996.

Herz, M. I. (ed.). Major advances in the treatment of schizophrenia. *Psychiatric Annals. 26:* 513–538, 1996.

Heston, L. L. Psychiatric disorders in foster home reared children of schizophrenic mothers. *British Journal of Psychiatry. 112:*819–825, 1966.

Hilger, T., P. Propping, and F. Haverkamp. Is there an increase of reproductive rates in schizophrenics? *Archiv für Psychiatrie und Nervenkrankheiten. 233:*177–186, 1983.

Hillger, L. A., and O. Koenig. Separable mechanisms in face processing: Evidence from hemispheric specialization. *Journal of Cognitive Neuroscience. 3:*42–58, 1991.

Hinde, R. A. (ed.). *Biological Bases of Human Social Behavior.* New York: McGraw-Hill, 1974.

Hinde, R. A. *Ethology.* New York: Oxford University Press, 1982.

Hinde, R. A. (ed.). *Primate Social Relationships.* Sunderland, MA: Sinauer, 1983.

Hinde, R. A. Was "the expression of the emotions" a misleading phrase? *Animal Behaviour. 33:*985–992, 1985.

Hinde, R. A. Developmental psychology in the context of other behavioral sciences. *Developmental Psychology. 28:*1018–1029, 1992.

Hinde, R. A., and Y. Spencer-Booth. Effects of brief separation from mother on rhesus monkeys. *Science. 173:*111–118, 1971.

Hinde, R. A., and J. Stevenson-Hinde (eds.). *Constraints on Learning: Limitations and Predispositions.* London: Academic Press, 1973.

Hirsch L. R., and L. Paul. Human male mating strategies: 2. Courtship tactics of the "quality" and "quantity" alternatives. *Ethology and Sociobiology. 17:*55–70, 1996.

Hirschfeld, R. M. A. (ed.). Longitudinal course of affective disorders. *Psychiatric Annals. 26:* 312–361, 1996.

Hirschfeld, R. M. A., G. L. Klerman, P. J. Clayton, M. B. Keller, P. McDonald-Scott, and B. H. Larkin. Assessing personality: Effects of the depressive state on trait measurement. *American Journal of Psychiatry. 140:*695–699, 1983.

Hirshleifer, J. Competition, cooperation, and conflict in economics and biology. *American Economic Association. 68:*238–243, 1978.

Hocking, J. E., and D. G. Leathers. Nonverbal indicators of deception: A new theoretical perspective. *Communication Monographs. 47:*119–131, 1980.

Hofer, M. A. Relationships as regulators: A psychobiologic perspective on bereavement. *Psychosomatic Medicine. 46:*183–197, 1984.

Hokanson, J. E., W. P. Sacco, S. R. Blumberg, and G. C. Landrum. Interpersonal behavior of depressive individuals in a mixed-motive game. *Journal of Abnormal Psychology. 89:* 320–332, 1980.

Hold-Cavell, B. C. L., and D. Borsutzky. Strategies to obtain high regard: Longitudinal study of a group of preschool children. *Ethology and Sociobiology. 7:*39–56, 1986.

Hollander, E., E. Schiffman, B. Cohen, M. A. Rivera-Stein, W. Rosen, J. M. Gorman, A. J. Fyer, L. Papp, and M. R. Liebowitz. Signs of central nervous system dysfunction in obsessive-compulsive disorders. *Archives of General Psychiatry. 47:*27–32, 1990.

Honer, W. G., A. S. Bassett, G. N. Smith, J. S. Lapointe, and P. Falkai. Temporal lobe abnormalities in multigenerational families with schizophrenia. *Biological Psychiatry. 36:*734–743, 1994.

Hopkins, W. D., and E. S. Savage-Rumbaugh. Vocal communication as a function of differential rearing experiences in Pan paniscus: A preliminary report. *International Journal of Primatology. 12:*559–583, 1991.

Howes, M. J., and J. E. Hokanson. Conversational and social responses to depressive interpersonal behavior. *Journal of Abnormal Psychology. 88:*625–634, 1979.

Howland, R. H., and M. E. Thase. Biological studies of dysthymia. *Biological Psychiatry. 30:* 283–304, 1991.

Hoyer, G., and E. Lund. Suicide among women related to number of children in marriage. *Archives of General Psychiatry. 50:*134–137, 1993.

Hudson, J. I., and H. G. Pope, Jr. Affective spectrum disorder: Does antidepressant response identify a family of disorders with a common pathophysiology? *American Journal of Psychiatry. 147:*552–564, 1990.

Humphreys, M. S., and W. Revelle. Personality, motivation, and performance: A theory of the relationship between individual differences and information processing. *Psychological Review. 91:*153–184, 1984.

Humphries, S. R. Munchausen syndrome. *British Journal of Psychiatry. 152:*416–417, 1988.

Hunter, R., and I. Macalpine. *Three Hundred Years of Psychiatry—1535–1860.* London: Oxford University Press, 1963.

Hutt, C., and C. Ounsted. The biological significance of gaze aversion with particular reference to the syndrome of infantile autism. *Behavior Science. 11:*346–356, 1966.

Hutt, S. J., and C. Hutt (eds.). *Behaviour Studies in Psychiatry.* Oxford: Pergamon Press, 1970.

Hyman, R. The psychology of deception. *Annual Review Psychology. 40:*133–154, 1989.

Igbal, N., W. Bajwa, and G. M. Asnis. The role of norepinephrine in depression. *Psychiatric Annals. 19:*354–359, 1989.

Igbal, N., and H. M. van Praag (eds.). *Biology and Treatment of Schizophrenia: Vols. 1, 2. Psychiatric Annals. 23:*105–215, 1993.

Innes, J. M., and S. Gilroy. The semantics of asking a favor: Asking for help in three countries. *Journal of Social Psychology. 110:*3–7, 1980.

Iny, L. J., J. Pecknold, B. E. Suranyi-Cadotte, B. Bernier, L. Luthe, N. P. V. Nair, and M. J. Meaney. Studies of a neurochemical link between depression, anxiety, and stress from [³H] imipramine and [³H] paroxetine binding on human platelets. *Biological Psychiatry. 36:*281–291, 1994.

Izard, C. E., E. A. Hembree, and R. R. Huebner. Infants' emotion expressions to acute pain: Developmental change and stability of individual differences. *Developmental Psychology. 23:*105–113, 1987.

Jablensky, A. Multicultural studies and the nature of schizophrenia: A review. *Journal of Royal Society Medicine. 80:*162–167, 1987.

Jackson, H. J., H. L. Whiteside, G. W. Bates, R. Bell, R. P. Rudd, and J. Edwards. Diagnosing personality disorders in psychiatric inpatients. *Acta Psychiatrica Scandinavica. 83:*206–213, 1991.

Jamieson, I. G. The functional approach to behavior: Is it useful? *American Naturalist. 127:* 195–208, 1986.

Janicki, M., and C. Crawford. Cognitive and perceptual biases in helping behavior. Paper presented at the Human Behavior and Evolution Society meeting, Albuquerque, NM, July 22–26, 1992.

Janicki, M. G. Detecting helpers and non-helpers: Their importance in reasoning about social exchange. Paper given at the Human Behavior and Evolution Society meeting, Santa Barbara, CA, June 28–July 2, 1995.

Johannsen, W. J. Responsiveness of chronic schizophrenics and normals to social and nonsocial feedback. *Journal of Abnormal Social Psychology. 62:*106–113, 1961.

John, E. R., L. S. Prichep, K. R. Alper, F. G. Mas, R. Cancro, P. Easton, and L. Sverdlov. Quantitative electrophysiological characteristics and subtyping of schizophrenia. *Biological Psychiatry. 36:*801–826, 1994.

Johnson, E. J., and A. Tversky. Affect, generalization, and the perception of risk. *Journal of Personality Social Psychology. 45:*20–31, 1983.

Jonas, A. D., and D. F. Jonas. The evolutionary mechanisms of neurotic behavior. *American Journal of Psychiatry. 131:*636–640, 1974.

Jonas, A. D., and D. F. Jonas. An evolutionary context for schizophrenia. *Schizophrenia Bulletin. 12:*33–41, 1975.

Jones, I. H. Ethology and psychiatry. Australia New Zealand *Journal of Psychiatry. 5:*258–263, 1971.

Jönsson, S. A. T., and H. Jönsson. Clusters in a cohort of untreated schizophrenia: Prognostic importance, atypical cases and the familial versus sporadic distinction. *Acta Psychiatrica Scandinavica. 86:*287–295, 1992.

Joyce, P. R., R. T. Mulder, and C. R. Cloninger. Temperament and hypercortisolemia in depression. *American Journal of Psychiatry. 151:*195–198, 1994.

Jung, C. G. *Memories, Dreams, Reflections.* New York: Random House, 1961.

Jung, C. G. *Two Essays on Analytical Psychology.* Princeton: Princeton University Press, 1966.

Jung, C. G. *Four Archetypes.* London: Routledge and Kegan Paul, 1972.

Kagan, J., J. S. Reznick, and N. Snidman. The physiology and psychology of behavioral inhibition in children. *Child Development. 58:*1459–1473, 1987.

Kagan, J., J. S. Reznick, and N. Snidman. Biological bases of childhood shyness. *Science. 240:* 167–171, 1988.

Kagan, J., and J. Schulkin. On the concepts of fear. *Harvard Review Psychiatry. 3:*1–5, 1995.

Kahn, R. S., H. M. van Praag, S. Wetzler, G. M. Asnis, and G. Barr. Serotonin and anxiety revisited. *Biological Psychiatry. 23:*189–208, 1988.

Kalin, N. H. The neurobiology of fear. *Scientific American. 268:*94–101, 1993.

Kalin, N. H., and S. E. Shelton. Defensive behaviors in infant rhesus monkeys: Environmental cues and neurochemical regulation. *Science. 243:*1718–1721, 1989.

Kaplan, H. I., and B. J. Sadock (eds.). *Comprehensive Textbook of Psychiatry,* Vol. 5. Baltimore: Williams and Wilkins, 1989a.

Kaplan, H. I., and B. J. Sadock. Typical signs and symptoms of psychiatric illness. In H. I. Kaplan and B. J. Sadock (eds.), *Comprehensive Textbook of Psychiatry,* Vol. 5. Baltimore: Williams and Wilkins, 1989b, pp. 468–475.

Kaplan, J. R., S. B. Manuck, T. B. Clarkson, F. M. Lusso, and D. M. Taub. Social status, environment and atherosclerosis in cynomolgus monkeys. *Arteriosclerosis. 2:*359–368, 1982.

Kaplan, J. R., S. B. Manuck, T. B. Clarkson, F. M. Lusso, D. M. Taub, and E. W. Miller. Social stress and atherosclerosis in normocholesterolemic monkeys. *Science. 220:*733–735, 1983.

Kapur, S., and J. J. Mann. Role of the dopaminergic system in depression. *Biological Psychiatry. 32:*1–17, 1992.

Kavanau, J. L. Conservative behavioral evolution, the neural substrate. *Animal Behaviour. 39:* 758–767, 1990.

Kay, S. R., F. Wolkenfeld, and L. M. Murrill. Profiles of aggression among psychiatric patients: 1. Nature and prevalence. *Journal of Nervous and Mental Disease. 176:*539–546, 1988.

Keil, F. C. Constraints on knowledge and cognitive development. *Psychological Review. 88:* 197–227, 1981.

Keller, E. F., and E. A. Lloyd. *Keywords in Evolutionary Biology.* Cambridge, MA: Harvard University Press, 1992.

Kellett, J. M. Evolutionary theory for the dichotomy of the functional psychoses. *Lancet. i:* 860–863, 1973.

Kendell, R. E. The concept of disease and its implications for psychiatry. *British Journal of Psychiatry. 127:*305–315, 1975.

Kendell, R. E. Reflections on psychiatric classification—For the architects of DSM-IV and ICD-10. *Integrative Psychiatry. 2:*43–47, 1984.

Kendell, R. E., and I. W. Kemp. Maternal influenza in the etiology of schizophrenia. *Archives of General Psychiatry. 46:*878–882, 1989.

Kendler, K. S., A. C. Heath, M. C. Neale, R. C. Kessler, and L. J. Eaves. Alcoholism and major depression in women. *Archives of General Psychiatry. 50:*690–698, 1993.

Kendler, K. S., N. Pedersen, L. Johnson, M. C. Neale, and A. A. Mathé. A pilot Swedish twin study of affective illness, including hospital- and population-ascertained subsamples. *Archives of General Psychiatry. 50:*699–706, 1993.

Kendon, A., R. M. Harris, and M. R. Key (eds.). *Organization of Behavior in Face-to-Face Interaction.* The Hague: Mouton, 1975.

Kenrick, D. T., and V. Sheets. Homicidal fantasies. *Ethology and Sociobiology. 14:*231–246, 1993.

Kessler, R. C., K. A. McGonagle, S. Zhao, C. B. Nelson, M. Hughes, S. Eshelman, H.-U. Wittchen, and K. S. Kendler. Lifetime and 12-month prevalence of DSM-III-R psychiatric disorders in the United States. *Archives of General Psychiatry. 51:*8–19, 1994.

Ketelaar, T., and G. C. Clore. Emotion as mental representations of fitness affordances: 2. Does anger make you more rational? Paper given at the Human Behavior and Evolution Society, Santa Barbara, CA, June 28–July 2, 1995.

Kim, K., P. K. Smith, and A. Palermiti. Conflict in childhood and reproductive development. *Evolution Human Behavior. 18:*109–143, 1997.

Kimura, D. Sex differences in the brain. *Scientific American. 267:*119–125, 1992.

Kiritz, S., and R. H. Moos. Physiological effects of social environments. *Psychosomatic Medicine. 36:*96–111, 1974.

Kirmayer, L. J. Cultural variations in the response to psychiatric disorders and emotional distress. *Social Science Medicine. 29:*327–339, 1989.

Kirmayer, L. J. The place of culture in psychiatric nosology: Taijin Kyofusho and DSM-III-R. *Journal of Nervous and Mental Disease. 179:*19–28, 1991.

Klein, D. F. Endogenomorphic depression. *Archives of General Psychiatry. 31:*447–454, 1974.

Klein, D. F. A proposed definition of mental illness. In R. L. Spitzer and D. F. Klein (eds.), *Critical Issues in Psychiatric Diagnosis.* New York: Raven, 1978, pp. 41–71.

Klein, D. F. What's new in DSM-IV? *Psychiatric Annals. 25:*461–474, 1995.

Klein, D. F., R. Gittelman, F. Quitkin, and A. Rifkin. *Diagnosis and Drug Treatment of Psychiatric Disorders: Adults and Children,* 2nd ed. Baltimore: Williams and Wilkins, 1980.

Klein, D. F., and H. M. Klein. The utility of the panic disorder concept. *European Archives Psychiatric Neurological Science. 238:*268–279, 1989.

Klerman, G. L. Depression and adaptation. In R. J. Friedman and M. M. Katz (eds.), *The Psychology of Depression: Contemporary Theory and Research.* New York: Wiley, 1974, pp. 127–145.

Klerman, G. L. Long term outcomes of neurotic depressions. In S. B. Sells, R. Crandall, M. Roff, J. S. Strauss, and W. Pollin (eds.), *Human Functioning in Longitudinal Perspectives: Studies of Normal and Psychopathic Populations.* Baltimore: Williams and Wilkins, 1980, pp. 58–70.

Klerman, G. L. Evaluating the efficacy of psychotherapy for depression: The USA experience. *European Archives Psychiatry Neurological Sciences. 19:*1–8, 1989.

Kligman, M., and C. M. Culver. An analysis of interpersonal manipulation. *Journal of Medicine Philosophy. 17:*173–197, 1992.

Kling, A. S. Neurological correlates of social behavior. In M. Gruter and R. Masters (eds.), *Ostracism: A Social and Biological Phenomenon.* New York: Elsevier, 1986, pp. 27–38.

Kling, A. S., and L. A. Brothers. The amygdala and social behavior. In J. Aggelton (ed.), *The Amygdala: Neurobiological Aspects of Emotion, Memory, and Mental Dysfunction.* New York: Wiley-Liss, 1992, pp. 353–377.

Kocsis, J. H. (ed.). Dsythymia and Chronic Depression States. *Psychiatric Annals. 23:*606–649, 1993.

Konner, M. Anthropology and psychiatry. In H. I. Kaplan and B. J. Saddock (eds.), *Comprehensive Textbook of Psychiatry,* Vol. 5. Baltimore: Williams and Wilkins, 1989, pp. 283–299.

Kosten, T. A., and T. R. Kosten. Pharmacological blocking agents for treating substance abuse. *Journal of Nervous and Mental Disease. 179:*583–592, 1991.

Kraemer, D. A., and W. T. McKinney, Jr. The overlapping territories of psychiatry and ethology. *Journal of Nervous and Mental Disease. 167:*3–22, 1979.

Kraemer, G. W. A psychobiological theory of attachment. *Behavioral Brain Science. 15:*493–541, 1992.

Kraemer, G. W., M. H. Ebert, C. R. Lake, and W. T. McKinney. Hypersensitivity to *d*-amphetamine several years after early social deprivation in rhesus monkeys. *Psychopharmacology. 82:*266–271, 1984.

Kraemer, G. W., M. H. Ebert, D. E. Schmidt, and W. T. McKinney. A longitudinal study of the effect of different social rearing conditions on cerebrospinal fluid norepinephrine and biogenic amine metabolites in rhesus monkeys. *Neuropsychopharmacology. 2:*175–189, 1989.

Kraemer, G. W., M. H. Ebert, D. E. Schmidt, and W. T. McKinney. Strangers in a strange land: A psychobiological study of infant monkeys before and after separation from real or inanimate mothers. *Child Development. 62:*548–566, 1991.

Krebs, J. R., and N. B. Davies. *Behavioral Ecology.* Sunderland, MA: Sinauer, 1978.

Krebs, J. R., and N. B. Davies. *An Introduction to Behavioural Ecology,* 2nd ed. Sunderland, MA: Sinauer, 1987.

Krone, K. P., J. A. Himle, and R. M. Nesse. A standardized behavioral group treatment program for obsessive-compulsive disorder: Preliminary outcomes. *Behavioral Research Therapy. 29:*627–631, 1991.

Kruesi, M. J. P., J. L. Rapoport, S. Hamburger, E. Hibbs, W. Z. Potter, M. Lenane, and G. L. Brown. Cerebrospinal fluid monoamine metabolites, aggression, and impulsivity in disruptive behavior disorders of children and adolescents. *Archives of General Psychiatry. 47:*419–426, 1990.

Kuhn, T. S. *The Structure of Scientific Revolutions,* 2nd ed. Chicago: University of Chicago Press, 1970.

Kurzban, R., J. Tooby, and L. Cosmides. Detecting coalitions: Evolutionary psychology and social categorization. Paper given at the Human Behavior and Evolution Society meeting, Santa Barbara, CA, June 28–July 2, 1995.

Lancaster, J. B., J. Altmann, A. S. Rossi, and L. R. Sherrod (eds.). *Parenting Across the Life Span.* New York: Aldine de Gruyter, 1987.

Lane, A., M. Byrne, F. Mulvany, A. Kinsella, J. L. Waddington, D. Walsh, C. Larkin, and E. O'Callaghan. Reproductive behaviour in schizophrenia relative to other mental disorders: Evidence for increased fertility in men despite decreased marital rate. *Acta Psychiatrica Scandinavica. 91:*222–228, 1995.

Lautenbacher, S., A. M. Pauls, F. Strian, K. Pirke, and J. Krieg. Pain sensitivity in anorexia nervosa and bulimia nervosa. *Biological Psychiatry. 29:*1073–1078, 1991.

Leak, G. K., and S. B. Christopher. Freudian psychoanalysis and sociobiology: A synthesis. *American Psychologist. 37:*313–322, 1982.

Leary, M. R., and R. M. Kowalski. Impression management: A literature review and two-component model. *Psychological Bulletin. 107:*34–47, 1990.

Leary, M. R., E. S. Tambor, S. K. Terdal, and D. L. Downs. Self-esteem as an interpersonal monitor: The sociometer hypothesis. *Journal of Personality and Social Psychology. 68:*518–530, 1995.

Lechin, F., B. van der Dijs, B. Orozco, A. E. Lechin, S. Báez, M. E. Lechin, I. Rada, E. Acosta, L. Arocha, V. Jiménez, G. León, and G. García. Plasma neurotransmitters, blood pressure and heart rate during supine resting, orthostasis, and moderate exercise in dysthymic depressed patients. *Biological Psychiatry. 37:*884–891, 1995.

Leckman, J. F., M. M. Weissman, B. A. Prusoff, K. A. Caruso, K. R. Merikangas, D. L. Pauls, and K. K. Kidd. Subtypes of depression. *Archives of General Psychiatry. 41:*833–838, 1984.

LeMarquand, D., R. O. Pihl, and C. Benkelfat. Serotonin and alcohol intake, abuse, and dependence: Findings of animal studies. *Biological Psychiatry. 36:*395–421, 1994.

Lenzenweger, M. F., and A. W. Loranger. Detection of familial schizophrenia using a psychometric measure of schizotypy. *Archives of General Psychiatry. 46:*902–907, 1989.

Lepola, U., U. Nousiainen, M. Puranen, P. Riekkinen, and R. Rimón. EEG and CT findings in patients with panic disorder. *Biological Psychiatry. 28:*721–727, 1990.

Leslie, A. M. The natural origins of understanding other minds. Paper given at the Human Behavior and Evolution Society meeting, Santa Barbara, CA, June 28–July 2, 1995.

Lewis, A. J. Melancholia: A clinical survey of depressive states. *Journal of Mental Science. 80:*277–378, 1936.

Lewis, B. P., D. E. Linder, and D. T. Kenrick. Arousal and attraction: Reproductive potential versus threat assessment. Paper given at the Human Behavior and Evolution Society meeting, Santa Barbara, CA, June 28–July 2, 1995.

Lewis, M. H., J. P. Gluck, A. J. Beauchamp, M. F. Keresztury, and R. B. Mailman. Long-term effects of early social isolation in Macaca mulatta: Changes in dopamine receptor function following apomorphine challenge. *Brain Research. 513:*67–73, 1990.

Lewontin, R. C. Sociobiology as an adaptionist program. *Behavioral Science. 24:*5–14, 1979.

Lichtenstein, S., P. Slovic, B. Fischhoff, M. Layman, and B. Combs. Judged frequency of lethal events. *Journal of Experimental Psychology. 4:*551–578, 1978.

Lieberman, P. *The Biology and Evolution of Language.* Cambridge: Harvard University Press, 1984.

Linn, L. Clinical manifestations of psychiatric disorders. In H. I. Kaplan and B. J. Sadock (eds.), *Comprehensive Textbook of Psychiatry,* Vol. 4. Baltimore: Williams and Wilkins, 1985, pp. 550–590.

Linnoila, M., M. Virkkunen, M. Scheinin, A. Nuutila, R. Rimon, and F. K. Goodwin. Low cerebrospinal fluid 5-hydroxyindoleacic acid concentration differentiates impulsive from nonimpulsive violent behavior. *Life Sciences. 33:*2609–2614, 1983.

Littlefield, C. H., and C. J. Lumsden. Gene-culture coevolution and the strategies of psychiatric healing. *Ethology and Sociobiology. 8:*151S–163S, 1987.

Lockard, J. S. Origins of self-deception. In J. S. Lockard and D. L. Paulhus (eds.), *Self-Deception: An Adaptive Mechanism?* Englewood Cliffs, NJ: Prentice-Hall, 1988, pp. 14–22.

Lockard, J. S., and D. L. Paulhus (eds.). *Self-Deception: An Adaptive Mechanism?* Englewood Cliffs, NJ: Prentice-Hall, 1988.

Lorenz, K. *Evolution and Modification of Behaviour.* Chicago: University of Chicago Press, 1965.

Low, B. S. Cross-cultural patterns in the training of children: An evolutionary perspective. *Journal of Comparative Psychology. 103:*311–319, 1989.

Low, B. S. Marriage systems and pathogen stress in human societies. *American Zoologist. 30:* 325–339, 1990a.

Low, B. S. Sex, power, and resources: Ecological and social correlates of sex differences. *International Journal of Contemporary Sociology. 27:*49–73, 1990b.

Low, B. S. Reproductive life in nineteenth century Sweden: An evolutionary perspective on demographic phenomena. *Ethology and Sociobiology. 12:*411–448, 1991.

Low, B. S., A. L. Clarke, and K. A. Lockridge. *Family Patterns in Nineteenth-Century Sweden: Variation in Time and Space.* Umea, Sweden: Tryckeri, 1991.

Low, B. S., A. L. Clarke, and K. A. Lockridge. Toward an ecological demography. *Population Development Review. 18:*1–31, 1992.

Lumsden, C. J., and E. O. Wilson. *Genes, Mind, and Culture.* Cambridge, MA: Harvard University Press, 1981.

Lydiard, R. B. (ed.). Social phobia. *Psychiatric Annals. 25:*544–576, 1995.

Lykken, D. T., T. J. Bouchard, Jr., M. McGue, and A. Tellegen. Heritability of interests: A twin study. *Journal of Applied Psychology. 78:*649–661, 1993.

Maas, J. W., and M. M. Katz. Neurobiology and psychopathological states: Are we looking in the right place? *Biological Psychiatry. 31:*757–758, 1992.

MacDonald, K. *Social and Personality Development: An Evolutionary Synthesis.* New York: Plenum Press, 1988a.

MacDonald, K. (ed.). *Sociobiological Perspectives on Human Development.* New York: Springer-Verlag, 1988b.

MacDonald, K. Warmth as a developmental construct: An evolutionary analysis. *Child Development. 63:*753–773, 1992.

MacDonald, K. Evolution, the five–factor model, and levels of personality. *Journal of Personality. 63:*525–567, 1995.

MacKay, D. M. Formal analysis of communicative processes. In R. A. Hinde (ed.), *Nonverbal Communication.* Cambridge, England: Cambridge University Press, 1972, pp. 3–25.

MacLean, P. D. Brain evolution relating to family, play and the separation call. *Archives of General Psychiatry. 42:*405–417, 1985.

MacLean, P. D. *The Triune Brain in Evolution.* New York: Plenum Press, 1990.

MacMillan, J., and L. Kofoed. Sociobiology and antisocial personality. *Journal of Nervous and Mental Disease. 172:*701–706, 1984.

Madsen, D. A biochemical property relating to power seeking in humans. *American Political Science Review. 79:*448–457, 1985.

Madsen, D. Power seekers are biochemically different: Further biochemical evidence. *American Political Science Review. 80:*261–269, 1986.

Madsen. D., and M. T. McGuire. Whole blood serotonin and the Type A behavior pattern. *Psychosomatic Medicine. 46:*546–548, 1984.

Maestripieri, D. First steps in the macaque world: Do rhesus mothers encourage their infants' independent locomotion? *Animal Behaviour. 49:*1541–1549, 1995.

Maher, B. A. Delusions: Contemporary etiological hypotheses. *Psychiatric Annals. 22:*260–268, 1992.

Mahowald, M. B. To be or not be a woman: Anorexia nervosa, normative gender roles, and feminism. *Journal of Medicine Philosophy. 17:*233–251, 1992.

Makara, G. B. Mechanisms by which stressful stimuli activate the pituitary-adrenal system. *Federation Proceedings. 44:*149–153, 1985.

Mandell, A. J. Nonlinear dynamics in brain processes. *Psychopharmacological Bulletin. 18:* 59–63, 1982.

Mandell, A. J., and K. A. Selz. Dynamical systems in psychiatry: Now what? *Biological Psychiatry. 32:*299–301, 1992.

Mann, J. J., P. A. McBride, R. P. Brown, M. Linnoila, A. C. Leon, M. DeMeo, T. Mieczkowski, J. E. Myers, and M. Stanley. Relationship between central and peripheral serotonin indexes in depressed and suicidal psychiatric inpatients. *Archives of General Psychiatry. 49:*442–446, 1992.

Mann, J. J., P. A. McBride, G. M. Anderson, and T. A. Mieczkowski. Platelet and whole blood serotonin content in depressed inpatients: Correlations with acute and life-time psychopathology. *Biological Psychiatry. 32:*243–257, 1992.

Manning, J. T., D. Scutt, G. H. Whitehouse, S. J. Leinster, and J. M. Walton Asymmetry and the menstrual cycle in women. *Ethology and Sociobiology. 17:*129–143, 1996.

Mannuzza, S., R. G. Klein, P. H. Konig, and T. L. Giampino. Hyperactive boys almost grown up: 4. Criminality and its relationship to psychiatric status. *Archives of General Psychiatry. 46:*1073–1079, 1989.

Mannuzza, S., R. G. Klein, N. Bonagura, P. Malloy, T. L. Giampino, and K. A. Addalli. Hyperactive boys almost grown up: 5. Replication of psychiatric status. *Archives of General Psychiatry. 48:*77–83, 1991.

Manschreck, T. C. Delusional (paranoid) disorders. In H. I. Kaplan and B. J. Sadock (eds.), *Comprehensive Textbook of Psychiatry,* Vol. 5. Baltimore: Williams and Wilkins, 1989, pp. 816–829.

Manschreck, T. C. Delusional disorders: Clinical concepts and diagnostic strategies. *Psychiatric Annals. 22:*241–251, 1992.

Marin, R. S. Apathy—Who cares? An introduction to apathy and related disorders of diminished motivation. *Psychiatric Annals. 27:*18–23, 1997.

Markow, T. A., and I. I. Gottesman. Fluctuating dermatoglyphic asymmetry in psychotic twins. *Psychiatry Research. 29:*37–43, 1989.

Markow, T. A., and J. Martin. Inbreeding and developmental stability in a small human population. *Annuals of Human Biology,* in press.

Marks, I. M. *Fears, phobias, and rituals.* Oxford, England: Oxford University Press, 1987.

Marks, I. M., and R. M. Nesse. Fear and fitness: An evolutionary analysis of anxiety disorders. *Ethology and Sociobiology. 15:*247–261, 1994.

Marmor, J. Systems thinking in psychiatry: Some theoretical and clinical implications. *American Journal of Psychiatry. 140:*833–838, 1983.

Martinot, J.-L., P. Hardy, A. Feline, J.-D. Huret, B. Mazoyer, D, Attar-Levy, S. Pappata, and A. Syrota. Left prefrontal glucose hypometabolism in the depressed state: A confirmation. *American Journal of Psychiatry. 147:*1313–1317, 1990.

Maslow, A. *The Further Reaches of Human Nature.* New York: Viking, 1971.

Masters, R., and M. T. McGuire (eds.). *The Neurotransmitter Revolution.* Carbondale: Southern Illinois University Press, 1994.

Mathis, J. L Psychiatric diagnoses: A continuing controversy. *Journal of Medicine Philosophy. 17:*253–261, 1992.

Maynard Smith, J. *Evolution and the Theory of Games.* Cambridge, England: Cambridge University Press, 1982.

Mayr, E. How to carry out the adaptionist program? *American Naturalist. 121:*324–334, 1983.

Mayr, E. *Towards a New Philosophy of Biology.* Cambridge, MA: Harvard University Press, 1988.

McDonald, P. W., and K. M. Prkachin. The expression and perception of facial emotion in alexithymia: A pilot study. *Psychosomatic Medicine. 52:*199–210, 1990.

McEwan, K. L., C. G. Costello, and P. J. Taylor. Adjustment to infertility. *Journal of Abnormal Psychology. 96:*108–116, 1987.

McGuffin, P., M. J. Owen, M. C. O'Donovan, A. Thapar, and I. I. Gottesman. *Seminars in Psychiatric Genetics.* London: Gaskell, 1994.

McGuire, M. T. *The St. Kitts Vervet.* Basel: Karger, 1974.

McGuire, M.T. An ethological approach to psychiatric disorders and treatment systems. *McLean Hospital Journal. 1:*21–33, 1976.

McGuire, M. T. A descriptive research-justified data base for psychiatry: Would it change things? *Perspectives in Biology Medicine. 21:*240–257, 1978.

McGuire, M. T. Sociobiology—Its potential contribution to psychiatry. *Perspectives in Biology Medicine. 23:*50–69, 1979a.

McGuire, M. T. Stephen Pepper, world hypotheses and the structure of metapsychology. *International Review of Psychoanalysis. 6:*217–230, 1979b.

McGuire, M. T. On the possibility of ethological explanations of psychiatric disorders. *Acta Psychiatrica Scandinavica. 77*(Suppl.):7–22, 1988.

McGuire, M. T., G. L. Brammer, and M. J. Raleigh. Resting cortisol levels and the emergence of dominant status among male vervet monkeys. *Hormones and Behavior. 20:*106–117, 1986.

McGuire, M. T., and S. M. Essock-Vitale. Psychiatric disorders in the context of evolutionary biology: A functional classification of behavior. *Journal of Nervous and Mental Disease. 169:*672–686, 1981.

McGuire, M. T., and S. M. Essock-Vitale. Psychiatric disorders in the context of evolutionary biology: The impairment of adaptive behaviors during the exacerbation and remission of psychiatric illnesses. *Journal of Nervous and Mental Disease. 170:*9–20, 1982.

McGuire, M. T., S. M. Essock-Vitale, and R. H. Polsky. Psychiatric disorders in the context of evolutionary biology: An ethological model of behavioral changes associated with psychiatric disorders. *Journal of Nervous and Mental Disease. 169:*687–704, 1981.

McGuire, M. T., and L. A. Fairbanks (eds.). *Ethological Psychiatry*. New York: Grune and Stratton, 1977.

McGuire, M. T., L. A. Fairbanks, S. R. Cole, R. Sbordone, F. M. Silvers, M. Richards, and J. Akers. The ethological study of four psychiatric wards: Patient, staff, and systems behaviors. *Journal of Psychiatric Research. 13:*211–224, 1977.

McGuire, M. T., F. I. Fawzy, J. E. Spar, R. W. Weigel, and A. Troisi. Altruism and mental disorders. *Ethology and Sociobiology. 15:*299–321, 1994.

McGuire, M. T., F. I. Fawzy, J. E. Spar, R. W. Weigel, and A. Troisi. Altruism and mental disorders. *Ethology and Sociobiology. 15:*299–321, 1994.

McGuire, M. T., and S. Lorch. Natural language conversation modes. *Journal of Nervous and Mental Disease. 146:*239–248, 1968.

McGuire, M. T., S. Lorch, and G. C. Quarton. Man-machine natural language exchanges based on selected features of unrestricted input: 2. The use of the time-shared computer as a research tool in studying dyadic communication. *Journal of Psychiatric Research. 5:*179–191, 1967.

McGuire, M. T., I. Marks, R. M. Nesse, and A. Troisi. Evolutionary biology: A basic science for psychiatry? *Acta Psychiatrica Scandinavica. 86:*89–96, 1992.

McGuire, M. T., and R. Polsky. Behavioral changes in hospitalized acute schizophrenics. *Journal of Nervous and Mental Disease. 167:*651–657, 1979.

McGuire, M. T., M. J. Raleigh, and G. L. Brammer. Adaptation, selection, and benefit-cost balances: Implications of behavioral-physiological studies of social dominance in male vervet monkeys. *Ethology and Sociobiology. 5:*269–277, 1984.

McGuire, M. T., M. J. Raleigh, and C. Johnson. Social dominance in adult male vervet monkeys: General considerations. *Social Science Information. 22:*89–123, 1983.

McGuire, M. T., and A. Troisi. Physiological regulation-deregulation and psychiatric disorders. *Ethology and Sociobiology. 8:*9S–12S, 1987a.

McGuire, M. T., and A. Troisi. Unrealistic wishes and physiological change. *Psychotherapy Psychosomatics. 47:*82–94, 1987b.

McGuire, M. T., and A. Troisi. Aggression. In H. I. Kaplan and B. J. Sadock (eds.), *Comprehensive Textbook of Psychiatry,* Vol. 5. Baltimore: Williams and Wilkins, 1989a, pp. 271–282.

McGuire, M. T., and A. Troisi. Anger: An evolutionary view. In R. Plutchik and H. Kellerman (eds.), *Emotion: Theory, Research, and Experience.* New York: Academic Press, 1989b, pp. 43–57.

McGuire, M. T., and A. Troisi. Deception. *International Journal of Contemporary Sociology. 27:*75–87, 1990.

McGuire, M. T., A. Troisi, and M. J. Raleigh. Depression in evolutionary context. In S. Baron-Cohen (ed.), *The Maladapted Mind: Classic Readings in Evolutionary Psychopathology*. London: Erlbaum, Taylor and Francis, in press.

McHugh, P. R., and P. R. Slavney. *The Perspectives of Psychiatry*. Baltimore: Johns Hopkins University Press, 1986.

McSorley, K. An investigation into the fertility rates of mentally ill patients. *Annals of Human Genetics. 27:*247–255, 1964.

Mealey, L. The sociobiology of sociopathy: An integrated evolutionary model. *Behavioral Brain Sciences. 18:*523–599, 1995.

Mealey, L., C. Daood, and M. Krage Enhanced memory for faces of cheaters. *Ethology and Sociobiology. 17:*119–128, 1996.

Meller, R. E., E. B. Keverne, and J. Herbert. Behavioural and endocrine effects of naltrexone in male talapoin monkeys. *Pharmacology Biochemistry Behavior. 13:*663–672, 1980.

Meltzer, H. Serotonergic dysfunction in depression. *British Journal of Psychiatry. 155*(Suppl.): 25–31, 1989.

Merikangas, K. R. Assortative mating for psychiatric disorders and psychological traits. *Archives of General Psychiatry. 39:*1173–1180, 1982.

Merikangas, K. R., E. J. Bromet, and D. G. Spiker. Assortative mating, social adjustment, and course of illness in primary affective disorder. *Archives of General Psychiatry. 40:*795–800, 1983.

Meyer, A. *Collected Papers of Adolph Meyer*, 4 vols. Baltimore: Johns Hopkins University Press, 1948–1952.

Miczek, K. A., J. Woolley, S. Schlisserman, and H. Yoshimura. Analysis of amphetamine effects on agonistic and affiliative behavior in squirrel monkeys (*Saimiri sciureus*). *Pharmacology Biochemistry Behavior. 14*(Suppl.):103–107, 1981.

Mikhailova, E. S., T. V. Vladimirova, A. F. Iznak, E. J. Tsusulkovskaya, and N. V. Sushko. Abnormal recognition of facial expression of emotions in depressed patients with major depression disorder and schizotypal personality disorder. *Biological Psychiatry. 40:*697–705, 1996.

Miller, P. A., and N. Eisenberg. The relation of empathy to aggressive and externalizing/antisocial behavior. *Psychological Bulletin. 103:*324–344, 1988.

Miller, P. McC., P. G. Surtees, N. B. Kreitman, J. G. Ingham, and S. P. Sashidharan. Maladaptive coping reactions to stress. *Journal of Nervous and Mental Disease. 173:*707–716, 1985.

Millon, T. *Disorders of Personality: DSM-III Axis II*. New York: Wiley, 1981.

Minami, E., N. Tsuru, and T. Okita. Effect of subject's family name on visual event-related potential in schizophrenia. *Biological Psychiatry. 31:*681–689, 1992.

Mineka, S. Evolutionary memories, emotional processing, and the emotional disorders. *Psychology Learning Motivation. 28:*161–206, 1992.

Mineka, S., and M. Cook. Immunization against the observational conditioning of snake fear in rhesus monkeys. *Journal of Abnormal Psychology. 95:*307–318, 1986.

Mitchell, R. W. A framework for discussing deception. In R. W. Mitchell and N. S. Thompson (eds.), *Deception: Perspectives on Human and Nonhuman Deceit*. Albany: State University of New York Press, 1986, pp. 3–40.

Mithen, S. From the Neanderthal to the modern mind (or how evolutionary psychology and human ecology need Palaeolithic archaeology). Paper presented at the Human Behavior and Evolution Society meeting, Santa Barbara, CA, June 28–July 2, 1995.

Møller, A. P., M. Soler, and R. Thornhill. Breast asymmetry, sexual selection, and human reproductive success. *Ethology and Sociobiology. 16:*207–220, 1995.

Monk, T. H., M. L. Moline, J. E. Fookson, and S. M. Peetz. Circadian determinants of subjective alertness. *Journal of Biological Rhythms. 4:*393–404, 1989.

Moskowitz, D. S. Cross-situational generality and the interpersonal circumplex. *Journal of Personality and Social Psychology. 66:*921–933, 1994.

Mullen, P. E. Jealousy: The pathology of passion. *British Journal of Psychiatry. 158:*593–601, 1991.

Murphy, D. L. Neuropsychiatric disorders and the multiple human brain serotonin receptor subtypes and subsystems. *Neuropsychopharmacology. 3:*457–471, 1990.

Murphy, D. L., J. Zohar, C. Benkelfat, M. T. Pato, T. A. Pigott, and T. R. Insel. Obsessive-compulsive disorder as a 5-HT subsystem-related behavioural disorder. *British Journal of Psychiatry. 155*(Suppl.):15–24, 1989.

Murphy, E. A. A geneticist's approach to psychiatric disease. *Psychological Medicine. 17:* 805–815, 1987.

Murphy, J. M. Psychiatric labeling in cross-cultural perspective. In M. H. Logan and E. E. Hunt, Jr. (eds.), *Health and the Human Condition.* North Scituate, MA: Duxbury Press, pp. 248–270, 1978.

Murphy, J. M., D. C. Oliver, R. R. Monson, A. M. Sobol, E. B. Federman, and A. H. Leighton. Depression and anxiety in relation to social status. *Archives of General Psychiatry. 48:* 223–229, 1991.

Näätänen, R. The role of attention in auditory information processing as revealed by event-related potentials and other brain measures of cognitive function. *Behavioral Brain Sciences. 13:*201–288, 1990.

Nathanson, D. L. Understanding emotion: New theories, new therapy. *Psychiatric Annals. 23:* 543–555, 1993.

Nelson, E. C., C. R. Cloninger, T. R. Przybeck, and J. G. Csernansky. Platelet serotonergic markers and tridimensional personality questionnaire measures in a clinical sample. *Biological Psychiatry. 40:*271–278, 1996.

Nesse, R. M. An evolutionary perspective on panic disorder and agoaophobia. *Ethology and Sociobiology. 8:*73S–83S, 1987a.

Nesse, R. M. An evolutionary perspective on senescence. *Philosophy Medicine. 26:*45–64, 1987b.

Nesse, R. M. Can we agree on standards of evidence for testing evolutionary hypotheses about human behavior? Paper given at the Human Behavior and Evolution Society meeting, Ann Arbor, MI, October 28–30, 1988a.

Nesse, R. M. Panic disorder: An evolutionary view. *Psychiatric Annals. 18:*478–483, 1988b.

Nesse, R. M. Evolutionary explanations of emotions. *Human Nature. 1:*261–289, 1990a.

Nesse, R. M. The evolutionary functions of repression and the ego defenses. *Journal of the American Academy Psychoanalysis. 18:*260–285, 1990b.

Nesse, R. M. Psychiatry. In M. Maxwell (ed.), *The Sociobiological Imagination.* Albany: State University of New York Press, 1991a, pp. 23–45.

Nesse, R. M. What good is feeling bad? *The Sciences. 31:*30–37, 1991b.

Nesse, R. M. An evolutionary perspective on substance abuse. *Ethology and Sociobiology. 15:* 339–348, 1994.

Nesse, R. M., G. C. Curtis, B. A. Thyer, D. S. McCann, M. J. Huber-Smith, and R. F. Knopf. Endocrine and cardiovascular responses during phobic anxiety. *Psychosomatic Medicine. 47:*320–332, 1985.

Nesse, R. M., and R. Klass. Risk perception by patients with anxiety disorders. *Journal of Nervous and Mental Disease. 182:*465–470, 1994.

Nesse, R. M., and A. T. Lloyd. The evolution of psychodynamic mechanisms. In J. Barkow, L. Cosmides, and J. Tooby (eds.), *The Adaptive Mind: Evolutionary Psychology and the Generation of Culture.* Oxford, England: Oxford University Press, 1992, pp. 601–624.

Nesse, R. M., and G. C. Williams. *Why We Get Sick.* New York: Random House, 1994.

Nierenberg, A. A., J. A. Pava, K. Clancy, J. F. Rosenbaum, and M. Fava. Are neurovegetative symptoms stable in relapsing or recurrent atypical depressive episodes? *Biological Psychiatry. 40:*691–696, 1996.

Nisbett, R. E., and T. D. Wilson. Telling more than we can know: Verbal reports on mental processes. *Psychological Review. 84:*231–259, 1977.

Norris, K. S., and C. R. Schilt. Cooperative societies in three-dimensional space: On the origins of aggregations, flocks, and schools, with special reference to dolphins and fish. *Ethology and Sociobiology. 9:*149–180, 1988.

Nyborg, H. *Hormones, Sex, and Society.* Westport, CT: Praeger, 1994.

Nygaard, J. A. Anorexia nervosa: Treatment and triggering factors. *Acta Psychiatrica Scandinavica. 82*(Suppl.):44–49, 1990.

Oatley, K., and W. Bolton. A social-cognitive theory of depression in reaction to life events. *Psychological Review. 92:*372–388, 1985.

O'Callaghan, E., P. Buckley, C. Madigan, O. Redmond, J. P. Stack, A. Kinsella, C. Larkin, J. T. Ennis, and J. L. Waddington. The relationship of minor physical anomalies and other putative indices of developmental disturbance in schizophrenia to abnormalities of cerebral structure on magnetic resonance imaging. *Biological Psychiatry. 38:*516–524, 1995.

Ødegard, Ø. Marriage rate and fertility in psychotic patients before hospital admissions and after discharge. *International Journal of Social Psychiatry. 6:*25–33, 1960.

Ødegard, Ø. Fertility of psychiatric first admissions in Norway, 1936–1975. *Acta Psychiatrica Scandinavica. 62:*212–220, 1980.

Ogata, S. N., K. R. Silk, S. Goodrich, N. E. Lohr, D. Westen, and E. M. Hill. Childhood sexual and physical abuse in adult patients with borderline personality disorder. *American Journal of Psychiatry. 147:*1008–1013, 1990.

Oliveric, A. F. Infanticide in Western cultures: A historical overview. In S. Parmigiani and F. S. vom Saal (eds.), *Infanticide and Parental Care.* Chur, Switzerland: Harwood, 1994, pp. 105–120.

Olivier, B., J. Mos, and P. F. Brain (eds.). *Ethopharmacology of Agonistic Behaviour in Animals and Humans.* Dordrecht: Martinus Nijhoff, 1987.

Packer, C. Constraints on the evolution of reciprocity: Lessons from cooperative hunting. *Ethology and Sociobiology. 9:*137–148, 1988.

Pam, A. A critique of the scientific status of biological psychiatry. *Acta Psychiatrica Scandinavica. 82*(Suppl.):1–35, 1990.

Panksepp, J., S. M. Siviy, and L. A. Normansell. Brain opioids and social emotions. In M. Reite and T. Field (eds.), *The Psychobiology of Attachment and Separation.* New York: Academic Press, 1985, pp. 3–49.

Pardo, J. V., P. J. Pardo, and M. E. Raichle. Neural correlates of self-induced dysphoria. *American Journal of Psychiatry. 150:*713–719, 1993.

Parker, G., P. Johnston, and L. Hayward. Parental "expressed emotion" as a predictor of schizophrenic relapse. *Archives of General Psychiatry. 45:*806–813, 1988.

Parnas, J., T. D. Cannon, B. Jacobsen, H. Schulsinger, F. Schulsinger, and S. A. Mednic. Lifetime DSM-III-R diagnostic outcomes in the offspring of schizophrenic mothers. *Archives of General Psychiatry. 50:*707–714, 1993.

Partridge, L., and P. H. Harvey. The ecological context of life history evolution. *Science. 241:* 1449–1455, 1988.

Paul, L., and L. R. Hirsch. Human male mating strategies: 2. Moral codes of "quality" and "quantity" strategies. *Ethology and Sociobiology. 17:*71–86, 1996.

Paykel, E. S., and M. W. Weissman. Social adjustment and depression. *Archives of General Psychiatry. 28:*659–663, 1973.

Peccei, J. S. The origin and evolution of menopause: The altriciality-lifespan hypothesis. *Ethology and Sociobiology,* in press.

Pedersen, J., M. F. Livoir-Petersen, and J. T. M. Schelde. An ethological approach to autism:

An analysis of visual behaviour and interpersonal contact in a child versus adult interaction. *Acta Psychiatrica Scandinavica. 80:*346–355, 1989.

Pedersen, J., J. T. M. Schelde, E. Hannibal, K. Behnke, B. M. Nielsen, and M. Hertz. An ethological description of depression. *Acta Psychiatrica Scandinavica. 78:*320–330, 1988.

Pedersen, N. L., R. Plomin, G. E. McClearn, and L. Friberg. Neuroticism, extraversion, and related traits in adult twins reared apart and reared together. *Journal of Personality and Social Psychology. 55:*950–957, 1988.

Pepper, S. C. *World Hypotheses.* Berkeley: University of California Press, 1942.

Peralta, V., and M. J. Cuesta. Negative symptoms in schizophrenia: A confirmatory factor analysis of competing models. *American Journal of Psychiatry. 152:*1450–1457, 1995.

Peralta, V., M. J. Cuesta, and J. deLeon. An empirical analysis of latent structures underlying schizophrenic symptoms: A four-syndrome model. *Biological Psychiatry. 36:*726–736, 1994.

Peralta, V., M. J. Cuesta, and J. deLeon. Positive and negative symptoms/syndromes in schizophrenia: Reliability and validity of different diagnostic systems. *Psychological Medicine. 25:*43–50, 1995.

Perris, C., M. Eisemann, L. von Knorring, and H. Perris. Personality traits in former depressed patients and in healthy subjects without past history of depression. *Psychopathology. 17:* 178–186, 1984.

Pezard, L., J.-L. Nandrino, B. Renault, F. E. Massioui, J.-F. Allilaire, J. Müller, F. J. Varela, and J. Martinerie. Depression as a dynamical disease. *Biological Psychiatry. 39:*991–999, 1996.

Phillips, K. A., J. G. Gunderson, R. M. A. Hirshfeld, and L. E. Smith. A review of the depressive personality. *American Journal of Psychiatry. 147:*830–837, 1990.

Phillips, W., J. C. Gómez, S. Baron-Cohen, V. Laá, and A. Rivière. Treating people as objects, agents, or "subjects": How young children with and without autism make requests. *Journal of Child Psychology Psychiatry. 36:*1383–1398, 1995.

Pickens, R. W., D. S. Svikis, M. McGue, D. T. Lykken, L. L. Heston, and P. J. Clayton. Heterogeneity in the inheritance of alcoholism. *Archives of General Psychiatry. 48:*19–28, 1991.

Pierson, A., R. Ragot, J. Van Hooff, A. Partiot, B. Renault, and R. Jouvent. Heterogeneity of information-processing alterations according to dimensions of depression: An event-related potentials study. *Biological Psychiatry. 40:*98–115, 1996.

Pinker, S. *Language Learnability and Language Development.* Cambridge, MA: Harvard University Press, 1994.

Pitchot, W., M. Ansseau, A. G. Moreno, M. Hansenne, and R. von Frenckell. Dopaminergic function in panic disorder: Comparison with major and minor depression. *Biological Psychiatry. 32:*1004–1011, 1992.

Pitman, R., B. Kolb, S. Orr, J. de Jong, S. Yadati, and M. M. Singh. On the utility of ethological data in psychiatric research: The example of facial behavior in schizophrenia. *Ethology and Sociobiology. 8*(Suppl.):111S–116S, 1987.

Pitman, R. K. Animal models of compulsive behavior. *Biological Psychiatry. 26:*189–198, 1989.

Plomin, R. The role of inheritance in behavior. *Science. 248:*183–188, 1990.

Plomin, R., and C. S. Bergeman. The nature of nurture: Genetic influence on "environmental" measures. *Behavioral Brain Sciences. 14:*373–427, 1991.

Plomin, R., and D. Daniels. Why are children in the same family so different from one another? *Behavioral Brain Sciences. 10:*1–60, 1987.

Plomin, R., M. J. Owen, and P. McGuffin. The genetic basis of complex human behaviors. *Science. 264:*1733–1739, 1994.

Plomin, R., and R. Rende. Human behavioral genetics. *Annual Review of Psychology. 42:*161–190, 1991.

Ploog, D. Human neuroethology of emotion. *Progress Neuro-Psychopharmacology Biological Psychiatry. 13*(Suppl.):S15–S22, 1989a.

Ploog, D. Psychopathology of emotions in view of neuroethology. In K. Davison and A. Kerr (eds.), *Contemporary Themes in Psychiatry.* London: Royal College of Psychiatrists, 1989b, pp. 441–458.

Ploog, D. Ethological foundations of biological psychiatry. In H. M. Emrich and M. Wiegand (eds.), *Integrative Biological Psychiatry.* Berlin: Springer-Verlag, 1992, pp. 3–35.

Ploog, D. W., and K. M. Pirke. Psychobiology of anorexia nervosa. *Psychological Medicine. 17*:843–859, 1987.

Plotkin, H. *Darwin Machines and the Nature of Knowledge.* Cambridge, MA: Harvard University Press, 1994.

Plutchik, R. *Emotion: A Psychoevolutionary Synthesis.* New York: Harper and Row, 1980.

Plutchik, R. Emotions: A general psychoevolutionary theory. In K. R. Scherer and P. Ekman (eds.), *Approaches to Emotion.* Hillsdale, NJ: Erlbaum, 1984a, pp. 197–219.

Plutchik, R. Emotions and imagery. *Journal of Mental Imagery. 8*:105–112, 1984b.

Plutchik, R. Emotions and evolution. In K. T. Strongman (ed.), *International Review of Studies of Emotion.* New York: Wiley, 1991, pp. 37–58.

Polsky, R. H., and M. T. McGuire. An ethological analysis of manic-depressive disorder. *Journal of Nervous and Mental Disease. 167*:56–65, 1979.

Polsky, R. H., and M. T. McGuire. Observational assessment of behavioral changes accompanying clinical improvement in hospitalized psychiatric patients. *Journal of Behavioral Assessment. 2*:207–223, 1980.

Polsky, R., and M. T. McGuire. Naturalistic observations of pathological behavior in hospitalized psychiatric patients. *Journal of Behavioral Assessment. 3*:59–81, 1981.

Popper, K. *Conjectures and Refutations.* London: Routledge and Kegan Paul, 1969.

Post, R. M. Transduction of psychosocial stress into the neurobiology of recurrent affective disorder. *American Journal of Psychiatry. 149*:999–1010, 1992.

Post, R. M., and S. R. B. Weiss. Endogenous biochemical abnormalities in affective illness: Therapeutic versus pathogenic. *Biological Psychiatry. 32*:469–484, 1992.

Powell, M., and D. R. Hemsley. Depression: A breakdown of perceptual defence? *British Journal of Psychiatry. 145*:358–362, 1984.

Power, M., and C. R. Brewin. From Freud to cognitive science: A contemporary account of the unconscious. *British Journal of Clinical Psychology. 30*:289–310, 1991.

Price, J. S. The dominance hierarchy and the evolution of mental illness. *Lancet. 7502*:243–246, 1967.

Price, J. S. Neurotic and endogenous depression. *British Journal of Psychiatry. 114*:119–126, 1969a.

Price, J. S. The ritualization of agonistic behaviour as a determinant of variation along the neuroticism stability dimension of personality. *Proceedings of the Royal Society of Medicine. 62*:37–40, 1969b.

Price, J., and R. Gardner, Jr. The paradoxical power of the depressed patient: A problem for the ranking theory of depression. *British Journal of Medical Psychology. 68*:193–206, 1995.

Price, J. S., and L. Sloman. Depression as yielding behavior: An animal model based on Schyelderup-Ebbe's pecking order. *Ethology and Sociobiology. 8*(Suppl.):85S–98S, 1987.

Price, J., L. Sloman, R. Gardner, Jr., P. Gilbert, and P. Rohde. The social competition hypothesis of depression. *British Journal of Psychiatry. 164*:309–315, 1994.

Prigerson, H. G., E. Frank, S. V. Kasl, C. F. Reynolds III, B. Anderson, G. S. Zubenko, P. R. Houck, C. J. George, and D. J. Kupfer. Complicated grief and bereavement-related depression as distinct disorders: Preliminary empirical validation in elderly bereaved spouses. *American Journal of Psychiatry. 152*:22–30, 1995.

Profet, M. The function of allergy: Immunological defense against toxins. *Quarterly Review of Biology. 66:*23–62, 1991.

Profet, M. Menstruation as a defense against pathogens transported by sperm. *Quarterly Review of Biology. 68:*335–386, 1993.

Pulver, A. E., K.-Y. Liang, C. H. Brown, P. Wolyniec, J. McGrath, L. Adler, D. Tam, W. T. Carpenter, and B. Childs. Risk factors in schizophrenia. *British Journal of Psychiatry. 160:*65–71, 1992.

Pyszczynski, T., and J. Greenberg. Self-regulatory perserveration and the depressive self-focusing style: A self-awareness theory of reactive depression. *Psychological Bulletin. 102:* 122–138, 1987.

Quarton, G. C., M. T. McGuire, and S. Lorch. Man-machine natural language exchanges based on selected features of unrestricted input: 1. The development of the time-shared computer as a research tool in studying dyadic communication. *Journal of Psychiatric Research. 5:* 165–177, 1967.

Raine, A., P. H. Venables, and M. Williams. Relationship between central and autonomic measures of arousal at age 15 years and criminality at age 24 years. *Archives of General Psychiatry. 47:*1003–1007, 1990.

Raleigh, M. J., G. L. Brammer, M. T. McGuire, and A. Yuwiler. Dominant social status facilitates the behavioral effects of serotonergic agonists. *Brain Research. 348:*274–282, 1985.

Raleigh, M. J., and M. T. McGuire. Female influences on male dominance acquisition in captive vervet monkeys, Cercopithecus aethiops sabaeus. *Animal Behaviour. 38:*59–67, 1989.

Raleigh, M. J., M. T. McGuire, G. L. Brammer, D. B. Pollack, and A. Yuwiler. Serotonergic mechanisms promote dominance acquisition in adult male vervet monkeys. *Brain Research. 559:*181–190, 1991.

Raleigh, M. J., M. T. McGuire, G. L. Brammer, and A. Yuwiler. Social and environmental influences on blood serotonin concentrations in monkeys. *Archives of General Psychiatry. 41:*405–410, 1984.

Rea, M. M., A. M. Strachan, M. J. Goldstein, I. Falloon, and S. Hwang. Changes in patient coping style following individual and family treatment for schizophrenia. *British Journal of Psychiatry. 158:*642–647, 1991.

Redican, W. K. An evolutionary perspective on human facial displays. In P. Ekman (ed.), *Emotions in the Human Face.* Cambridge, England: Cambridge University Press, 1982, pp. 264–289.

Reich, J. H., and A. I. Green. Effect of personality disorders on outcome of treatment. *Journal of Nervous and Mental Disease. 179:*74–82, 1991.

Reiss, D., R. Plomin, and E. M. Hetherington. Genetics and psychiatry: An unheralded window on the environment. *American Journal of Psychiatry. 148:*283–291, 1991.

Reiss, S., and S. Havercamp. The sensitivity theory of motivation: Implications for psychopathology. *Behavioral Research Therapy. 8:*621–632, 1996.

Reite, M., K. Kaemingk, and M. L. Boccia. Maternal separation in bonnet monkey infants: Altered attachment and social support. *Child Development. 60:*473–480, 1989.

Reite, M., R. Short, C. Seiler, and J. D. Pauley. Attachment, loss and depression. *Journal of Child Psychology Psychiatry. 22:*141–169, 1981.

Rescorla, R. A. Pavlovian conditioning. *American Psychologist. 43:*151–160, 1988.

Reznek, L. *The Nature of Disease.* London: Routledge and Kegan Paul, 1987.

Rice, J., T. Reich, N. C. Andreasen, P. W. Lavori, J. Endicott, P. J. Clayton, M. B. Keller, R. M. A. Hirschfeld, and G. L. Klerman. Sex–related differences in depression. *Journal of Affective Disorders. 71:*199–210, 1984.

Richer, J. M., and R. G. Coss. Gaze aversion in autistic and normal children. *Acta Psychiatrica Scandinavica. 53:*193–210, 1976.

Richerson, P. J., and R. Boyd. A dual inheritance model of the human evolutionary process:

1. Basic postulates and a simple model. *Journal of Social Biological Structures. 1:*127–154, 1978.

Rimé, B., B. Boulanger, P. Laubin, M. Richir, and K. Stroobants. The perception of interpersonal emotions originated by patterns of movement. *Motivation and Emotion. 9:*241–260, 1985.

Ritvo, E. R., B. J. Freeman, A. Mason-Brothers, A. Mo, and A. M. Ritvo. Concordance for the syndrome of autism in 40 pairs of afflicted twins. *American Journal of Psychiatry. 142:* 74–77, 1985.

Robins, E., and S. B. Guze. Establishment of diagnostic validity in psychiatric illness: Its application to schizophrenia. *American Journal of Psychiatry. 126:*983–987, 1970.

Rose, M. R. *Evolutionary Biology of Aging.* New York: Oxford University Press, 1991.

Rose, R. J., M. Koskenvuo, J. Kaprio, S. Sarna, and H. Langinvainio. Shared genes, shared experiences, and similarity of personality: Data from 14,288 adult Finnish co-twins. *Journal of Personality and Social Psychology. 54:*161–171, 1988.

Rose, R. J., J. Z. Miller, M. F. Pogue-Geile, and G. F. Cardwell. Twin-family studies of common fears and phobias. Twin Research 3: Intelligence, Personality, and Development, pp. 169–174, 1981.

Rose, S. The rise of neurogenetic determinism. *Nature. 373:*380–382, 1995.

Rosen, A. J., K. T. Mueser, S. Sussman, and J. M. Davis. The effects of neuroleptic drugs on the social interactions of hospitalized psychotic patients. *Journal of Nervous and Mental Disease. 169:*240–243, 1981.

Rosen, A. J., S. E. Tureff, J. S. Lyons, and J. M. Davis. Pharmacotherapy of schizophrenia and affective disorders: Behavioral assessment of psychiatric medications. *Journal of Behavioral Assessment. 3:*133–148, 1981.

Rosen, L., S. Targum, M. Terman, M. J. Bryant, H. Hoffman, S. F. Kasper, J. R. Hamovit, J. P. Docherty, B. Welch, and N. E. Rosenthal. Prevalence of seasonal affective disorder at four latitudes. *Journal of Psychiatry Research. 31:*131–144, 1990.

Rosenthal, D. *Genetic Theory and Abnormal Behavior.* New York: McGraw-Hill, 1970.

Rosenthal, N. E., D. A. Sack. J. C. Gillin, A. J. Lewy, F. K. Goodwin, Y. Davenport, P. S. Mueller, D. A. Newsome, and T. A. Wehr. Seasonal affective disorder. *Archives of General Psychiatry. 41:*72–80, 1984.

Roth, M. (ed.). Treatment and outcome of phobic and related disorders. *Psychiatric Annals. 21:* 326–381, 1991.

Rothstein, S. I., and R. Pierotti. Distinctions among reciprocal altruism, kin selection, and cooperation and a model for the initial evolution of beneficent behavior. *Ethology and Sociobiology. 9:*189–210, 1988.

Rowan, J. *Subpersonalities.* London: Routledge, 1990.

Roy, A., F. Karoum, and S. Pollack. Marked reduction in indexes of dopamine metabolism among patients with depression who attempt suicide. *Archives of General Psychiatry. 49:* 447–450, 1992.

Roy, A., N. L. Segal, B. S. Centerwall, and C. D. Robinette. Suicide in twins. *Archives of General Psychiatry. 48:*29–32, 1991.

Roy, A., N. L. Segal, and M. Sarchiapone. Attempted suicide among living co-twins of twin suicide victims. *American Journal of Psychiatry. 152:*1075–1076, 1995.

Roy-Byrne, P. P., H. Weingartner, L. M. Bierer, K. Thompson, and R. M. Post. Effortful and automatic cognitive processes in depression. *Archives of General Psychiatry. 43:*265–267, 1986.

Sachdev, P., and A. M. Aniss. Slowness of movement in melancholic depression. *Biological Psychiatry. 35:*253–262, 1994.

Sadler, J. Z., O. P. Wiggins, and M. A. Schwartz. *Philosophical Perspectives on Psychiatric Diagnostic Classification.* Baltimore: Johns Hopkins University Press, 1994.

Salter, F. K. *Emotions in Command.* Oxford, England: Oxford University Press, 1995.

Salzen, E. Why there are eight primary emotions. *Across Species Comparison and Psychopathology. 8:*8–10, 1995.

Salzen, E. A. On the nature of emotion. *International Journal of Comparative Psychology. 5:* 47–88, 1991.

Sapolsky, R. M. Individual differences in cortisol secretory patterns in the wild baboon: Role of negative feedback sensitivity. *Endocrinology. 113:*2263–2267, 1983.

Sapolsky, R. M. Hypercortisolism among socially subordinate wild baboons originates at the CNS level. *Archives of General Psychiatry. 46:*1047–1051, 1989.

Sapolsky, R. M. Adrenocortical function, social rank, and personality among wild baboons. *Biological Psychiatry. 28:*862–878, 1990a.

Sapolsky, R. M. Stress in the wild. *Scientific American. 262:*116–123, 1990b.

Savin-Williams, R. C. *Adolescence: An Ethological Perspective.* New York: Springer-Verlag, 1987.

Sawaguchi, T. Correlations of cerebral indices for "extra" cortical parts and ecological variables in primates. *Brain Behavior Evolution. 32:*129–140, 1988.

Scadding, J. G. Diagnosis: The clinician and the computer. *Lancet. 2:*877–882, 1967.

Scadding, J. G. Health and disease: What can medicine do for philosophy? *Journal of Medical Ethics. 14:*118–124, 1988.

Scarr, S. Developmental theories for the 1990s: Development and individual differences. *Child Development. 63:*1–19, 1992.

Schatzberg, A. F., J. A. Samson, K. L. Bloomingdale, P. J. Orsulak, B. Gerson, P. P. Kizuka, J. O. Cole, and J. J. Schildkraut. Towards a biochemical classification of depressive disorders. *Archives of General Psychiatry. 46:*260–268, 1989.

Scheflen, A. E. Communication and regulation in psychotherapy. *Psychiatry. 26:*126–136, 1963.

Scheflen, A. E. The significance of posture in communication systems. *Psychiatry. 27:*316–331, 1964.

Schelde, J. T. M., J. Pedersen, E. Hannibal, K. Behnke, B. M. Nielsen, and M. Hertz. An ethological analysis of depression: Comparison between ethological recording and Hamilton rating of five endogenously depressed patients. *Acta Psychiatrica Scandinavica. 78:* 331–340, 1988.

Schelde, T. Characteristic, nonverbal behaviors of endogenous depression, remission, and normal psychic state. Paper presented at the Scandinavian Society for Biological Psychiatry, Oslo, November 20–22, 1992.

Schelde, T. Ethological research in psychiatry. *Ethology and Sociobiology. 15:*349–368, 1994.

Schildkraut, J. The catecholamine hypothesis of affective disorders: A review of supporting evidence. *American Journal of Psychiatry. 122:*509–522, 1965.

Schino, G., and A. Troisi. Opiate receptor blockade in juvenile macaques: Effect on affiliative interactions with their mothers and group companions. *Brain Research. 576:*125–130, 1992.

Schneider, M. L., and C. L. Coe. Repeated social stress during pregnancy impairs neuromotor development of the primate infant. *Developmental Behavioral Pediatrics. 14:*81–87, 1993.

Schore, A. N. *Affect Regulation and the Origin of the Self.* Hillsdale, NJ: Erlbaum, 1994.

Schroeder, M. L., J. A. Wormworth, and W. J. Livesley. Dimensions of personality disorder and the five-factor model of personality. In P. T. Costa and T. A. Widiger (eds.), *Personality Disorders and the Five-Factor Model of Personality.* Washington, DC: American Psychological Association, 1994, pp. 117–127.

Schulsinger, H. Clinical outcome of ten years of follow-up of children of schizophrenic mothers. In S. B. Sells, R. Crandall, M. Roff, J. S. Strauss, and W. Pollin (eds.), *Human Functioning in Longitudinal Perspectives.* Baltimore: Williams and Wilkins, 1980, pp. 33–43.

Schultz, D. P. *Sensory Restriction.* New York: Academic Press, 1965.

Schwartz, B. L., R. B. Rosse, C. Veazey, and S. I. Deutsch. Impaired motor skill learning in schizophrenia: Implications for corticostriatal dysfunction. *Biological Psychiatry. 39:* 241–248, 1996.

Scott, J. P. The social behavior of dogs and wolves: An illustration of sociobiological systematics. *Annals of the New York Academy of Sciences. 51:*1009–1021, 1950.

Segal, N. L. Cooperation, competition, and altruism within twin sets: A reappraisal. *Ethology and Sociobiology. 5:*163–177, 1984.

Segal, N. L., and S. A. Blozis Bereavement in monozygotic and dizygotic twins: An evolutionary perspective. Paper given at the Human Behavior and Evolution Society meeting, Santa Barbara, CA, June 28–July 2, 1995.

Sharma, R. P., and P. G. Janicak (eds.). Treatment of schizophrenia. *Psychiatric Annals. 26:* 67–104, 1996.

Shawcross, C. R., and P. Tyrer. Influence of personality on response to monoamine oxidase inhibitors and tricyclic antidepressants. *Journal of Psychiatric Research. 19:*557–562, 1985.

Shea, M. T., P. A. Pilkonis, E. Beckham, J. F. Collins, I. Elkin, S. M. Sotsky, and J. P. Docherty. Personality disorders and treatment outcome in the NIMH treatment of depression collaborative research program. *American Journal of Psychiatry. 147:*711–718, 1990.

Siever, L. J., and K. L. Davis. Overview: Toward a dysregulation hypothesis of depression. *American Journal of Psychiatry. 142:*1017–1031, 1985.

Sifneos, P. E., R. Apfel-Savitz, and F. H. Frankel. The phenomenon of "alexithymia." *Psychotherapy Psychosomatics. 28:*47–57, 1977.

Sigman, M., S. E. Cohen, L. Beckwith, R. Asarnow, and A. H. Parmelee. Continuity in cognitive abilities from infancy to 12 years of age. *Cognitive Development. 6:*47–57, 1991.

Sigman, M., M. A. MacDonald, C. Neumann, and N. Bwibo. Prediction of cognitive competence in Kenyan children from toddler nutrition, family characteristics and abilities. *Journal of Child Psychology Psychiatry. 32:*307–320, 1990.

Sigvardsson, S., M. Bohman, A.-L. von Knorring, and C. R. Cloninger. Symptom patterns and causes of somatization in men: 1. Differentiation of two discrete disorders. *Genetic Epidemiology. 3:*153–169, 1986.

Silberg, J. L., A. C. Heath, R. Kessler, M. C. Neale, J. M. Meyer, L. J. Eaves, and K. S. Kendler. Genetic and environmental effects on self-reported depressive symptoms in a general population twin sample. *Journal of Psychiatric Research. 24:*197–212, 1990.

Silberman, E. K., H. Weingartner, M. Laraia, S. Byrnes, and R. M. Post. Processing of emotional properties of stimuli by depressed and normal subjects. *Journal of Nervous and Mental Disease. 171:*10–14, 1983.

Silberman, E. K., H. Weingartner, and R. M. Post. Thinking disorder in depression. *Archives of General Psychiatry. 40:*775–780, 1983.

Silverman, J. S., J. A. Silverman, and D. A. Eardley. Do maladaptive attitudes cause depression? *Archives of General Psychiatry. 41:*28–30, 1984.

Singh, D. Female health, attractiveness, and desirability for relationships: Role of breast asymmetry and waist-to-hip ratio. *Ethology and Sociobiology,* in press.

Singh, D., and R. K. Young. Body weight, waist-to-hip ratio, breasts, and hips: Role in judgments of female attractiveness and desirability for relationships. *Ethology and Sociobiology,* in press.

Singh, M. M., S. R. Kay, and R. K. Pitman. Territorial behavior of schizophrenics. *Journal of Nervous and Mental Disease. 169:*503–512, 1981.

Slater, E., E. H. Hare, and J. S. Price. Marriage and fertility of psychiatric patients compared with national data. *Social Biology. 18*(Suppl.):560–573, 1971.

Slavin, M. O. The origins of psychic conflict and the adaptive function of repression: An evolutionary biological view. *Psychoanalysis Contemporary Thought. 8:*407–440, 1985.

Slavin, M. O., and D. Kriegman. *The Adaptive Design of the Human Psyche.* New York: Guilford Press, 1992.

Slavney, P. R., and G. Rich. Variability of mood and the diagnosis of hysterical personality disorder. *British Journal of Psychiatry. 136:*402–404, 1980.

Sloman, L. The role of neurosis in phylogenetic adaptation, with particular reference to early man. *American Journal of Psychiatry. 133:*543–547, 1976.

Sloman, L., M. Konstantareas, and D. W. Dunham. The adaptive role of maladaptive neurosis. *Biological Psychiatry. 14:*961–972, 1979.

Sloman, L., and J. S. Price. Losing behavior (yielding subroutine) and human depression: Proximate and selective mechanisms. *Ethology and Sociobiology. 8:*99S–109S, 1987.

Sloman, L., J. Price, P. Gilbert, and R. Gardner. Adaptive function of depression: Psychotherapeutic implications. *American Journal of Psychotherapy. 48:*1–16, 1994.

Sloman, S., and L. Sloman. Mate selection in the service of human evolution. *Journal of Social Biological Structures. 11:*457–468, 1988.

Smith, W. J. *The Behavior of Communicating.* Cambridge, MA: Harvard University Press, 1977.

Smith-Gill, S. J. Developmental plasticity: Developmental conversion versus phenotypic modulation. *American Zoologist. 23:*47–55, 1983.

Smuts, B. B., and R. W. Smuts. Male aggression and sexual coercion of females in nonhuman primates and other mammals: Evidence and theoretical implications. In P. J. B. Slater, J. S. Rosenblatt, C. T. Snowden and M. Milinski (eds.), *Advances in the Study of Behavior,* Vol. 22. New York: Academic Press, 1993, pp. 1–63.

Snaith, R. N. Measurement in psychiatry. *British Journal of Psychiatry. 159:*78–82, 1991.

Sober, E. *The Nature of Selection.* Cambridge, MA: MIT Press, 1987.

Soldz, S., S. Budman, A. Demby, and J. Merry. Representation of personality disorders in circumplex and five-factor space: Explorations with a clinical sample. *Psychological Assessment. 5:*41–52, 1993.

Sorensen, G., and A. Randrup. Possible protective value of severe psychopathology against lethal effects of an unfavorable milieu. *Stress Medicine. 2:*103–105, 1986.

Spalletta, G., A. Troisi, M. Saracco, N. Ciani, and A. Pasini. Symptom profile, Axis II comorbidity and suicidal behaviour in young males with DSM-III-R depressive illnesses. *Journal of Affective Disorders. 39:*141–148, 1996.

Spitz, R. A. Hospitalism. *The Psychoanalytic Study of the Child,* Vol. 1. New York: International Universities Press, 1945, pp. 53–74.

Squire, L. *Memory and Brain.* New York: Oxford University Press, 1987.

Stack, S. The effects of marital dissolution on suicide. *Journal of Marriage and the Family. 42:*83–91, 1980.

Staddon, J. E. R. *Adaptive Behavior and Learning.* Cambridge, England: Cambridge University Press, 1983.

Stampfer, H. G., and G. A. German. The neuroscience imperative in psychiatry. *Integrative Psychiatry. 6:*152–164, 1988.

Staub, E. Helping a distressed person: Social, personality, and stimulus determinants. In L. Berkowitz (ed.), *Advances in Experimental Social Psychology,* Vol. 7. New York: Academic Press, 1974, pp. 293–342.

Steadman, L. B., and C. T. Palmer. Religion as an identifiable traditional behavior subject to natural selection. *Journal of Social and Evolutionary Systems. 18:*149–164, 1995.

Stearns, S. C. *The Evolution of Life Histories.* Oxford, England: Oxford University Press, 1992.

Stein, D. J., and K. O. Jobson (eds.). Pharmacotherapy algorithms for anxiety disorders. *Psychiatric Annals. 26:*189–232, 1996.

Stein, D. J., and M. Stanley. Serotonin and suicide. In R. D. Masters and M. T. McGuire (eds.), *The Neurotransmitter Revolution*. Carbondale: Southern Illinois Press, 1994, pp. 47–60.

Stein, M. B., M. E. Tancer, C. S. Gelernter, B. J. Vittone, and T. W. Uhde. Major depression in patients with social phobia. *American Journal of Psychiatry. 147:*637–639, 1990.

Stenhouse, D. *The Evolution of Intelligence.* New York: Barnes and Noble, 1973.

Stevens, A., and J. Price. *Evolutionary Psychiatry.* New York: Routledge, 1996.

Stone, V. E. The evolution of status and dominance: Implications for nonverbal signals. Paper given at the Human Behavior and Evolution Society Meeting, Los Angeles, August 1990.

Strauss, J. S., R. F. Kokes, W. T. Carpenter, Jr., and B. A. Ritzler. The course of schizophrenia as a developmental process. In L. C. Wynne, R. L. Cromwell, and S. Matthysse (eds.), *The Nature of Schizophrenia.* New York: Wiley, 1978. pp. 617–630.

Strongman, K. T., P. E. Wookey, and R. E. Remington. Elation. *British Journal of Psychology. 62:*481–492, 1971.

Sturt, E., N. Kumakura, and G. Der. How depressing life is—Life-long morbidity risk for depressive disorder in the general population. *Journal of Affective Disorders. 7:*109–122, 1984.

Suddath, R. L., G. W. Christison, E. F. Torrey, M. F. Casanova, and D. R. Weinberger. Anatomical abnormalities in the brains of monozygotic twins discordant for schizophrenia. *New England Journal of Medicine. 322:*789–794, 1990.

Sullivan, E. V., P. K. Shear, K. O. Lim, R. B. Zipursky, and A. Pfefferbaum. Cognitive and motor impairments are related to gray matter volume deficits in schizophrenia. *Biological Psychiatry. 39:*234–240, 1996.

Sullivan, E. V., P. K. Shear, R. B. Zipursky, H. J. Sagar, and A. Pfefferbaum. A deficit profile of executive memory and motor functions in schizophrenia. *Biological Psychiatry. 36:* 641–653, 1994.

Sulloway, F. J. *Freud, Biologist of the Mind: Beyond the Psychoanalytic Legend.* New York: Basic Books, 1979.

Sulloway, F. J. Birth order and evolutionary psychology: A meta-analytic overview. Paper given at the Human Behavior and Evolution Society meeting, Santa Barbara, CA, June 28–July 2, 1995.

Sulser, F. Serotonin-norepinephrine receptor interactions in the brain: Implications for the pharmacology and pathophysiology of affective disorders. *Journal of Clinical Psychiatry. 48*(Suppl.):12–18, 1987.

Surbey, M. K. Anorexia nervosa, amenorrhea, and adaptation. *Ethology and Sociobiology. 8*(Suppl.):47S–62S, 1987.

Suzuki, T., Y. Yoshino, K. Tsukamoto, and K. Abe. Concordant factors of depression rating scales and dimensionality of depression. *Biological Psychiatry. 37:*253–258, 1995.

Svrakic, D. M., C. Whitehead, T. R. Przybeck, and C. R. Cloninger. Differential diagnosis of personality disorders by the seven-factor model of temperament and character. *Archives of General Psychiatry. 50:*991–999, 1993.

Symons, D. *The Evolution of Human Sexuality.* New York: Oxford University Press, 1979.

Symons, D. Adaptiveness and adaption. *Ethology and Sociobiology. 11:*427–444, 1990.

Takei, N., P. B. Mortensen, U. Klaening, R. M. Murray, P. C. Sham, E. O. O'Callaghan, and P. Munk-Jørgensen. Relationship between in utero exposure to influenza epidemics and risk of schizophrenia in Denmark. *Biological Psychiatry. 40:*817–824, 1996.

Talbot, J. D., S. Marrett, A. C. Evans, E. Meyer, M. C. Bushnell, and G. H. Duncan. Multiple representations of pain in human cerebral cortex. *Science. 251:*1355–1358, 1991.

Tantam, D. Lifelong eccentricity and social isolation: I. Psychiatric, social, and forensic aspects. *British Journal of Psychiatry. 153:*777–782, 1988.

Tardiff, K. The current state of psychiatry in the treatment of violent patients. *Archives of General Psychiatry. 49:*493–499, 1992.

Tellegen, A., D. T. Lykken, T. J. Bouchard, Jr., K. J. Wilcox, N. L. Segal, and S. Rich. Personality similarity in twins reared apart and together. *Journal of Personality and Social Psychology. 54:*1031–1039, 1988.

Thayer, J. F., B. H. Friedman, and T. D. Borkovec. Autonomic characteristics of generalized
anxiety disorder and worry. *Biological Psychiatry. 39:*255–266, 1996.

Thayer, R. E. *The Biopsychology of Mood and Arousal.* New York: Oxford University Press,
1989.

Thiessen, D., and B. Gregg. Human assortative mating and genetic equilibrium: An evolutionary perspective. *Ethology and Sociobiology. 1:*111–140, 1980.

Thorngate, W. Efficient decision heuristics. *Behavior Science. 25:*219–225, 1980.

Thornhill, N. W. The comparative method of evolutionary biology in the study of the societies
of history. *International Journal of Contemporary Sociology. 27:*7–27, 1990.

Thornhill, N. W., and R. Thornhill. The evolution of psychological pain. In R. W. Bell and
N. B. Bell (eds.), *Sociobiology and the Social Sciences.* Lubbock: Texas Tech University
Press, 1989, pp. 73–103.

Thornhill, N. W., and R. Thornhill. An evolutionary analysis of psychological pain following
rape: 1. The effects of victim's age and marital status. *Ethology and Sociobiology. 11:*
155–176, 1990a.

Thornhill, N. W., and R. Thornhill. An evolutionary analysis of psychological pain following
rape: 2. The effects of stranger, friend, and family-member offenders. *Ethology and Sociobiology. 11:*177–193, 1990b.

Thornhill, N. W., and R. Thornhill. An evolutionary analysis of psychological pain following
rape: 3. Effects of force and violence. *Aggressive Behavior. 16:*297–320, 1990c.

Thornhill, R., S. W. Gangestad, and R. Comer. Human female orgasm and mate fluctuating
asymmetry. *Animal Behaviour. 50:*1601–1615, 1995.

Tierney, A. J. The evolution of learned and innate behavior: Contributions from genetics and
neurobiology to a theory of behavioral evolution. *Animal Learning Behavior. 14:*339–348,
1986.

Tiger, L. *Men in Groups.* New York: Random House, 1969.

Tiger, L. *The Pursuit of Pleasure.* Boston: Little, Brown, 1992.

Tiger, L., and R. Fox. *The Imperial Animal.* New York: Holt, Rinehart and Winston, 1971.

Tiihonen, J., J. Kuikka, H. Viinämäki, J. Lehtonen, and J. Partanen. Altered cerebral blood
flow during hysterical paresthesia. *Biological Psychiatry. 37:*134–135, 1995.

Tinbergen, E. A., and N. Tinbergen. Early childhood autism—An ethological approach. Z.
Tierpsychologie. 10(Suppl.):1–53, 1972.

Tinbergen, N. *The Study of Instinct.* New York: Oxford University Press, 1951.

Tinbergen, N. Ethology and stress diseases. *Science. 185:*20–27, 1974.

Tooby, J., and L. Cosmides. Evolutionary psychology and the generation of culture: 1. Theoretical considerations. *Ethology and Sociobiology. 10:*29–50, 1989.

Tooby, J., and L. Cosmides. On the universality of human nature and the uniqueness of the
individual: The role of genetics in adaptation. *Journal of Personality. 58:*17–67, 1990a.

Tooby, J., and L. Cosmides. The past explains the present: Emotional adaptions and the structure of ancestral environments. *Ethology and Sociobiology. 11:*375–424, 1990b.

Tooby, J., and L. Cosmides. Statistical reasoning in a multimodular mind: 1. An evolutionary
approach to the study of judgment under uncertainty. Paper given at the Human Behavior
and Evolution Society meeting, Albuquerque, NM, July 22–26, 1992.

Tooby, J., and L. Cosmides. The evolution of memory, modularity, and information integrity.
Paper given at the Human Behavior and Evolution Society meeting, Santa Barbara, CA,
June 28–July 2, 1995.

Torgersen, S. Hereditary differentiation of anxiety and affective neuroses. *British Journal of
Psychiatry. 146:*530–534, 1985.

Torrey, E. F. Prevalence studies in schizophrenia. *British Journal of Psychiatry, 150:*598–608, 1987.

Townsend, J. M. Sexuality and partner selection: Sex differences among college students. *Ethology and Sociobiology. 14:*305–330, 1993.

Tracy, J. I. Nonspecific abnormalities in psychiatry: Do they reflect an ordered pathway of brain dysfunction? *Biological Psychiatry. 38:*207–209, 1995.

Trivers, R. L. The evolution of reciprocal altruism. *Quarterly Review of Biology. 46:*35–57, 1971.

Trivers, R. L. Parental investment and sexual selection. In B. Campbell (ed.), *Sexual Selection and the Descent of Man 1871–1971.* Chicago: Aldine, 1972, pp. 136–179.

Trivers, R. L. Parent-offspring conflict. *American Zoologist. 14:*249–264, 1974.

Trivers, R. L. *Social Evolution.* Menlo Park, CA: Benjamin-Cummings, 1985.

Troisi, A., F. R. D'Amato, A. Carnera, and L. Trinca. Maternal aggression by lactating group-living Japanese macaque females. *Hormones and Behavior. 22:*444–452, 1988.

Troisi, A., and M. Marchetti. Epidemiology. In M. Hersen, R. T. Ammerman, and L. A. Sisson (eds.), *Handbook of Aggressive and Destructive Behavior in Psychiatric Patients.* New York: Plenum Press, 1994, pp. 95–112.

Troisi, A., and M. T. McGuire. Deception and somatizing disorders. In C. N. Stefanis, A. D. Rabavilas, and C. R. Soldatos (eds.), *Psychiatry: A World Perspective,* Vol. 3. Amsterdam: Excerpta Medica, 1991, pp. 973–978.

Troisi, A., and M. T. McGuire. Evolutionary biology and life events research. *Archives of General Psychiatry. 49:*501–502, 1992.

Troisi, A., A. Pasini, G. Bersani, M. Di Mauro, and N. Ciani. Negative symptoms and visual behavior in DSM-III-R prognostic subtypes of schizophreniform disorder. *Acta Psychiatrica Scandanivica. 83:*391–394, 1991.

Troisi, A., A. Pasini, G. Bersani, A. Grispini, and N. Ciani. Ethological predictors of amitriptyline response in depressed outpatients. *Journal of Affective Disorders. 17:*129–136 (submitted, 1997).

Troisi, A., A. Pasini, G. Bersani, A. Grispini, and N. Ciani. Ethological assessment of DSM-III subtyping of unipolar depression. *Acta Psychiatrica Scandinavica. 81:*560–564, 1990.

Trower, P., and P. Chadwick. Pathways to defense of the self: A theory of two types of paranoia. *Clinical Psychological Science Practice. 2:*263–278, 1995.

Tsuang, M. T. A thirty-five year follow-up study of schizophrenia, mania, and depression. In S. B. Sells, R. Crandall, M. Roff, J. S. Strauss, and W. Pollin (eds.), *Human Functioning in Longitudinal Perspectives.* Baltimore: Williams and Wilkins, 1980, pp. 46–52.

Tsuang, M. T., and R. F. Woolson. Mortality in patients with schizophrenia, mania, depression, and surgical conditions. *British Journal of Psychiatry. 130:*162–166, 1977.

Tsuang, M. T., and R. G. Woolson. Excess mortality in schizophrenia and affective disorders. *Archives of General Psychiatry. 35:*1181–1185, 1978.

Tsuang, M. T., R. F. Woolson, and J. A. Fleming. Premature deaths in schizophrenia and affective disorders. *Archives of General Psychiatry. 37:*979–983, 1980.

Tuomi, J., T. Hakala, and E. Haukioja. Alternative concepts of reproductive effort, costs of reproduction, and selection in life-history evolution. *American Zoologist. 23:*25–34, 1983.

Turke, P. Which humans behave adaptively, and why does it matter? *Ethology and Sociobiology, 11:*305–340, 1990.

Turke, P. A hypothesis: Menopause discourages infanticide and encourages continued investment by agnates. *Ethology and Sociobiology,* in press.

Ungerleider, L. G. Functional brain imaging studies of cortical mechanisms for memory. *Science. 270:*769–774, 1995.

Vaccarino, F. J., S. H. Kennedy, E. Ralevski, and R. Black. The effects of growth hormone-releasing factor on food consumption in anorexia nervosa patients and normals. *Biological Psychiatry. 35:*446–451, 1994.

Vaillant, G. E. Theoretical hierarchy of adaptive ego mechanisms. *Archives of General Psychiatry. 24:*107–118, 1971.

Vaillant, G. E. Natural history of male psychological health. *Archives of General Psychiatry. 33:*535–545, 1976.

Vaillant, G. E. Natural history of male psychological health. *New England Journal of Medicine. 301:*1249–1254, 1979.

Vaillant, G. E., and C. O. Vaillant. Natural history of male psychological health: 12. A 45-year study of predictors of successful aging at age 65. *American Journal of Psychiatry. 147:* 31–37, 1990.

Vallejo, J., C. Gasto, R. Catalan, A. Bulbena, and J. M. Menchon. Predictors of antidepressant treatment outcome in melancholia: Psychosocial, clinical and biological indicators. *Journal of Affective Disorders. 21:*151–162, 1991.

van den Berghe, P. L. The family and the biological base of human sociality. In E. E. Filsinger (ed.), *Biosocial Perspectives on the Family.* Newbury Park, CA: Sage, 1988, pp. 39–60.

van de Rijt-Plooij, H. H. C., and F. X. Plooij. Growing independence, conflict, and learning in mother-infant relations in free-ranging chimpanzees. *Behaviour. 101:*1–86, 1987.

van Praag, H. M. Moving ahead yet falling behind. *Neuropsychobiology. 22:*181–193, 1989.

van Praag, H. M., R. S. Kahn, G. M. Asnis, C. Z. Lemus, and S. L. Brown. Therapeutic indications for serotonin potentiating compounds: A hypothesis. *Biological Psychiatry. 22:* 205–212, 1987.

van Praag, H. M., R. S. Kahn, G. M. Asnis, S. Wetzler, S. L. Brown, A. Bleich, and M. L. Korn. Denosologization of biological psychiatry or the specificity of 5-HT disturbances in psychiatric disorders. *Journal of Affective Disorders. 13:*1–8, 1987.

Vaughn, C. E., K. S. Snyder, S. Jones, W. B. Freeman, and I. R. H. Falloon. Family factors in schizophrenic relapse. *Archives of General Psychiatry. 41:*1169–1177, 1984.

Viewig, W. V. R. Behavioral approaches to polydipsia. *Biological Psychiatry. 34:*125–127, 1993.

Virkkunen, M., R. Rawlings, R. Tokola, R. E. Poland, A. Guidotti, C. Nemeroff, G. Bissette, K. Kalogeras, S.-L. Karonen, and M. Linnoila. CSF biochemistries, glucose metabolism, and diurnal activity rhythms in alcoholic, violent offenders, fire setters, and healthy volunteers. *Archives of General Psychiatry. 51:*20–27, 1994.

Vogt, R., G. Bürckstümmer, L. Ernst, K. Meyer, and M. von Rad. Differences in phantasy life of psychosomatic and psychoneurotic patients. *Psychotherapy Psychosomatics. 28:*98–105, 1977.

Voland, E., and R. Voland. Evolutionary biology and psychiatry: The case of anorexia nervosa. *Ethology and Sociobiology. 10:*223–240, 1989.

Volkart, R., A. Dittrich, T. Rothenfluh, and W. Paul. Eine kontrollierte Untersuchung über psychopathologische Effekte der Einzelhaft. In H. Huber (ed.), *Revue Suisse de Psychologie Pure et Appliquée.* Bern: Verlag, 1983, pp. 1–24.

Von Hartmann, E. *Philosophy of the Unconscious* (1868). Trans. W. C. Coupland. New York: Macmillan, 1884.

Wake, M. H. The evolution of integration of biological systems: An evolutionary perspective through studies on cells, tissues, and organs. *American Zoologist. 30:*897–906, 1990.

Wakefield, J. C. The concept of mental disorder. *American Psychologist. 47:*373–388, 1992.

Walker, E., and R. J. Lewine. Prediction of adult-onset schizophrenia from childhood home movies of the patients. *American Journal of Psychiatry. 147:*1052–1056, 1990.

Walsh, A. Love styles, masculinity/femininity, physical attractiveness, and sexual behavior: A test of evolutionary theory. *Ethology and Sociobiology. 14:*25–38, 1993.

Walters, E. E., and K. S. Kendler. Anorexia nervosa and anorexic-like syndromes in a population-based female twin sample. *American Journal of Psychiatry. 152:*64–71, 1995.

Wang, X. T. Evolutionary hypotheses of risk-sensitive choice: Age differences and perspective change. *Ethology and Sociobiology. 17:*1–17, 1997.

Warburton, D. M. Stress and distress in response to change. In H. Box (ed.), *Primate Responses to Environmental Change.* London: Chapman-Hall, 1990, pp. 337–356.

Weintraub, M., and J. Frankel. Sex differences in parent-infant interaction during freeplay, departure, and separation. *Child Development. 48:*1240–1249, 1977.

Weissman, M. M., and G. L. Klerman. Sex differences and the epidemiology of depression. *Archives of General Psychiatry. 34:*98–111, 1977.

Weissman, M. M., P. J. Leaf, C. E. Holzer III, J. K. Myers, and G. L. Tischler. The epidemiology of depression. *Journal of Affective Disorders. 7:*179–188, 1984.

Wenegrat, B. *Sociobiology and Mental Disorder.* Menlo Park, CA: Addison-Wesley, 1984.

Wenegrat, B. *Sociobiological Psychiatry.* Lexington, MA: Lexington Books, 1990.

Wenegrat, B., L. Abrams, E. Castillo-Yee, and I. Jo Romine. Social norm compliance as a signaling system: 1. Studies of fitness-related attributions consequent on everyday norm violations. *Ethology and Sociobiology. 17:*403–416, 1996.

Wenegrat, B., E. Castillo-Yee, and L. Abrams. Social norm compliance as a signaling system: 2. Studies of fitness-related attributions consequent on a group norm violation. *Ethology and Sociobiology. 17:*417–431, 1996.

West-Eberhard, M. J. Adaptation: Current Usages. In E. F. Keller and E. A. Lloyd (eds.), *Keywords in Evolutionary Biology.* Cambridge, MA: Harvard University Press, 1992, pp. 13–18.

Westen, D. Motivation and affect regulation: A psychodynamic-cognitive-evolutionary model. Paper given at the Human Behavior and Evolution Society meeting, Santa Barbara, CA, June 28–July 2, 1995.

White, N. F. Ethology and psychiatry. In N. F. White (ed.), *Ethology and Psychiatry.* Toronto: University of Toronto Press, 1974, pp. 3–25.

Whiten, A., and R. W. Byrne. Tactical deception in primates. *Behavioral Brain Sciences. 11:* 233–273, 1988.

Whybrow, P. C., H. S. Akiskal, and W. T. McKinney, Jr. *Mood Disorders.* New York: Plenum Press, 1984.

Wiederman, M. W., and E. R. Allgeier. Gender differences in mate selection criteria: Sociobiological or socioeconomic explanation? *Ethology and Sociobiology. 13:*115–124, 1992.

Wiggins, J. S. Personality structure. *Annual Review of Psychology. 19:*293–350, 1968.

Wiggins, O. P., and M. A. Schwartz. The limits of psychiatric knowledge and the problem of classification. In J. Z. Sandler, O. P. Wiggins, and M. A. Schwartz (eds.), *Philosophical Perspectives on Psychiatric Diagnostic Classification.* Baltimore: Johns Hopkins University Press, 1994, pp. 89–103.

Wilhelm, K., and G. Parker. Is sex necessarily a risk factor to depression? *Psychological Medicine. 19:*401–413, 1989.

Wilkinson, G. S. Reciprocal altruism in bats and other mammals. *Ethology and Sociobiology. 9:*85–100, 1988.

Williams, G. C. Pleiotropy, natural selection, and the evolution of senescence. *Evolution. 11:* 398–411, 1957.

Williams, G. C. *Adaption and Natural Selection.* Princeton: Princeton University Press, 1966.

Williams, G. C. *Natural Selection: Domains, Levels, and Challenges.* New York: Oxford University Press, 1992.

Williams, G. C., and R. M. Nesse. The dawn of Darwinian medicine. *Quarterly Review of Biology. 66:*1–22, 1991.

Wilson, D. S. Adaptive genetic variation and human evolutionary psychology. *Ethology and Sociobiology. 15:*209–236, 1994.

Wilson, E. O. *Sociobiology—The New Synthesis.* Cambridge, MA: Harvard University Press, 1975.

Wilson, M., and M. Daly. The evolutionary psychology of mate-guarding and the risk of homicide in *Homo sapiens.* Paper given at the Human Evolution and Behavior Society meeting, Albuquerque, NM, July 22–26, 1992.

Wilson, R. S. Synchronies in mental development: An epigenetic perspective. *Science. 202:* 939–948, 1978.

Wolfner, G. D., and R. J. Gelles. A profile of violence toward children: A national study. *Child Abuse and Neglect. 17:*197–212, 1993.

Wolpe, J. The renascence of neurotic depression: Its varied dynamics and implications for outcome research. *Journal of Nervous and Mental Disease. 176:*607–613, 1988.

Woods, B. T., D. Yurgelun-Todd, J. M. Goldstein, L. J. Seidman, and M. T. Tsuang. MRI brain abnormalities in chronic schizophrenia: One process or more? *Biological Psychiatry. 40:*585–596, 1996.

Wrangham, R. W. Evolution of social structure. In B. B. Smuts, D. L. Cheney, R. M. Seyfarth, R. W. Wrangham, and T. T. Struhsaker (eds.), *Primate Societies.* Chicago: University of Chicago Press, 1987, pp. 282–296.

Wright, S. The genetical theory of natural selection. *Journal of Heredity. 21:*349–356, 1930.

Wu, J. C., M. S. Buchsbaum, T. G. Hershey, E. Hazlett, N. Sicotte, and J. C. Johnson. PET in generalized anxiety disorder. *Biological Psychiatry. 29:*1181–1199, 1991.

Yatham, L. N. Prolactin and cortisol responses to fenfluramine challenge in mania. *Biological Psychiatry. 39:*285–288, 1996.

Yuwiler, A. Diagnosis and the hunt for etiology. *Biological Psychiatry. 37:*1–3, 1995.

Yuwiler, A., and D. X. Freedman. Neurotransmitter research in autism. In E. Schopler and G. B. Mesibov (eds.), *Neurobiological Issues in Autism.* New York: Plenum Press, 1987, pp. 263–284.

Zeki, S. The visual image in mind and brain. *Scientific American. 267:*68–77, 1992.

Zilboorg, C. A. *A History of Medical Psychology.* New York: Norton, 1941.

Zimmerman, M., and W. Coryell. DSM-III personality disorder diagnoses in a nonpatient sample. *Archives of General Psychiatry. 46:*682–689, 1989.

Zitrin, C. M., and D. C. Ross. Early separation anxiety and adult agoraphobia. *Journal of Nervous and Mental Disease. 176:*621–625, 1988.

Zoccolillo, M., and C. R. Cloninger. Somatization disorder: Psychological symptoms, social disability, and diagnosis. *Comprehensive Psychiatry. 27:*65–73, 1986.

Zohar, J., T. R. Insel, K. F. Berman, E. B. Foa, J. L. Hill, and D. R. Weinberger. Anxiety and cerebral blood flow during behavioral challenge. *Archives of General Psychiatry. 46:* 505–510, 1989.

Zohar, J., E. A. Mueller, T. R. Insel, R. C. Zohar-Kadouch, and D. L. Murphy. Serotonegergic responsivity in obsessive-compulsive disorder. *Archives of General Psychiatry. 44:*946–951, 1987.

Zola-Morgan, S., and L. R. Squire. Neuroanatomy of memory. *Annual Review of Neuroscience. 16:*547–563, 1993.

Zukerman, M., M. S. Buchsbaum, and D. L. Murphy. Sensation seeking and its biological correlates. *Psychological Bulletin. 88:*187–214, 1980.

Index